IMPLEMENTATION OF THE CLEAN AIR ACT NATIONAL AMBIENT AIR QUALITY STANDARDS (NAAQS) REVISIONS FOR OZONE AND PARTICULATE MATTER

JOINT HEARING

BEFORE THE

SUBCOMMITTEE ON HEALTH AND ENVIRONMENT

AND THE

SUBCOMMITTEE ON OVERSIGHT AND INVESTIGATIONS

OF THE

COMMITTEE ON COMMERCE
HOUSE OF REPRESENTATIVES

ONE HUNDRED FIFTH CONGRESS

FIRST SESSION

OCTOBER 1, 1997

Serial No. 105–62

Printed for the use of the Committee on Commerce

U.S. GOVERNMENT PRINTING OFFICE

43–841CC WASHINGTON : 1998

For sale by the U.S. Government Printing Office
Superintendent of Documents, Congressional Sales Office, Washington, DC 20402
ISBN 0-16-056087-X

COMMITTEE ON COMMERCE

TOM BLILEY, Virginia, *Chairman*

W.J. "BILLY" TAUZIN, Louisiana	JOHN D. DINGELL, Michigan
MICHAEL G. OXLEY, Ohio	HENRY A. WAXMAN, California
MICHAEL BILIRAKIS, Florida	EDWARD J. MARKEY, Massachusetts
DAN SCHAEFER, Colorado	RALPH M. HALL, Texas
JOE BARTON, Texas	RICK BOUCHER, Virginia
J. DENNIS HASTERT, Illinois	THOMAS J. MANTON, New York
FRED UPTON, Michigan	EDOLPHUS TOWNS, New York
CLIFF STEARNS, Florida	FRANK PALLONE, Jr., New Jersey
BILL PAXON, New York	SHERROD BROWN, Ohio
PAUL E. GILLMOR, Ohio	BART GORDON, Tennessee
Vice Chairman	ELIZABETH FURSE, Oregon
SCOTT L. KLUG, Wisconsin	PETER DEUTSCH, Florida
JAMES C. GREENWOOD, Pennsylvania	BOBBY L. RUSH, Illinois
MICHAEL D. CRAPO, Idaho	ANNA G. ESHOO, California
CHRISTOPHER COX, California	RON KLINK, Pennsylvania
NATHAN DEAL, Georgia	BART STUPAK, Michigan
STEVE LARGENT, Oklahoma	ELIOT L. ENGEL, New York
RICHARD BURR, North Carolina	THOMAS C. SAWYER, Ohio
BRIAN P. BILBRAY, California	ALBERT R. WYNN, Maryland
ED WHITFIELD, Kentucky	GENE GREEN, Texas
GREG GANSKE, Iowa	KAREN McCARTHY, Missouri
CHARLIE NORWOOD, Georgia	TED STRICKLAND, Ohio
RICK WHITE, Washington	DIANA DeGETTE, Colorado
TOM COBURN, Oklahoma	
RICK LAZIO, New York	
BARBARA CUBIN, Wyoming	
JAMES E. ROGAN, California	
JOHN SHIMKUS, Illinois	

JAMES E. DERDERIAN, *Chief of Staff*
CHARLES L. INGEBRETSON, *General Counsel*
REID P.F. STUNTZ, *Minority Staff Director and Chief Counsel*

SUBCOMMITTEE ON HEALTH AND ENVIRONMENT

MICHAEL BILIRAKIS, Florida, *Chairman*

J. DENNIS HASTERT, Illinois,	SHERROD BROWN, Ohio
Vice Chairman	HENRY A. WAXMAN, California
JOE BARTON, Texas	EDOLPHUS TOWNS, New York
FRED UPTON, Michigan	FRANK PALLONE, Jr., New Jersey
SCOTT L. KLUG, Wisconsin	PETER DEUTSCH, Florida
JAMES C. GREENWOOD, Pennsylvania	ANNA G. ESHOO, California
NATHAN DEAL, Georgia	BART STUPAK, Michigan
RICHARD BURR, North Carolina	GENE GREEN, Texas
BRIAN P. BILBRAY, California	TED STRICKLAND, Ohio
ED WHITFIELD, Kentucky	DIANA DeGETTE, Colorado
GREG GANSKE, Iowa	RALPH M. HALL, Texas
CHARLIE NORWOOD, Georgia	ELIZABETH FURSE, Oregon
TOM COBURN, Oklahoma	JOHN D. DINGELL, Michigan,
RICK LAZIO, New York	(Ex Officio)
BARBARA CUBIN, Wyoming	
TOM BLILEY, Virginia,	
(Ex Officio)	

SUBCOMMITTEE ON OVERSIGHT AND INVESTIGATIONS

JOE BARTON, Texas, *Chairman*

CHRISTOPHER COX, California	RON KLINK, Pennsylvania
Vice Chairman	HENRY A. WAXMAN, California
JAMES C. GREENWOOD, Pennsylvania	PETER DEUTSCH, Florida
MICHAEL D. CRAPO, Idaho	BART STUPAK, Michigan
RICHARD BURR, North Carolina	ELIOT L. ENGEL, New York
BRIAN P. BILBRAY, California	THOMAS C. SAWYER, Ohio
GREG GANSKE, Iowa	JOHN D. DINGELL, Michigan,
TOM COBURN, Oklahoma	(Ex Officio)
TOM BLILEY, Virginia,	
(Ex Officio)	

(II)

CONTENTS

IMPLEMENTATION OF THE CLEAN AIR ACT NATIONAL AMBIENT AIR QUALITY STANDARDS (NAAQS) REVISIONS FOR OZONE AND PARTICULATE MATTER

WEDNESDAY, OCTOBER 1, 1997

House of Representatives,
Committee on Commerce,
Subcommittee on Health and Environment, joint
with Subcommittee on Oversight and Investigations,
Washington, DC.

The subcommittees met, pursuant to notice, at 10:10 a.m, in room 2123, Rayburn House Office Building, Hon. Michael Bilirakis (chairman of the Subcommittee on Health and Environment) presiding.

Members present, Subcommittee on Health and the Environment: Representatives Bilirakis, Barton, Upton, Greenwood, Deal, Burr, Bilbray, Ganske, Norwood, Coburn, Cubin, Bliley (ex officio), Brown, Waxman, Pallone, Deutsch, Eshoo, Stupak, Green, Strickland, Hall, Furse, and Dingell (ex officio).

Members present, Subcommittee on Oversight and Investigations: Representatives Barton, Greenwood, Crapo, Burr, Bilbray, Ganske, Coburn, Bliley (ex officio), Klink, Waxman, Deutsch, Stupak, Sawyer, and Dingell (ex officio).

Staff present: Charles Ingebretson, general counsel; Robert Meyers, majority counsel; Joseph Stanko, majority counsel; Stephen Sayle, majority counsel, and Bill Tyndall, minority counsel.

Mr. BILIRAKIS. The subcommittees will come to order.

I very much would like to limit opening statements. Without belittling the significance of opening statements, in the interest of time, there will be a number of votes toward the tailend of our session today on the Floor. There will be a number of suspension votes, as many as 14—we don't know—which will take quite a long period of time. I'm not sure that we can possibly finish up our hearing by the time we need to vote on the Floor, but it certainly would be nice if we could.

So I would like to limit opening statements to 5 minutes; hopefully, others, less than 3 minutes. The gavel will come down in 3 minutes' time, and when we go into the questioning, in 5 minutes the gavel will come down.

I'd like to thank Administrator Browner and the other witnesses for their appearance here today, some of which have traveled from out of town to be with us.

This is the sixth hearing we've held on the new ozone and particulate matter standards, but this is the first hearing that we have held since the standards became legally effective on September 16. Thus, our attention naturally has shifted from the scientific basis of the standards, which many consider to be still in dispute, to how the administration proposes to implement the new standards over the next 20 years.

As one member who was on this committee during the 1990 Clean Air Act Amendments, I remember the debate and controversy that initially surrounded the 1990 amendments and the bipartisan cooperation which finally led to their enactment. As the Administrator herself often points out, in 1990, a Democrat Congress and a Republican White House worked together to establish national policy and to craft black letter environmental law.

However, as I reviewed the Administrator's written statement last night—and I might add, as others will also say, we received that information somewhat tardily—I'm not so sure that a line has not been crossed. The implementation plan either stretches the Clean Air Act far beyond its original intent or seeks, without proper authority, to make new law.

The administration's plan indicates that we won't require State implementation plans for 6 to 11 years from now. Compliance activities will not be completed for 15 to 20 years. And, yet, the Clean Air Act provides that this work be done "as expeditiously as practicable."

The administration's plan will also defer requirements designed to protect clean areas from becoming dirty. Somehow non-binding guidance will overrule clear statutory language.

All this is being done in the name of flexibility and common sense, but I fear that EPA walks a dangerous path when it chooses to both insulate itself from judicial review and not seek legislative authority for new programs. And I have to ask, Why? Why would EPA run the risk of lawsuits overturning major elements of its implementation plan? What's the purpose? Why not build a firm legal foundation to support flexible, common-sense implementation of the Clean Air Act? Why risk years of litigation, delay, and uncertainty?

We have time. Administrator Browner herself said that sources would not be legally required to reduce their pollution until 2009. So we have the opportunity. We can learn from the successful implementation of the 1990 act. And we have the means, the ability in the time available, to craft legislation which can withstand judicial review.

But I seriously question what rationale with respect to the implementation plan would favor years of uncertainty and the law of unintended consequences. To me, that is not responsible environmental policy or responsible public policy.

Perhaps it is not surprising then that, although legislative authority for the Clean Air Act expires in 45 days, all that we have heard from the EPA in this regard is silence.

Mr. Brown, for an opening Statement?

Mr. BROWN. Thank you, Mr. Chairman. I would like to thank Chairmen Bilirakis and Barton for holding today's hearing, and thank Administrator Browner and the others for appearing before us today.

Following hours of testimony, after reading reams of scientific evidence on these proposed standards, and meeting with numerous public health experts and residents in my district in northeast Ohio, I've come to the conclusion that these new air quality standards will significantly improve the quality of life for millions of Americans and save countless lives. Improved air quality will especially benefit our most vulnerable citizens: children with asthma and seniors with lung reduction diseases.

This is a public health issue.

In my State, the Lung Association, the Heart Association, pediatricians, and other advocates for children and the elderly have lauded these new regulations. A recent study showed significantly higher infant mortality rates in northeast Ohio in part because of particulate matter.

I genuinely appreciate, however, the concerns of businesses, especially small business owners, who fear these standards will saddle them with additional regulatory costs which they say will stifle our current 6-year economic expansion. I believe, however, that American industry, as in the past, will find innovative and cost-effective ways to meet these new standards.

For instance, in reducing acid rain, industry years ago estimated it would cost between $1,000 and $1,500 for every ton of sulphur dioxide emissions reduced. The estimated cost today for reducing one ton of SO_2 is $78. American industry remains the leader in developing cutting-edge environmental technology, allowing industry to grow while reducing emissions and being environmentally responsible.

I applaud Administrator Browner's work on establishing a flexible implementation plan to meet these new standards. The ozone implementation plan will use a regional State-sponsored plan to address the long-distance transport of ozone. Areas that don't meet the new ozone standard will be able to avoid burdensome measures associated with compliance by being given the new designation of transitional.

The PM implementation plan will allow EPA to conduct another full scientific review of the health effects of fine particulates before making any non-attainment designations. EPA will allow 5 years to gather and analyze data; then use its discretion under the Clean Air Act to allow another 3 years for areas not in compliance to submit new air quality plans.

Many have questioned the administration's commitment to this flexible implementation plan. However, based on Administrator Browner's honest and straightforward work with Congress to craft bipartisan legislation to improve the Safe Drinking Water Act, and other important environmental issues, I have faith in her commitment to honor this flexible implementation plan and work with interested parties to address the unique needs of many of the communities we represent.

On a personal note, several months ago, members of the Amish community in my district in northeast Ohio expressed concerns that these new standards would restrict or lead to an outright ban of their use of wood-burning stoves. Since the Amish do not use electricity, their stoves are critical to their way of life. In response to these concerns, I spoke with Administrator Browner to seek her assurance that these new standards would not affect my constituents adversely. Within a short period of time, she was able to provide me with written assurance that the residents of the Amish community would not lose their stoves.

Clearly, Congress produces the best policy when we work together in a bipartisan manner. While we have our differences over these standards, I remain hopeful that Congress and the Clinton administration can find common ground which balances the benefits of cleaner air with the economic health and well-being of the local communities we represent.

Having said that, I look forward to Administrator Browner's testimony today on implementation of these proposed standards, which I believe will make a positive difference in the everyday lives of our constituents. Thank you, Mr. Chairman.

Mr. BILIRAKIS. I thank the gentleman. I will now recognize the chairman of the Oversight and Investigations Subcommittee and co-chairman of today's hearing, Mr. Barton.

Mr. BARTON. Thank you, Mr. Chairman.

Administrator Browner, it's always good to have you before us.

Mr. Chairman, I'm going to submit my opening statement for the record. I'll just very briefly summarize it.

I think we need to look at the proposed implementation plan. I think that the Administrator has made some judgment calls. I don't agree with her judgments. I think the ozone standard's too tight. I think it should be 90 parts per billion instead of the standard that she set it at. I don't think there's enough information to set a PM standard. I think we need more data collection on that.

I think the President agrees with me, because, as I understand the implementation strategy that's been laid out, they don't plan to enforce any of these standards on PM for anywhere from 7 to 10 years. So I don't see how setting a standard now that has no effect is any different than waiting until we have scientific evidence to set the standard.

I don't think you can show enough flexibility to make these payments. EPA's own numbers are it's going to cost $47 billion a year. Those are your numbers; those aren't our numbers. Forty-seven billion dollars is quite a bit of paying in Texas where I come from.

To be a little bit positive, I think you are trying to show some flexibility, and that's a good thing. And if we ever have a chance to ask questions at this hearing, just to put you on notice, I'm going to ask some questions about some statements that one of your officials has made down in Texas at a hearing that Congressman Frost had.

Again, Thank you for being here today.

Mr. BILIRAKIS. I thank the gentleman.

Mr. Dingell, for an opening statement.

Mr. DINGELL. Mr. Chairman, I ask unanimous to have my statement entered into the record.

Mr. BILIRAKIS. Without objection, the ranking member's opening statement and all the other opening statements will be made a part of the record.

Mr. DINGELL. Mr. Chairman, this is a very vital subject, one which will affect the entire future of the United States in terms of its economics and in terms of its environment.

We are here to inquire into a number of hard questions. As Mayor Daly of Chicago recently said about this matter, "Setting tougher pollution standards will not give us cleaner air. Indeed, good standards, if carried out badly, can adversely affect public health and the environment." I would note the air is getting cleaner. I would note the air is getting cleaner because the Clean Air Act is working. I would note the regulations relative to ozone and relative to particulates are working splendidly. I would note that we have before us the curious situation where EPA now seeks to impose new standards, and I would note that at the same time, through a number of addenda and other changes which they seek to make in the way the statute would be implemented by these new regulations, it appears that we have on our hands a situation where we may be well observing the strong possibility of extralegal and improper behavior which will be challenged successfully in court, with severe and drastic consequences to the regulatory efforts at EPA.

These are only a part of the questions that need to be addressed today. I'm delighted to see Ms. Browner before us today. Thank you for being here. We have some splendid questions for you, which I know will be enjoyed by all concerned.

Mr. Chairman, I thank you.

[The prepared statement of Hon. John D. Dingell follows:]

PREPARED STATEMENT OF HON. JOHN D. DINGELL, A REPRESENTATIVE IN CONGRESS FROM THE STATE OF MICHIGAN

I want to thank the Chairmen for holding this hearing on this very vital subject.

As Mayor Daley of Chicago said shortly after the Administration announced its decision to adopt stringent new air quality standards: "Setting tougher pollution standards will not give us cleaner air. Indeed, good standards, if carried out badly, can adversely affect public health and the environment."

Put another way: New, tight standards may grab the headlines, but they do not remove one molecule of ozone or a single fine particle from the air.

With this in mind, we must consider what progress we will make in cleaning the air under the new standards.

In this sense, this is no longer a public health question. While EPA has adopted stringent new air quality standards, the value of these standards to public health is determined by the timing and amount of actual emissions reductions. If EPA's implementation of these standards causes confusion, uncertainty, delays, and disruptions, then the standards, while impressive on paper, do not provide much relief to the breathing public.

Adding to the importance of this inquiry is the reality—which Ms. Browner has affirmed repeatedly—that we are currently making progress cleaning the air and will continue to do so simply by continuing to implement the Clean Air Act Amendments of 1990.

Indeed, the real question before us is what is the best way to make additional progress on cleaning the air? The Administration has now spoken on that issue. Faced with legitimate issues regarding the scientific record and the economic impact of the proposed now standards, the Administration responded by concocting a complicated and convoluted scheme that delays implementation of the standards while trying to assure that further review of the standards occurs before they are implemented.

As for fine particles, the Administration seeks to provide assurances that no areas will be designated for nonattainment—nor will any new controls be implemented—until EPA completes a new scientific review of the fine particle standard. In this, the Administration is perfectly in sync with H.R. 1984. The difference is that the implementation plan leaves State and local governments and industry subject to the whims of federal courts and citizen suits. H.R. 1984 ensures that EPA can and will keep its word on this.

For ozone, the implementation plan seeks to avoid—in EPA's word—the "stigma" of nonattainment for the hundreds of new nonattainment areas through an aggressive bending of the law. With a new "transitional" classification, EPA expects these areas to avoid—again, in EPA's own words—"burdensome new local planning requirements and restrictions on economic growth." This is a worthy goal, but I am not sure that the sudden discovery of new authority to relax nonattainment requirements for some parts of the country is legal or equitable.

In fact, all EPA is trying to do is delay the impact of the new ozone standard and reduce the number of counties that must be classified as nonattainment. After five years of continued implementation of the Clean Air Act Amendments of 1990 and implementation of whatever transport controls the States agree to, ozone levels will fall and most communities will be in compliance with even a new, stringent ozone standard.

Again, H.R. 1984 will guarantee this extra time. The Administration's Implementation Plan guarantees only that EPA, the States, and industry will be held hostage to every citizen group with access to a first year law student.

So I welcome this discussion of the implementation of these new standards. We owe the American people clean air. It is high time to look at the question of how we continue to make the additional progress that we know our constituents want and deserve.

Mr. BILIRAKIS. I thank the gentleman. I will now recognize Dr. Coburn for an opening statement. Please limit it to 3 minutes or less, if you would, Tom.

Mr. COBURN. Thank you, Mr. Chairman. I will limit it to less than that.

I would welcome the Administrator. I would just simply say that I continue to be concerned, since I have been in Washington, at the examples of what I see in terms of incestuous science, relationships that are based on grant-making to former employees, and a lack of clarity and true science in terms of how we make decisions. And that's not anything new; I've said that a long time in terms of the studies that have been made.

I am interested to hear the Secretary tell us when they're going to enforce these regulations, how they're going to enforce these regulations, and in fact inquire when real science comes about to say what we have in the area of particulate matter, whether or not the EPA is going to be flexible, once we really do know the answers, instead of guessing in the dark.

And I yield back.

Mr. BILIRAKIS. I thank the gentleman. Mr. Klink?

Mr. KLINK. Thank you, Mr. Chairman. I look forward to today's hearing.

It has been nearly 5 months since Administrator Browner testified before these subcommittees about the then-proposed ozone and fine particulate standards, and a lot has happened since then, but one thing has remained the same: No one at the EPA or the administration has agreed to sit down and have discussions with those of us in Congress who have had problems with these new proposed standards.

Over 40 members of this very committee signed a letter asking for some time with the President. Over 80 Democrats signed onto one of various letters sent to the President voicing our concern. We didn't just write; we called; we wrote; we talked to everyone we could. I offered to sit down and speak with the President, the Vice President, Administrator Browner. In fact, on May 15, after our last hearing on this issue, Administrator Browner approached the desk up here, apologized for canceling a meeting that we had set up the previous day, and promised to reschedule soon. Administrator, I'm still waiting for your call. The EPA and the administration has found time for everyone except those members of the President's own party who have serious concerns about the effect that these standards will have on our districts and our Nation.

We need to finish the work that we started under the old standards. The Ozone Transport Assessment Group, which began its work under the old standards, and based its conclusions, which they just issued this past summer, on the old standards—these findings have not yet been implemented. Pennsylvania and many other States have State implementation plans that are not completely in place yet from the last revisions, and there are serious questions about the credibility of the studies that the EPA has used in justifying these standards.

My question is: Why don't we slow down? Why don't we take 5 years and give EPA the time and give them the resources to resolve the scientific and the technical controversy surrounding these standards? Why not ensure the clean air progress over the next 5 years will continue through implementation of ozone-fighting provisions of the Clean Air Act Amendments of 1990? Why not provide the implementation of the monitoring and research programs that the EPA advocates for fine particles want?

I've introduced a bill, H.R. 1984, which does just that, and if 1984 is enacted, I would say that, even if we're wrong, if those of us 200 Republicans and Democrats who have introduced this bill, even if we're wrong, we will not delay by 1 day the implementation plan that EPA has suggested. The EPA will have had the chance to conduct more studies on fine particles. They will have been able to buy and set up monitors, so they can enforce these fine particle standards. The EPA will be able to focus resources on stopping the exceedances of the old standards, and they will be able to review the fine particle and ozone standards and put those standards into effect. That would be 7 years before any cleanup would be required under the EPA's own plan. Administrator Browner testified before the House Agriculture Committee that no one would be legally required to reduce pollution under the new standards until 2009; that we would beat that by at least 7 years.

Now what happens if the EPA's assumptions about the new standards are incorrect and we don't pass 1984? First of all, many States are going to lose billions of dollars in State highway funds. People will lose access to affordable health care due to job loss or jobs that aren't created. Over 600 counties will be thrown into nonattainment. Companies will find it difficult to justify expanding and locating in most U.S. cities. Job creation and urban revitalization will be put on hold. Building permits will not be issued. Many provisions of the EPA's implementation plan will, indeed, violate the Clean Air Act, and I would stress the citizens' lawsuit provision of the Clean Air Act would cause a lot of the assurances that members feel they've gotten from the EPA to fall like a deck of cards.

I simply say that we don't want to have a rush to judgment. We have nearly 200 co-sponsors, both Republicans and Democrats, to H.R. 1984. It is a practical, bipartisan solution proposed by the elected representatives rather than some bureaucratic solution proposed by an ambitious political appointee.

Now the questions are these: Is there anything wrong with the approach we've taken in H.R. 1984? What are the administration's and EPA's problems with this bill? I will tell you, Mr. Chairman, we don't know because no one from the EPA or the administration has agreed to meet with us or to talk with us. The only way we can get answers to these questions is to drag them before this committee. So I look forward to today's hearing and their answers to these questions.

Mr. BILIRAKIS. I thank the gentleman. Dr. Norwood, 3 minutes or less, please, sir.

Mr. NORWOOD. Thank you, Mr. Chairman. I thank both of you for having these hearings. I look forward to trying to get an answer or two from Ms. Browner that I've tried to get in the last meeting that we had here, and I would say so far there's good news and bad news. I've already lost a company moving into one of my counties down in Georgia, just because that county is potentially a nonattainment county. So it is already beginning to affect us in our districts.

But, second, there's good news here, too. This, of all things I know, of a Federal agency ramming it down the throats of the American people is certainly going to keep regulatory reform live and well in the future Congresses.

And with that, I'll submit the balance of mine for the record, please, sir.

Mr. BILIRAKIS. I thank the gentleman. Let's see, Mr. Waxman, 3 minutes or less, please.

Mr. WAXMAN. Thank you very much. Mr. Chairman, I welcome Administrator Browner. I understand this is her 11th appearance on the Hill on this subject, 25th hearing in the Congress on this subject. It's hard to think that you're not coming here and meeting enough with us about this matter.

The United States has a proud history of working toward clean air. For the last 27 years, the Clean Air Act has employed cost-effective, common-sense measures to move the country steadily toward the goal of ensuring that all Americans are breathing healthy air. Unfortunately, this progress has not always been achieved cooperatively. Each time we act to control air pollution, industry brings out the same tired predictions of economic disaster. Those predictions have been proven wrong time and time again. Industry, generally, has risen to the challenge when the challenge has been placed before them.

The idea that we have to choose between a good economy and a clean environment is a false choice. We can have both. These clean air standards are popularly supported by the American people and Members of Congress. It's time everyone realizes that the new standards are in place, and they're not going to go away.

Now we need to work with the States, local governments, and business, and get past the fear that opponents of the standards have worked so hard to generate. In fact, the estimate I've seen is that there's been a $30 million media campaign designed to show that the clean air standards lack either scientific or popular support. They're wrong.

We have time to figure out a cost-effective, common-sense way to implement these standards. We have found solutions to pollution problems before, and we can find them again. I hope this joint committee will exercise the leadership necessary to get the job done.

Let us have all the meetings that are necessary for us to work together. Let's get on with this important work.

I yield back the balance of my time.

Mr. BILIRAKIS. I thank the gentleman. Mr. Upton, for an opening statement, 3 minutes or less, please.

Mr. UPTON. Thank you, Mr. Chairman. I'm going to submit my full statement for the record.

Mr. BILIRAKIS. Without objection.

[The prepared statement of Hon. Fred Upton follows:]

PREPARED STATEMENT OF HON. FRED UPTON, A REPRESENTATIVE IN CONGRESS FROM THE STATE OF MICHIGAN

Mr. Chairman, thank you for convening today's hearing on the Administration's plan for implementing the Environmental Protection Agency's (EPA) newly revised standards for ozone and particulate matter. I have a number of concerns and questions about the assumptions behind this implementation plan, about its viability if challenged in court, and about the extent to which it really addresses some of the very serious problems posed by the new standards, particularly for those areas of the country, such as the area I represent, which are especially hard hit by transient pollution.

I think it is interesting that the implementation plan was developed by an inter-agency Administration group, not by the EPA alone, and that President Clinton has *directed* Ms. Browner to implement this interagency plan. I wish that President Clinton had directed Ms. Browner to respond to the many concerns about the new standards raised by other agencies and departments within the Administration when the EPA proposed these standards last November. As I have pointed out in previous hearings, during the interagency review of the proposed standards, very serious concerns and reservations were raised by the Departments of Agriculture and Transportation and the Small Business Administration, in particular. Those concerns, however, were not addressed. Nor were the very serious concerns raised by many of us in Congress, who were hearing from our states, our cities, our local governments, and our businesses about the enormous potential economic and social costs of the standard. Instead, the EPA insisted on its proposed standards, and the President signed off on the final rule.

I welcome the commitment to the virtues of "flexibility" and "common sense" and "cost effectiveness" President Clinton makes in his forward to the implementation plan guidance document. But I wish he had come a little earlier in the process to this commitment to these virtues and required the EPA to re-examine its proposed standards and respond to the real scientific and economic concerns raised about them.

Had such virtues entered the process earlier, we would not now be confronted with an implementation plan that is certainly imaginative, but that may not work in reality and that may not stand the test of a court challenge. The prime example is the plan to create "transition areas" to address the problem faced by many areas of the country affected by transient ozone. As a result of transient ozone, many areas of the country which met the previous ozone standard, including my district in Southwestern Michigan, would be thrust into non-attainment status under the revised standard, with all that means in lost jobs, economic growth, and personal life-style changes.

Under the new standard, for example, Cass County in my district, which has more hogs than people, no four-lane roads, and one stop light in the county seat, will not be in attainment under the new standards because of pollution blowing in from Chicago, Gary, Indiana, and Milwaukee. Under the implementation plan, Cass County may qualify us a "transition area," an EPA created designation that may or may not be within a very, very broad reading of the Clean Air Act. Transition areas, the implementation plan states, are areas that would in all likelihood be in compliance with the new ozone standard when the Ozone Transport Assessment Group's (OTAG) recommendations for regional ozone controls are in place. "Transition areas" would avoid the stigma and some of the regulatory burden that come with designation as a non-attainment area.

Sounds good, but there are some catches. Transition area designation for Cass County and similar areas hinges on the state's submission of a plan to implement the OTAG strategy. If Michigan decides against participating in the OTAG plan, areas such as Cass County cannot be designated as transition areas. This just doesn't make sense. The problem in Cass County isn't pollution from Michigan sources, it's pollution from Chicago, Gary, and Milwaukee. If anything, transition designation should hinge in these cities' participation in the OTAG plan, not Michigan's. Transition area designation seems less a means of addressing the problem of Cass County and similar areas under the revised standard than a means of forcing states such as Michigan to participate in the OTAG plan.

Second, the implementation plan assumes that pollution reductions under the OTAG plan will result in Cass County and similar areas coming into attainment. But that assumption is questionable. Suppose the EPA's modeling is wrong?

Third, under the Clean Air Act as I read it, you are either in attainment or not in attainment. I don't see any mention of "transition areas." Will this artful designation stand up under a court challenge?

As I think these examples of the problems with the "transition area" plan reveal, the Administration's implementation plan is not a model of common sense, flexibility, and cost-effectiveness. It is, instead, a rather imaginative but highly questionable and probably ultimately unworkable way to address the real problems inherent in the new standards. The common-sense, flexible, and cost-effective thing to do at this point is to put the new standards on hold and take hard second look at them.

That is what H.R. 1984, legislation I have introduced with Congressmen Klink and Boucher, will do. It doesn't amend the Clean Air Act. It doesn't change the new ozone or particulate matter standard. It doesn't undermine our continuing clean air progress under the 1990 Clean Air Act Amendments. It permits the EPA to re-promulgate its new ozone and particulate matter standards after a four-year review, or to propose another standard, as required under the Clean Air Act. Further, H.R. 1984 would *codify* and protect from court challenge exactly what the implementation plan proposes for the new particulate matter standard. It provides the EPA with the *statutory* authority and resources to immediately install fine particulate matter monitors, develop a data base, do the further research necessary to determine whether, in fact, a 2.5 micron standard or another standard is appropriate, and promulgate a standard based on that determination.

H.R. 1984, not the flawed Administration implementation plan, is the common-sense, flexible, cost-effective approach to take to clean air regulation.

Mr. UPTON. I just want to echo Mr. Klink's concerns that there are a number of us that have felt that we've been ignored, particularly with some of the concerns that we raised at the hearing last spring when you were here. Specifically, we do have a number of questions, and I'm looking forward to the question-and-answer session later on. I hope we don't have too many votes which would delay that. But, specifically, I hope you can address this in your testimony on the implementation of the new standard.

You will remember that I raised questions with regard to the transient air coming across Lake Michigan to Michigan, from centers like Gary, Indiana, Milwaukee, and Chicago, who are, in essence, 95 percent responsible for pollution in the air that we breathe in southwest Michigan. I have a county two counties in from Lake Michigan, Cass County, that has four times as many hogs than people; doesn't have a four-lane road; has one traffic light, and this county will not be in attainment. There is nothing they can do to get out from underneath that until Gary, Chicago, and Milwaukee get their acts cleaned up.

And with regard to the implementation plan, no one has any idea whether or not this will stand up in court, when it's taken there. If it doesn't stand up, they're just caught, and it's very upsetting to a lot of folks, not only on this side of the aisle, but on the other as well.

And I yield back the balance of my time.

Mr. BILIRAKIS. I thank the gentleman. Mr. Stupak, 3 minutes or less, please, sir.

Mr. STUPAK. Thank you, Mr. Chairman. Thus far, so many words have been spoken, but so little have been said. I don't want to add to this futility, so I'm going to waive my opening statement and look forward to hearing from the witness.

Mr. BILIRAKIS. Mr. Greenwood?

Mr. GREENWOOD. Pass, Mr. Chairman.

Mr. BILIRAKIS. Mr. Greenwood passes. Ms. Cubin?

Ms. CUBIN. I'll put my opening comments in the record.

Mr. BILIRAKIS. I thank the young lady.

Mr. Sawyer?

Mr. SAWYER. Thank you, Mr. Chairman. We've heard a number of comments this morning, some deeply sympathetic to the message that the Administrator brings to us; others only somewhat less deeply sympathetic to the message that the Administrator brings to us. I'd like to associate myself with the remarks of my colleagues.

The common theme that runs through all of this is implementation, implementation, and so I'll forego my opening statement and just say that, clearly, all of us are looking forward to your clarifications on those questions. Thank you.

Mr. BILIRAKIS. I thank the gentleman. Let's see, Dr. Ganske?

Mr. GANSKE. I'll submit, Mr. Chairman. Thank you.

Mr. BILIRAKIS. I thank you, doctor.

[The prepared statement of Hon. Greg Ganske follows:]

PREPARED STATEMENT OF HON. GREG GANSKE, A REPRESENTATIVE IN CONGRESS FROM THE STATE OF IOWA

Thank you Mr. Chairman. This morning we have an opportunity to review the Administration's plan for the implementation of the of the newly revised ozone and particulate matter standards.

Many members in the House, from both sides of the aisle, have attempted to offer a compromise to address the revised standards. The compromise would have allowed four years for EPA to review the science before implementing regulations.

Both EPA and CASAC have recommended additional research be conducted to better examine the relationship between fine particles and our health. Both EPA and CASAC have also acknowledged a lack of fine particle matter data. Yet, the EPA claimed the science was sound enough to implement the revised standards, and on September 16 the revised standards became effective.

The Administration has issued a memorandum regarding the implementation of the new standards. The planned implementation calls for the EPA to collect particle matter data for five years and complete a scientific review of the particulate matter standards.

Does it not still make sense to delay the implementation of this standard? If the science was so strong why are we waiting so long? Doesn't the decision to delay implementation confirm the concerns expressed by the members of Congress, scientists, and the American public?

Also, I will be very interested in learning why the implementation plan was issued as a memorandum instead of a regulation? I am not sure we should subject state and local governments to an implementation plan that can be modified with little notice, no opportunity for public comment, and no judicial review.

State and local governments should have the opportunity for long term planning. These revised air quality standards would be burdensome even if the standards were associated with a definite time line.

There is substantial doubt as to whether the EPA has the legal authority to delay enforcement of the new particle matter and ozone standards. If EPA is mistaken and a civil action filed by a private citizen could force immediate implementation of the air quality standards, states, localities, and small businesses will all be the losers.

I look forward to learning more about these issues today and I yield back the balance of my time.

Mr. BILIRAKIS. Let's see, who was next? Mr. Green?

Mr. GREEN. Mr. Chairman, I'll submit my full statement, but let me just say——

Mr. BILIRAKIS. Without objection.

Mr. GREEN. [continuing] that after sitting through the five earlier meetings and the sixth today, I didn't realize we had 25 or so on the Hill. I guess one of my frustrations is trying to establish a meeting or discussion on these new standards with the potential for implementation, and that did not have an impact, even though we had lots of hearings, I'm sure, before Congress, but to try to problem-solve.

I co-signed the bill 1984 in September, so to speak, to give the EPA some breathing room. And I would hope in your remarks you would deal with how we can, without correcting it statutorily, not have a citizen's right to sue to overcome what EPA may say is an implementation schedule that's flexible, and I would hope we would hear that today. As someone who supported the citizens' right to sue, I also know that we want to make sure that we realize that EPA's not forced to do something that today or the last 6 months you have been telling us that you will give us that delay on the implementation schedule.

Again, Mr. Chairman, I look forward to the testimony.

Mr. BILIRAKIS. I thank the gentleman. Mr. Crapo, 3 minutes or less, please, sir.

Mr. CRAPO. Mr. Chairman, I'll submit my opening statement.

[The prepared statement of Hon. Michaed D. Crapo follows:]

PREPARED STATEMENT OF HON. MICHAEL D. CRAPO, A REPRESENTATIVE IN CONGRESS FROM THE STATE OF IDAHO

Thank you, Mr. Chairman.

I want to express my appreciation to you for holding this important hearing on the implementation plan proposed by the Environmental Protection Agency (EPA) for its new National Ambient Air Quality Standards (NAAQS) on $PM_{2.5}$ and ozone. This hearing is particularly timely as Members of Congress have been hearing from states and localities about the virtual impossibility of their meeting the new standards for the foreseeable future.

In the dozens of committee hearings Congress has held since the standards were drafted last fall, members have repeatedly called for an implementation system that reflects the needs of localities. Therefore, I also want to applaud the EPA for finally acknowledging that a flexible and locality-specific approach is necessary in implementing the new air quality standards.

Unfortunately, while the EPA talks about working toward a flexible implementation plan, its actions tell another story. So far, the only documentary evidence of the implementation scheme appears to be an interagency memorandum between the White House and EPA. Of course, this memorandum cannot be considered either official policy or regulation. Legally, the document is exactly what it says it is, a memo.

This fact makes the EPA's absolute refusal to consider legislation codifying the agency's flexible implementation plan all the more confusing. Although EPA Administrator Carol Browner has conceded that the so-called "Klink-Upton Moratorium" legislation would in no way hinder the agency's proposed implementation of the NAAQS standards, she has repeatedly opposed passage of the bill. I am hopeful that the Administrator will be able to enlighten us today about her reasons for why this legislation would undermine the EPA's implementation scheme.

As the representative of a largely rural, agriculture-based region, I am especially interested to hear how Administrator Browner intends to modify the implementation plan to reflect her oft-repeated assurances that the NAAQS standards are not directed at rural cities or farming communities. It would appear difficult to create a Washington-based, national scheme that is also flexible enough to accommodate the unique characteristics of the West. If these new standards are indeed primarily designed to mitigate air quality problems in the East, then concrete efforts must be made by the Administrator to address our concerns.

I also look forward to hearing the various panels' views on the legal defenses available to the EPA should this undocumented implementation scheme be challenged. It is easy to envision a legal action being brought against the agency for making regulatory decisions not permitted under the Clean Air Act. Moreover, there doesn't appear to be evidence that the EPA will defend itself any more vigorously than it did in the suit brought by the American Lung Association which created all of the new standards.

It is important for this subcommittee to hear the comments of those who will be in the forefront of carrying out the EPA's new NAAQS mandates. Members of Congress have tried in recent months to make the Administration aware of the American people's concerns about the new standards. This hearing will provide a forum for the state and local representatives here today to identify problems with the EPA's proposed implementation plan. I look forward to a lively and enlightening discussion.

Thank you, Mr. Chairman.

Mr. BILIRAKIS. Mr. Pallone, 3 minutes or less, please.

Mr. PALLONE. Thank you, Mr. Chairman. I'll submit my statement for the record, but I'd just like to summarize.

First of all, I do want to welcome Administrator Browner back, and I want to commend her for the job she's doing to protect the health of our children and families. And because of the efforts made by States and industries under the Clean Air Act, the quality of air we breathe has shown steady improvement for the past three decades. But, most important, the history of the act shows that the economy can thrive in tandem with environmental protection.

During the past 25 years, six major pollutants and their precursors have dropped by nearly 30 percent, while the Gross Domestic Product has increased by more than 90 percent. And under the act, the EPA sets the standards based solely upon human health considerations, and rightly so.

However, because of EPA's authority to consider cost implementing these standards, I've stated all along that I believe the agency should do all that it can to implement the standards for the least possible cost in terms of both money and effort. And I'm pleased that the White House memo on implementation of the standards clearly makes this point. I believe that the Administrator will do that, and I hope we in Congress will work with the agency and the States to achieve that goal.

I just wanted to say, I think that it's very important that we implement these standards fairly. We must take steps to reduce pollution in a way that fully recognizes the fact that these pollutants are transported from one region to another. NO_x from industrial sources and power plants in other parts of the country continue to be a key contributor to elevated levels of ozone pollution in the Northeast, including my home State of New Jersey, and we can't continue to force businesses, industry, and individuals in the Northeast to unfairly shoulder the burden of cleaning the air simply because that pollution is being blown in from other States or regions.

And I'm, therefore, pleased that the White House memo recognizes the need to address transportation of these pollutants in implementing the new standards. I think the most cost-effective way to implement these standards is to move toward a cap-and-trade system for utilities that are emitting NO_x and fine particles. The cap-and-trade for the utility industry would provide fairness to utilities, balancing those who pollute more against those who pollute less within the industry. Moreover, by tackling the big polluters in this industry first, I believe that we would relieve the burden of compliance for small businesses and other industry, including automobile companies.

In general, I believe that the Clean Air Act's process for setting standards is an appropriate and fair mechanism for making public health decisions. Throughout the existence of the law, this process has led to many different outcomes, including the relaxing of ozone standards. Yet, regardless of the outcome of each individual review, administrators of both parties have made the decisions based upon the scientific knowledge.

In the end, Mr. Chairman, I believe that if we move forward carefully, we can protect and improve the health and economic well-being of all Americans, and I know that the Administrator is trying to accomplish this goal, and I commend her for doing that.

Mr. BILIRAKIS. Ms. Eshoo, for an opening statement.

Ms. ESHOO. I don't have a formal opening statement, Mr. Chairman. My thanks to you for holding this hearing, and a warm welcome to Administrator Browner, and I'm looking forward to hearing her remarks on implementation.

In my view, I think that this is a set of bookends that can and should and will stand. One, the public health being protected, and the other bookend being the economic health of our Nation. Is it difficult to do? Of course. Is it our responsibility to implement, to make it work? Yes. So I'm glad that we're having the hearing today, and I look forward to the Administrator's comments. Thank you.

Mr. BILIRAKIS. I thank the gentlelady. Ms. Furse, for an opening statement, 3 minutes or less, please.

Ms. FURSE. Thank you, Mr. Chairman. I'll introduce a longer statement. But I'm very pleased to have the opportunity for this hearing.

I know that everyone in this room—everyone—given a choice, would choose clean air over dirty air. Certainly, the American people have said they will. But many citizens and businesses rightly have concerns about how we get to cleaner air. Now I believe that if we all apply the same energy and commitment to implementation that we did to the fight over whether or not to have new standards, we could easily meet the challenge for the cleanest air possible. And, Mr. Chairman, I believe that our children and our grandchildren deserve no less.

I look forward to hearing from these qualified witnesses, and, in particular, to Administrator Browner. It is always a pleasure to meet with you, Administrator.

Thank you, Mr. Chairman.

Mr. BILIRAKIS. I thank the gentlelady. Mr. Hall has just come in. Do you have a very short opening statement, less than 3 minutes?

Mr. HALL. No. No, thank you, Mr. Chairman, I won't have an opening statement. I'm really sorry that I missed all the good opening statements that were given.

Mr. BILIRAKIS. And we're sorry to miss yours.

Any further opening statements?

[No response.]

Apparently not. Good.

[Additional statements submitted for the record follow:]

PREPARED STATEMENT OF HON. BRIAN P. BILBRAY, A REPRESENTATIVE IN CONGRESS FROM THE STATE OF CALIFORNIA

Thank you, Mr. Chairman. I'd like to first thank the Chairmen of both the Oversight Subcommittee and the Health and Environment Subcommittee, Mr. Barton and Mr. Bilirakis, for convening this important hearing today. This is an important issue, and in order to see it done correctly, in order to properly protect the public health, there are a number of questions which must be clarified beforehand.

I'd also like to thank Administrator Browner and our other witnesses for agreeing to be with us here today, to help us tackle the complicated questions which have arisen from the President's July guidance memo on implementation of the new NAAQS. Ms. Browner, I look forward to your comments, and eagerly anticipate your answers and clarifications. Before going further, however, I want to make something very clear.

All of us before you here today stand in support of cleaner air, and tough new standards. We all want the American people to breathe the cleanest air possible. That is not in question, and that is not the issue. For myself, I very much appreciate the President's and your efforts to be flexible in how we implement these new NAAQS, and to do so in the most common-sense and practical ways possible. Without question, it is with the utmost practicality that we must proceed, in order to best protect the public health.

But we are here today to seek your help. We need for you to help us understand how this complex implementation program is intended to work, as it is overlaid on to the preexisting infrastructure of state and local pollution control strategies now in place all around the country. I know I have questions and concerns of a very specific nature about how San Diego will be affected—these are legitimate questions to which answers are not readily apparent in the Administration's guidance memo, which does not carry the force of regulation or judicial review.

Here are just a few examples of the concerns which local public health officials have raised to me:

* Local pollution control districts will be making considerable investments in time and resources in order to implement the plan for these new NAAQS. What assurances can they have that the rug will not be pulled from beneath them, as a result of a court interpretation, or a change in policy by EPA? What level of confidence can we provide them as they set this plan in motion?
* Will local pollution control districts be vulnerable to outside lawsuits, because their preexisting permits do not expressly fall under the new NAAQS?

I do not mean to be alarmist, nor do my colleagues. However, it is a formidable plan which you have set before us, and one that cannot be lightly taken up. In order for this to be effective, we have to instill some confidence into those who will be implementing it on the front lines. It would be a disservice to those whose health we claim to represent, to enter into this process without the assurances needed to ensure that it will be effective. Ms. Browner, I'm sure you share these concerns. Before proceeding any further, I would like to have entered into the record, along with my statement, a letter and enclosure from EPA (dated August 7, 1997) responding to a series of questions which I had asked of EPA following our most recent NAAQS hearing May. At that point in time, some of the questions which we will ask here today had started to emerge, and I anticipate that the information which EPA has provided in response to my questions will be useful to the dialogue we will have here today and in subsequent days.

There are two points which I would like to specifically address in the time remaining. The first of these has to do with the issue of ozone transport. In EPA's response to Question #2 from the above mentioned list, it states the following:

"As San Diego prepares a State implementation plan (SIP) to address the revised NAAQS, it will have to account for the seasonal and daily variability of emissions, for ozone and its precursors transported from outside its boundaries, and for variations in industrial and public activities. This accounting will be done through the technical analyses required in the SIP development process in order to demonstrate attainment of the NAAQS. San Diego's ozone problem may be complicated by transported emissions from Mexico. The EPA will consider the impact of transport on international border areas, such as San Diego, in developing the implementation strategy for the revised ozone standard."

I will digress briefly by saying that I appreciate the acknowledgement that San Diego's ozone problem "may be complicated" by Mexico. I look forward to elaborating further on this topic when the Health and Environment Subcommittee holds a field hearing later this fall on my bill to reduce cross-border smog (H.R. 8), and here again, I would hope that I can count on the support of EPA in this regard.

Back to the question at hand—right now, under the current attainment deadlines, San Diego is scheduled to come into attainment in 1999, roughly 10 years before Los Angeles is to reach attainment. Without calling a new SIP, as the implementation plan would require for San Diego, I can tell you that the numbers show very clearly that transported ozone from Los Angeles comprises the bulk of San Diego's ozone nonattainment problem.

So my question here is—how would EPA advise San Diego to proceed, under the new plan, to continue to maintain the high quality of air we now enjoy while transitioning to the new standard? It seems clear that it will be a number of years, according to EPA's own timetable, before attainment with the new NAAQS will be required. In the interim, what will be San Diego's status, and how will the external factors of Mexico and Los Angeles be mitigated?

Pursuing this further, EPA appears to have focused in on "regional control strategies" as an effective means to reach attainment of the new NAAQS. While this might make sense for the 37 OTAG states, it would seem to have limited application to certain areas of the West, such as California, which have fewer stationary sources. If a state chooses not to participate in this regional strategy, will EPA's new "transitional" designation still be allowed to apply to the state?

I would close with a repeat warning on a topic which the Administrator and I have discussed in past hearings, that of precisely where the burden of new source reductions will fall as the new NAAQS are implemented. We have already talked about the concepts of emissions trading and regional control strategies, and while they may be good ideas in and of themselves, I continue to be concerned with what I see as the Administration's emphasis on further reductions from utilities and power plants. The Administrator has stated in previous testimony that further reductions in ozone will come from new controls on utilities.

Mr. Chairman, I have to say this again—we can only get so much more out of our stationary sources. Given the great volume of reductions we hope to achieve over time from these new NAAQS, it is only fair that we are realistic about where these gains will come. Because our stationary sources are so fully tapped now, at least in California, we have to recognize that our mobile sources are going to bear an increasing share of the burden. Because this means that older, more polluting vehicles are going to be singled out as a result, the demographic impact will be greater on lower-middle class working neighborhoods, many of which are going to be disproportionately minority.

If this is the price we as a society must pay for cleaner air, then so be it. But we owe to those we represent to be honest about the cost. It is not realistic to proceed under the assumption that power plants and other stationary sources can continue to deliver the reductions we seek. We have to accept that the cost of clean air is going to begin to involve, on a greater scale, consumer choices and lifestyle decisions.

I will end my statement here, with one final request. I have a number of additional technical questions for the Administrator which I put into letter format (dated September 26, 1997) and sent last week. I would like to have this letter included in the hearing record, and ask that EPA's responses to these questions also be included in the record at the appropriate point. I would also ask that this letter to me from the San Diego Chamber of Commerce, dated today, be included in the hearing record.

With that, Mr. Chairman, I thank you once again for holding this hearing.

GREATER SAN DIEGO CHAMBER OF COMMERCE
October 1, 1997

The Honorable BRIAN BILBRAY
United States House of Representatives
1530 Longworth HOB
Washington, D.C. 20515

RE: New National Ambient Air Quality Standards (NAAQS)

DEAR CONGRESSMAN BILBRAY, On behalf of the 4,200 members of the Greater San Diego Chamber of Commerce, I would like to thank you and your staff for your continued efforts to make some sense of the EPA's new NAAQS and their implementation. Clearly, we here in San Diego and our counterparts throughout the State of California have gone to substantial lengths in attempting to improve our air quality. In fact, the residents and businesses of California, as you are well aware, have had to not only comply with the federal Clean Air Act, we have also had to implement some of the toughest and most comprehensive air quality standards in the country.

With this in mind, the Chamber has met with your staff and staff from our local air quality district to gain a better understanding of the new standards, what they'll mean to San Diego and some sense of the implementation schedule. Unfortunately, the questions were many and the answers were few, especially as it relates to the science used by the EPA in generating these new standards. As your letter to Administrator Browner states, prior to the implementation of any new standards, we here in San Diego would like to have "a more robust understanding of the science involved" and any identifiable health benefits of the NAAQS especially as they relate to the new $PM_{2.5}$ standard as well as the final implementation schedule.

What concerns us here in San Diego is that large scientific gaps exist in the evidence, which has been provided by the EPA to justify the need for the higher standards. Other concerns include the extent to which "transport" will be considered by the EPA in determining an area's attainment status and the severity of its non-attainment problem. As you have been instrumental in proving, San Diego has demonstrated that a great many of its non-attainment days are caused by transport of smog from Los Angeles. The relationship between the modeling evidence developed by the San Diego County APCD and the new EPA standards must be clarified.

Additionally, the tools which the EPA will consider acceptable in analyzing transport and other air quality issues in California should also be identified, including any funding the EPA plans to provide in the implementation of these new standards.

We are in receipt of your letter of comment and concern to Administrator Browner (Dated 9/27/97) and feel that it is directly on point. The EPA must answer these questions if the regulated community, at least in Southern California, is to have any comfort level in working with the new NAAQS. We are also in receipt of the EPA's response to your last letter (dated August 7, 1997). Although the comments help illuminate some of the rationale behind the EPA's thinking, many, if not all, of the answers remain too general and anecdotal to clarify any of our concerns. We encourage your office to continue to pursue specific technical and scientific answers to these questions, including those presented in your letter of September 26, 1997.

In closing, the EPA has yet to demonstrate the immediate need for these new standards. To suggest anecdotally that the public's health is at risk without them is simply political rhetoric that could cost billions of dollars to implement without any health benefit whatsoever. California has achieved a great deal in cleaning its air and is still improving through tough standards on both stationary and mobile source emissions. Unfortunately, the EPA has not seen fit to allow the current efforts to fully mature in order to determine how much further we can reasonably go to provide for additional clean air benefits.

The ultimate answer for Congress is to urge caution. Our U.S. Chamber has pushed for delay in implementation of these new standards. This proposal has a great deal of merit considering the lack of specific answers from the EPA. We do request, in the interim, answers to these specific questions, so the regulated community can better determine which position is appropriate for us to take.

Thank you again for your efforts in this matter. As always, your office has proven to be pragmatic in dealing with these very complex issues. We look forward to continuing to work with you to clarify all of our mutual concerns related to the new NAAQS.

Sincerely,

CRAIG S. BENEDETTO
Vice President

Mr. BILIRAKIS. Madam Administrator, I know it's traditional that you have, or the administration witness have, 10 minutes, and so we'll adhere to that. Obviously, the shorter, the better.

TESTIMONY OF HON. CAROL M. BROWNER, ADMINISTRATOR, ENVIRONMENTAL PROTECTION AGENCY; ACCOMPANIED BY JON CANNON, COUNSEL; ROB BRENNER; RICHARD OSSIAS; BILL HARNETT, AND DICK WILSON

Ms. BROWNER. Yes, I agree. Mr. Chairman, Mr. Chairman, members of the committee——

Mr. BILIRAKIS. I beg your pardon. Our committee doesn't make a practice of swearing in, but the Oversight and Investigations Subcommittee does.

Ms. BROWNER. Do you want to swear in the technical people at the same time, in case——

Mr. BILIRAKIS. I think it might be a good idea if they're going to be possibly testifying.

[Witnesses sworn.]

Mr. BILIRAKIS. Each of you is now under oath. Thank you.

Ms. BROWNER. Thank you. Again, Mr. Chairman, members of the committee, thank you for inviting me here today to discuss EPA's newly updated public health air quality standards for ozone and particulate matter. Joining me today at the table is Jon Cannon, our general counsel for the agency.

These standards, which were announced by the President in June and which I signed in July, are the most significant step we have taken in a generation to protect the American people, most particularly our children, from the health hazards of air pollution. Together, they will protect 125 million Americans, including 35 million children, from the adverse health effects of breathing polluted air. They will prevent approximately 1,500 premature deaths, about 350,000 aggravated asthma attacks, and nearly a million cases of significantly decreased lung function in children. Clearly, the best available science shows that the previous standards were not adequately protecting Americans from the hazards of breathing polluted air. Revising these standards will bring enormous health benefits to the Nation. That is why we took this action on clean air.

Mr. Chairman, these proposed standards, now finalized, are based on the most thorough, the most extensive, scientific review ever conducted by the EPA for any rulemaking. The review of particulate matter has taken nearly a decade, 10 years, and it has been nearly 20 years, two decades, since the last time the ozone standard was thoroughly reviewed. These updated standards are based on more than 250 of the latest, best available studies on ozone and PM, all of them published, all of them peer-reviewed, all of them fully debated, thoroughly analyzed, by an independent, scientific committee, CASAC. We're literally talking peer review of peer review of peer review. It is good science; it is solid science.

Mr. Chairman, EPA has worked very hard to make itself available to Congress to keep Congress apprised of the job we are doing. We have now responded on this issue alone to 258 congressional letters; we have provided 17,400 pages of documents. The requests continue, and we will continue to comply with all of those requests. We have appeared at more than a dozen congressional hearings. I think I heard one member say the count, as of today, is 25 or 26. We will continue to appear at hearings, where requested.

We welcome your interest, and most particularly, we welcome the oversight of this body. We will continue to do whatever we can to satisfy your request.

Let me turn now to EPA's common-sense plan for implementing these new standards and achieving cleaner air in our communities. The plan is designed to give States, local governments, businesses, the flexibility they will need to meet these standards in a reasonable, affordable, cost-effective way. Notice I said "flexibility" to meet these standards. I am not suggesting, nor have I ever suggested, that these new, more protective air quality standards can be achieved effortlessly. We have a lot of work ahead of us, and by "we," I don't mean just the EPA; I mean State, local government, industries, and Congress.

No doubt some of you have heard the allegation that I've been going around the country and disingenuously promising industry after industry that these updated standards are not about them or will have no effect on them. That is simply not the case.

From the beginning, what I have said time and time again on implementation of these standards in hearings before this committee, in meetings with industry officials, in speeches and letters, on the telephone, and so on, has carried essentially the same message:

First, we will continue on the path of progress toward meeting the previous air quality standards, respecting agreements that have already been reached with communities, with States, with businesses—respecting the agreements that we worked hard to develop in partnership with all of these parties. It is not in our interest to undo our hard work of the last 5, 6, 7 years.

Second, no non-attainment designations will be made for particulate matter until the completion of another 5-year review of the health effects of fine particles. No designations; ample opportunity to review the science, as the law passed by this Congress, reaffirmed by President Bush, President Carter, promises the American people.

Third, the necessary reductions in ozone-forming emissions will come from large industrial sources such as major power plants, not from small businesses, from large, industrial sources, from large industrial smokestacks, from those who perhaps have not done as much as others to reduce their pollution of the public's air. This is where we will focus because, quite frankly, this is where the lion's share of the problem is coming from. This is where we can achieve a cost-effective solution and provide the public with cleaner air. This is where we stand to gain the most in our efforts to reduce the regional, the transported ozone that so many of you have asked about and spoken to. We now understand, the States understand, that the transport of ozone is a major part of the small problem in many metropolitan areas, and it is by far the most cost-effective way to address the problem of ozone.

Our next phase of implementation in this effort is a regional strategy. It was developed collectively by 37 States. This is not EPA's strategy; it is the strategy of your States. They worked for 2 years. They brought us this idea.

It is designed to target major utilities for pollution reductions through a market-based cap-and-trade program. Soon EPA will ask the States to submit their plans for achieving the necessary ozone reductions through their strategy, through their regional approach. Once this plan is given a chance to work, we believe that the vast majority of cities that based on current data, would not meet a more protective health standard, will be able to do so through this State-designed strategy without any additional, new, local pollution controls or measures.

In fact, we are so confident that this will happen that we will not be treating as traditional non-attainment the States, the areas that fail to meet the new, updated standard for ozone, so long as they participate in the regional program they designed and asked us to adopt. This is the way government should work, partnerships—

Federal, State, local government—ideas, flexibility, innovation, ingenuity. This is what you have impressed upon me over the last 4½ years, and we now bring it back to you. We have done precisely what you asked us to do in looking at how to find better ways to protect the public's health, to reduce the pollution in our air, our water, on our land.

The States will receive a transitional classification, one that has been carefully crafted under the authority provided by the law on the books today, the Clean Air Act, as amended by this body in 1990. This classification will enable them to avoid undue local planning requirements and the restrictions on economic growth, which many of you have properly asked about. This plan is responsive to those concerns, to those questions.

So what will this mean for electricity consumers in the Midwest and elsewhere? A question we have often heard. Well, let me tell you what a utility said. Don't take my word for it; let me tell you what a utility said. They issued a report this week.

In a recent study conducted by Public Service Electric and Gas, a New Jersey-based utility company, the cost of complying with the cap-and-trade program for ozone-causing emissions will be only a small fraction, 3 to 5 percent of the amount of money the utilities will save—of the amount of money utilities will save—through deregulation. In other words, the consumers can see a savings in their utility bill and cleaner air. Now that's a good deal.

The fact is that the technology to reduce ozone-causing emissions exists. It can be done cost-effectively. It is now available.

Let me just give you one other utility as an example.

Mr. BILIRAKIS. Your time has expired, Madam Administrator. Please finish your statement, though. Hopefully, with all of these questioners, you're going to be able——

Ms. BROWNER. Last time I think I stayed until 5. I'm happy to stay as long as you need me.

Mr. BILIRAKIS. Maybe 5 or later.

Ms. BROWNER. Was it later than 5?

Mr. BILIRAKIS. Yes, I think it was. Please finish.

Ms. BROWNER. If I can just give one more utility example—this is not EPA; this is utilities out there in your States. One of the Nation's largest coal-fired utilities, based in Ohio, last week announced, that it will make major reductions in its emissions in advance of any trading requirement. This is a utility that has recognized the cost advantages of newly available technologies, as well as the importance of a regional approach to reducing ozone-causing emissions.

The point is, this is not pie in the sky. This is a sensible implementation plan that will allow us to do what all of us agree needs to be done: Protect the public's health through common-sense, cost-effective solutions.

[The prepared statement of Hon. Carol M. Browner follows:]

PREPARED STATEMENT OF HON. CAROL M. BROWNER, ADMINISTRATOR, ENVIRONMENTAL PROTECTION AGENCY

Messrs. Chairmen, Members of the Subcommittees, thank you for inviting me to discuss implementation plans for the Environmental Protection Agency's (EPA's) revisions to the national ambient air quality standards for ground-level ozone and particulate matter.

As you know, the Clean Air Act directs EPA to set national standards for certain air pollutants to protect public health and the environment. For each of these pollutants, Congress directed EPA to set what are known as "primary" standards to protect public health without consideration of cost. Under the Act, Congress directs EPA to review these standards every five years to determine whether the latest scientific research indicates a need to revise the standards.

In July of this year, I set new standards for ozone and particulate matter that will be a major step forward in public health and welfare protection. Each year, these updated standards have the potential to prevent as many as 15,000 premature deaths, and hundreds of thousands of cases of significantly decreased lung function in children and cases of aggravated asthma.

Numerous other public health and welfare benefits will result from implementation of the new standards. Additional public health benefits would include: reduced respiratory illnesses, reduced acute health effects, reduced cancer from air toxics reductions, and various other public health benefits. Public welfare benefits will include: reduced adverse effects on vegetation, forests, and natural ecosystems, improved visibility, and protection of sensitive waterways and estuaries from deposition of airborne nitrogen that can cause algal blooms, fish kills, and loss of aquatic vegetation. Estimated total monetized health and public welfare benefits associated with the new standards are enormous, ranging in the tens of billions of dollars annually. Many additional potentially large benefit categories, such as reduced chronic respiratory damage, infant mortality, and other health and welfare benefit categories, cannot be monetized.

The new ozone and particulate matter standards are based on an extensive scientific and public review process. Congress directs EPA to consult with an independent scientific advisory board, the Clean Air Scientific Advisory Committee (CASAC). In conducting these reviews, EPA analyzed thousands of peer-reviewed scientific studies that had been published in well-respected scientific journals. These studies were then synthesized and, along with a recommendation on whether the existing standards were adequately protective, presented to CASAC. After three-and-a-half years of work, 11 meetings totaling more than 125 hours of public discussion, and based on 250 of the most relevant studies, the CASAC panel concluded that EPA's air quality standards for ozone and particulate matter should be revised. CASAC unanimously supported changing the ozone standards from a 1-hour averaging period to an 8-hour average to reflect increasing concern over prolonged exposure to ozone. CASAC also supported adding a fine particle standard. Fine particles are inhaled deeply into the lungs and are more strongly associated with serious health effects and visibility impairment than larger particles.

Based on scientific evidence reviewed by EPA and CASAC, EPA proposed revised standards and conducted an extensive public comment process, receiving approximately 57,000 comments at public hearings across the country and through written, telephone and E-mail message communications.

After carefully considering the results of this extensive process, and with the support of the President, I issued a final rule updating the ozone standard from 0.12 parts per million (ppm) of ozone measured over one hour to a standard of 0.08 ppm measured over eight hours, with the three-year average of the annual fourth highest concentration determining whether an area is out of compliance. The new standard will reduce "flip-flopping" in and out of attainment by changing it from an "expected exceedance" to a "concentration-based" form.

For particulate matter, EPA has added new standards for particles smaller than 2.5 micrometers in diameter (known as "$PM_{2.5}$" or fine particles). The fine particle standard has two components: an annual standard, set at 15 micrograms per cubic meter and a 24-hour standard, set at 65 micrograms per cubic meter. EPA has also changed the form of the current 24-hour PM_{10} standard; this will provide some additional stability and flexibility to states in meeting that standard.

We believe it is critical to move forward with these standards now. The American public deserves to know whether its air is healthy or not. The standards we have set serve as an essential benchmark for people to use in understanding whether the air they are breathing is safe. In addition, the implementation plan for the standards will encourage early action to help reduce adverse health effects as soon as possible. By setting the standards, states will now be able to proceed with the monitoring and planning requirements over the next several years. $PM_{2.5}$ areas can begin to develop inventories and characterize the nature of their $PM_{2.5}$ problem. As I will now discuss, the President has directed EPA to undertake an implementation strategy that has been developed through an extensive interagency consultative process to assure that concerns of state and local governments and affected industries, such as transportation and agriculture, are addressed. This strategy will allow states and local areas the time they need to implement these standards in a cost-effective and reasonable way.

Implementation of the Revised Air Standards

In evaluating the changes to these air quality standards, I believed that it was critical to develop a common sense implementation plan. Because of the vital importance to states, cities, and industry, we believed that early implementation guidance should be given and that implementation be done in a flexible, cost-effective way. In the interagency process leading up to the issuance of these standards, EPA worked with other federal agencies to develop a strategy for implementing the standards. In a memorandum signed July 16, 1997, President Clinton set forth several general principles for implementing the standards, and directed EPA to follow the interagency implementation strategy. I would like to summarize the principal features of that strategy for you today.

Achieving the air quality benefits of the updated standards requires a common sense, cost-effective means for communities and businesses to meet the standards. We believe it is important that these standards be implemented in the most flexible, reasonable and least burdensome manner. The President's implementation package has four basic features, all of which can be carried out under existing legal authority:

"1. Implementation of the air quality standards is to be carried out to maximize common sense, flexibility, and cost effectiveness;

2. Implementation shall ensure that the Nation continues its progress toward cleaner air by respecting the agreements already made by States, communities, and businesses to clean up the air, and by avoiding additional burdens with respect to the beneficial measures already underway in many areas. Implementation also shall be structured to reward State and local governments that take early action to provide clean air to their residents; and to respond to the fact that pollution travels hundreds of miles and crosses many State lines;

3. Implementation shall ensure that the Environmental Protection Agency ('Agency') completes its next periodic review of particulate matter, including review by the Clean Air Scientific Advisory Committee, within 5 years of issuance of the new standards, as contemplated by the Clean Air Act. Thus, by July 2002, the Agency will have determined, based on data available from its review, whether to revise or maintain the standards. This determination will have been made before any areas have been designated as 'nonattainment' under the $PM_{2.5}$ standards and before imposition of any new controls related to the $PM_{2.5}$ standards; and

4. Implementation is to be accomplished with the minimum amount ofpaperwork and shall seek to reduce current paperwork requirements wherever possible."

Strategy for Meeting the Revised Ozone Standard

Ozone and ozone precursors travel great distances and it is increasingly important to address them as a regional problem. For the past two years, EPA has been working with the 37 most eastern states through the Ozone Transport Assessment Group (OTAG) in the belief that reducing interstate pollution will help all areas in the OTAG region attain the NAAQS. A regional approach can reduce compliance costs and allow areas to avoid most traditional local nonattainment planning requirements. The OTAG was an effort sponsored by the Environmental Council of States, with the objective of assessing ozone transport and recommending strategies for mitigating interstate pollution.

The OTAG completed its work in June 1997 and forwarded recommendations to EPA. Based on these recommendations, EPA will soon propose a rule requiring states in the OTAG region that are significantly contributing to nonattainment, or interfering with maintenance of attainment, in downwind states to submit state implementation plans (SIPs) to reduce their interstate pollution. EPA will issue the final rule by September 1998.

EPA will encourage and assist the states to develop and implement a cap and trade program for nitrogen oxides (NO_X), including developing a model program with the states. A regional emissions cap and trade system, similar to the current acid rain program, is expected to achieve cost-effective reductions for meeting the new standards. Most important, based on EPA's review of the latest modeling, a regional approach, coupled with the implementation of other already existing state and Federal Clean Air Act requirements, will allow the vast majority of areas that currently meet the 1-hour standard but would not otherwise meet the new 8-hour ozone standard to achieve healthful air quality without further local controls.

Areas in the OTAG region that would still exceed the new standard after the regional strategy, including areas that do not meet the current 1-hour standard, will benefit as well, because the regional NO_X program will reduce the extent of additional local measures needed to achieve the 8-hour standard. In many cases these regional reductions may be adequate to meet CAA progress requirements for a number of years, allowing areas to defer additional local controls.

Phase-out of 1-hour Ozone Standards

EPA's revised ozone standard will replace the current 1-hour standard with an 8-hour standard. However, the 1-hour standard will continue to apply to areas not attaining it for an interim period to ensure an effective transition to the new 8-hour standard.

As you know, the Clean Air Act includes provisions (Subpart 2 of part D of Title I) that address requirements for different nonattainment areas that do not meet the 1-hour standard (i.e., those classified as marginal, moderate, serious, severe and extreme). These requirements include such items as mandatory control measures, annual rate of progress requirements for emission reductions and emission offset requirements for new sources. All of these requirements have contributed significantly to the improvements in air quality since 1990. Although EPA initially proposed an interpretation of the Clean Air Act that would have been more flexible in how these provisions applied to existing ozone nonattainment areas after promulgation of a new ozone standard, based on comments received, EPA has reconsidered its interpretation and EPA has concluded that these provisions should continue to apply as a matter of law for the purpose of achieving attainment of the 1-hour standard. Once an area attains the 1-hour standard, those provisions and the 1-hour standard will no longer apply to that area. An area's implementation of the new 8-hour standard would then be governed by the provisions of Subpart 1 of Part D of Title I.

The purpose of retaining the current standard is to ensure a smooth transition to the new standard. It is important not to disrupt the controls that are currently in place as well as those that are underway to meet the current ozone standard. These controls will continue to be important for reaching the new 8-hour standard.

General Time Line for Meeting the Ozone Standard

Following promulgation of a revised NAAQS, the Clean Air Act provides up to three years for state governors to recommend and EPA to designate areas according to their most recent air quality. In addition, states will have up to three years from designation to develop and submit SIPs to provide for attainment of the new standard. Under this approach, areas would be designated as nonattainment for the 8-hour standard by July 2000 and would be required to submit their nonattainment SIP by July 2003. The Act allows up to 10 years plus two 1-year extensions from the date of designation for areas to attain the revised NAAQS.

Transitional Classification

For areas that attain the 1-hour standard but not the new 8-hour standard, EPA will follow a flexible implementation approach that encourages cleaner air sooner, responds to the fact that ozone is a regional as well as local problem, and eliminates unnecessary planning and regulatory burdens for state and local governments. A primary element of the plan will be the establishment under Section 172(a)(1) of the CAA of a special "transitional" classification for areas that participate in a regional strategy and that opt to submit early plans to attain the new 8-hour standard. Because many areas will need little or no additional new local emission reductions to reach attainment, beyond those reductions that will be achieved through the regional control strategy, and will come into attainment earlier than they otherwise might be required, EPA will exercise its discretion under the law to eliminate unnecessary local planning requirements for such areas. EPA will revise its rules for new source review (NSR) and conformity so that states will be able to comply with only minor revisions to their existing programs in areas classified as transitional. During this rulemaking, EPA will also reexamine the NSR requirements applicable to existing nonattainment areas, in order to deal with issues of fairness among existing and new nonattainment areas. The transitional classification would be available for any area attaining the 1-hour standard but not attaining the 8-hour standard as of the time EPA promulgates designations for the 8-hour standard. In terms of process, areas would follow the approaches described below based on their status.

(1) Areas attaining the 1-hour standard, but not attaining the 8-hour standard, that would attain the 8-hour standard through the implementation of the regional NO_X transport strategy for the East.

Based on the OTAG analyses, areas in the OTAG region that would reach attainment through implementation of the regional transport strategy would not be required to adopt and implement additional local measures. When EPA designates these areas under section 107(d), it will place them in the new transitional classification if they would attain the standard through implementation of the regional transport strategy and are in a state that by 2000 submits an implementation plan that includes control measures to achieve the emission reductions required by EPA's rule for states in the OTAG region. This is three years earlier than an attainment SIP would otherwise be required. We anticipate that we will be able to determine whether such areas will attain the revised ozone standard based on the OTAG and other regional modeling and that no additional local modeling would be required.

(2) Areas attaining the 1-hour standard but not attaining the 8-hour standard for which a regional transport strategy is not sufficient for attainment of the 8-hour standard.

To encourage early planning and attainment for the 8-hour standard, EPA will make the transitional classification available to areas not attaining the 8-hour standard that will need additional local measures beyond the regional transport strategy, as well as to areas that are not affected by the regional transport strategy, provided they meet certain criteria. To receive the transitional classification, these areas must submit an attainment SIP prior to the designation and classification process in 2000. The SIP must demonstrate attainment of the 8-hour standard and provide for the implementation of the necessary emissions reductions on the same time schedule as the regional transport reductions.

(3) Areas not attaining the 1-hour standard and not attaining the 8-hour standard.

The majority of areas not attaining the 1-hour standard have made substantial progress in evaluating their air quality problems and developing plans to reduce emissions of ozone-causing pollutants. These areas would be eligible for the transitional classification provided that they attain the 1-hour standard by the year 2000 and comply with EPA's regional transport rule, as applicable.

Areas not Eligible for the Transitional Classification

Existing nonattainment areas which cannot attain the 1-hour standards by 2000 will not be eligible for the transitional classification. However, their work on planning and control programs to meet the 1-hour standard by their current attainment date will take them a long way toward meeting the 8-hour standard. While areas will need to submit an implementation plan for achieving the 8-hour standard within three years of designation as nonattainment for the new standard, such a plan can rely in large part on measures needed to attain the 1-hour standard. For virtually all of these areas, no additional local control measures beyond those needed to meet the requirements of Subpart 2 and needed in response to the regional transport strategy would be required to be implemented prior to their applicable attainment date for the 1-hour standard. This approach allows them to make continued progress toward attaining the 8-hour standard throughout the entire period without requiring new additional local controls for attaining the 8-hour standard until the 1-hour standard is attained.

Implementing the New Particulate Matter Standards

Implementing the new particulate matter standards will require a different path from the one I just discussed for ozone. As required under the Act, within the next 5 years EPA will complete the next periodic review of the particulate matter criteria and standards, including review by the CASAC. As with all NAAQS reviews, the purpose is to update the pertinent scientific and technical information and to determine whether it is appropriate to revise the standards in order to protect the public health with an adequate margin of safety or to protect the public welfare. EPA has concluded that the current scientific knowledge provides a strong basis for the revised PM_{10} and new $PM_{2.5}$ standards. We, along with the Departments of Transportation, Health and Human Services, Energy, and others, will continue to sponsor research to better understand the causes and mechanisms of fine particles effects on human health, and the species and sources of $PM_{2.5}$. EPA will also promptly initiate a new review of the scientific criteria on the effects of airborne particles on human health and the environment. By July 2002, we will have determined, based on data available from its review, whether to revise or maintain the standards. This determination will have been made before any areas have been designated nonattainment under the $PM_{2.5}$ standards and before imposition of any new controls related to the $PM_{2.5}$ standards.

Implementation of New $PM_{2.5}$ NAAQS

The first priority for implementing the new $PM_{2.5}$ standard is establishing a comprehensive monitoring network to determine ambient fine particle concentrations across the country. The monitoring network will help EPA and the states determine which areas do not meet the new air quality standards, what the major sources of

PM$_{2.5}$ in various regions are, and what action is needed to clean up the air. EPA and the states will consult with affected stakeholders on the design of the network and will then establish the network, which will consist of approximately 1,500 monitors. All monitors will provide for limited "speciation," or analysis of the chemical composition, of the particles measured. At least 50 of the monitors will provide for a more comprehensive speciation of the particles. EPA will work with states to deploy the PM$_{2.5}$ monitoring network. Our intent will be to work with states to ensure that monitors are placed in urban areas with the most significant population exposures and generally not in agricultural areas. The EPA will fund the cost of purchasing the monitors, as well as the cost of analyzing particles collected at the monitors to determine their chemical composition.

Because we are establishing standards for a new indicator for particulate matter (i.e., PM$_{2.5}$), it is critical to develop the best information possible before attainment and nonattainment designation decisions are made. Three calendar years of federal reference monitoring data will be used to determine whether an area does or does not attain the new PM$_{2.5}$ standards. In view of the time needed to establish the network and collect data, EPA expects that three years of PM$_{2.5}$ monitoring data will not be available until between 2001 and 2004, depending on when monitors are installed in a given locality. Therefore, actual designations of attainment or nonattainment will not take place until between 2002 and 2005. The Clean Air Act, however, requires that EPA make designation determinations (i.e., attainment, nonattainment, or unclassifiable) within two to three years of revising a NAAQS. To fulfill this requirement, in 1999 EPA will issue "unclassifiable" designations for PM$_{2.5}$ These designations will not trigger the nonattainment planning or control requirements of Title I of the Act.

When EPA designates nonattainment areas for PM$_{2.5}$ pursuant to the governors' recommendations beginning in 2002, areas will be allowed three years to develop and submit to EPA pollution control plans showing how they will meet the new standards. As for ozone, areas will have up to 10 years from the date of being redesignated as nonattainment until they will have to attain the PM$_{2.5}$ standards. In addition, two 1-year extensions are possible.

In developing strategies for attaining the PM$_{2.5}$ standards, it will be important to focus on measures that decrease emissions that contribute to regional pollution. Available information indicates that nearly one-third of the areas projected to not meet the new PM$_{2.5}$ standards, primarily in the Eastern U.S., could come into compliance as a result of the regional SO2 emission reductions already mandated under the Clean Air Act's acid rain program, which will be fully implemented between 2000 and 2010. Similarly, the Grand Canyon Visibility Transport Commission, consisting of western states and tribes, committed to reductions in regional emissions of PM$_{2.5}$ precursors (sulfates, nitrates, and organics) to improve visibility across the Colorado Plateau.

As detailed PM$_{2.5}$ air quality data and data on the chemical composition of PM$_{2.5}$ in different areas become available, EPA will work with the states to analyze regional strategies that could reduce PM$_{2.5}$ levels. If further cost-effective regional reductions help areas meet the new standard, EPA will encourage states to work together to use a cap and trade approach similar to that used to curb acid rain. The acid rain program delivered environmental benefits at a greatly reduced cost than originally anticipated by EPA and industry.

Given the regional dimensions of the PM$_{2.5}$ problem, local governments and local businesses should not be required to undertake unnecessary planning and local regulatory measures when the problem requires action on a regional basis. Therefore, as long as the states are doing their part to carry out regional reduction programs, the areas that would attain the PM$_{2.5}$ standards based on full implementation of the acid rain program will not face new local requirements. Early identification of other regional strategies could also assist local areas in completing their programs to attain the PM$_{2.5}$ standards after those areas have been designated nonattainment.

The EPA will also encourage states to coordinate their PM$_{2.5}$ control strategy development and efforts to protect regional visibility. Visibility monitoring and data analysis will support both PM$_{2.5}$ implementation and the visibility program.

Implementation of Revised PM$_{10}$ NAAQS

In its rule, EPA revised the current set of PM$_{10}$ standards. Given that health effects from coarse particles are still of concern, the overall goal during this transition period is to ensure that PM$_{10}$ control measures remain in place to maintain the progress that has been achieved toward attainment of the current PM$_{10}$ NAAQS (and which provides benefits for PM$_{2.5}$) and protection of public health. To ensure that this goal is met, the existing PM$_{10}$ NAAQS will continue to apply until actions by EPA, and by states and local agencies, are taken to sustain the progress already made.

Cost-Effective Implementation Strategies

Consistent with states' ultimate responsibility to attain the standards, EPA will encourage the states to design strategies for attaining the particulate matter and ozone standards that focus on getting low cost reductions and limiting the cost of control to under $10,000 per ton for all sources, which is the high end of the range of reasonable cost to impose on sources. Market-based strategies can be used to reduce compliance costs. EPA will work with states to develop a model program for a Clean Air Investment Fund, which would allow sources facing control costs higher than $10,000 a ton for any of these pollutants to pay a set annual amount per ton to fund cost-effective emissions reductions from non-traditional and small sources. Compliance strategies like this will likely lower the costs of attaining the standards through more efficient allocation, minimize the regulatory burden for small and large pollution sources, and serve to stimulate technology innovation as well.

Future Activities

In accordance with the President's July 16th directive, to ensure that the final details of the implementation strategy are practical, incorporate common sense, and provide for appropriate steps toward cleaning the air, input is needed from many stakeholders including representatives of state and local governments, industry, environmental groups, and Federal agencies. EPA will continue seeking advice from a range of stakeholders and, after evaluating their input, propose the necessary guidance to make these approaches work. In particular, EPA will continue working with the Subcommittee on Implementation of Ozone, Particulate Matter and Regional Haze Rules which EPA established to help develop innovative, flexible and cost-effective implementation strategies. Moreover, EPA will continue to work with a number of Federal agencies to ensure that those agencies comply with these new standards in cost-effective, common sense ways. EPA plans to issue all guidance and rules necessary for this implementation strategy by the end of 1998.

EPA will continue to work with the Small Business Administration (SBA) because small businesses are particularly concerned about the potential impact resulting from future control measures to meet the revised PM and ozone standards. EPA, in partnership with SBA, will work with the states to include in their SIPs flexible regulatory alternatives which minimize the economic impact and paperwork burden on small businesses to the greatest possible degree consistent with public health protection.

Legal Authority for Implementation

Messrs. Chairmen, you asked that I address the legal basis for the Administration's implementation strategy for these revised standards. This implementation strategy was subject to careful legal review by EPA and EPA believes that it is properly based within the authority Congress has provided under the Clean Air Act, and general principles of administrative law and statutory construction. As you are aware, Title I of the Clean Air Act establishes the steps that EPA and States must take once a new national ambient air quality standard is established. Under section 107 of the Clean Air Act, EPA must designate areas two years after promulgation of a new or revised standard, but may take an additional year if there is insufficient information to promulgate the designations. For areas designated as nonattainment, the general planning provisions of Subpart 1 of Part D of the Act apply. In particular, section 172(b) requires EPA to establish a deadline not later than 3 years from the date an area is designated for states to submit plans showing how they will meet the new or revised standards. Subpart 1 provides ample room for flexibility in implementation. EPA has historically developed policy and guidance documents to further delineate these and other planning requirements for states.

As discussed earlier, part of the Administrations's approach to implementing the new 8-hour ozone standard will include a "transitional" classification. Section 172(a) provides EPA with discretion to create classifications for areas designated as nonattainment. Classifications may be created for the purpose of establishing attainment dates or for other purposes. In developing such a classification scheme, EPA has the authority under the Clean Air Act and generally applicable legal principles to interpret and apply the provisions of Subpart 1, including new source review and transportation conformity, in a way that recognizes the particular circumstances of areas. As previously explained, areas that have attained the 1-hour ozone standard, but that have air quality that violates the 8-hour standard, will be eligible for this transitional classification.

Conclusions

In summary, EPA believes that the new ozone and particulate matter standards will provide important new health protection and will improve the lives of Americans in coming years. Our implementation strategy will ensure that these new standards are implemented in a common sense, cost-effective and flexible manner. We intend to work closely with state and local governments, other Federal agencies and all other interested parties to accomplish this goal.

Messrs. Chairmen, this concludes my written statement. I will be happy to answer any questions that you might have.

Mr. BILIRAKIS. Thank you. Thank you, Madam Administrator.

I'll start off the questioning. I applaud your recognition that increased flexibility and relief from certain onerous requirements are a common-sense—I believe were your words and mine—approach to implementation. The Clean Air Act, however, contains numerous—and this is very, very significant, I think—contains numerous, mandatory requirements and deadlines that might preclude EPA from exercising such flexibility. When you were questioned on this issue before the Agriculture Committee on September 16, you stated that EPA had, and again I quote, "had structured a legal argument with the Justice Department" regarding the implementation plan. I'm not certain what a structured legal argument is, but, according to the information provided to the committee, the Justice Department has not supplied you with a written opinion, nor has EPA developed a written legal opinion addressing this matter.

So I guess the question is, how can States and Congress evaluate your statements regarding EPA's legal authority if EPA's has not even taken the time to put these arguments in writing?

Ms. BROWNER. The process that we use is probably similar to a process you would use in crafting legislation. You don't simply come up with an idea and then go to the lawyers. The lawyers are part and parcel of crafting the plan. They are the keepers of the law. In this case, I personally met with our lawyers for I don't know how many hours. They had meetings with the Justice Department. It wasn't simply a question of one side of the agency coming up with an idea and then saying to the other side of the agency, "What do you think?" We did this as a team—our lawyers, our technical experts, our air experts—and that is the way government should function. It's common sense.

Mr. BILIRAKIS. Is much of this documented in papers back and forth, containing that sort of legal opinion, et cetera?

Ms. BROWNER. We're more than happy to provide a list of the various meetings.

Mr. BILIRAKIS. We requested that.

Ms. BROWNER. The meetings that we had? Oh, we're more than——

Mr. BILIRAKIS. Documents.

Ms. BROWNER. Mr. Chairman, we will give you any document you want. We have had a lengthy process within the agency, across the administration, including the Justice Department, to look at how, within the confines of the law on the books, to honor our commitment to flexible implementation. That is what we have done, and we have involved all of the experts within our agency and at the Justice Department in doing that, and we've already provided 17,000 pages of documents. We're happy to provide more.

Mr. BILIRAKIS. Well, Madam Administrator, you say that, but, as I understand it, that has not taken place. Now you've said that this is the biggest and most important EPA rule of your administration, of our generation. So the question is: Isn't it big enough to have some of your lawyers prepare a legal opinion? Now we're very much concerned about that. In fact, we're almost in the position here of trying to help you as far as this implementation is concerned.

Ms. BROWNER. But, Mr. Chairman——

Mr. BILIRAKIS. But there has to be legal authority.

Ms. BROWNER. And there is, and we——

Mr. BILIRAKIS. But not on the basis—forgive me—but not on the basis of your saying that there's legal authority. I don't think you have to be a lawyer to realize that these things have to be in writing.

Ms. BROWNER. Well, we have done this to whatever degree we have been asked, and we do have some requests pending which came in very late, which we are seeking to respond to as quickly as possible—we're more than happy to go through and say to you, for example, this legal authority, this section of the Clean Air Act, is where we draw our basis for this flexibility. We're more than happy to do that.

Mr. BILIRAKIS. Well, as I understand it, it was 1 week ago, and I guess we will say that maybe that was a little late, but we sent you a three-page letter containing four basic questions regarding EPA's legal authority for the implementation plan—pretty basic and awfully foundational. And your Associate Administrator said that EPA could not give us answers to those four questions because a response, "requires consultation with the Department of Justice, and EPA is engaged in that consultation process now." So if EPA has already worked out its legal authority for the plan, why does it take more than 1 week to answer four questions?

Ms. BROWNER. We have been working with your staff to ensure that we give you a thorough response, and you will have it. There's nothing going on here except that you want us to, I presume, talk to all of the parties that were part of crafting this plan, of looking within the law, finding the flexibility within the law. That's all we're doing.

Mr. BILIRAKIS. Well, my time's up. I'm just going to ask for any legal opinions from the EPA lawyers and from the Department of Justice supporting your implementation plan. Now when might we receive those documents?

Ms. BROWNER. We will provide you all of the documents you have sought, but I really—I feel very strongly about making a point here, with all due respect, Mr. Chairman. If I hadn't had the lawyers meeting with me in some instances on a daily basis for numerous hours as we looked at how best to structure the implementation plan, I would be here and you would be criticizing me for not having them. Now you're criticizing me in some way for having them. You know, we tried to do this in a full, comprehensive manner——

Mr. BILIRAKIS. I think it's basic—forgive me—I don't think I've been really all that critical. You tell us that you have legal justification and you have your legal opinions, and what this committee is saying is, Can you furnish us with copies of all this?

Ms. BROWNER. We'll furnish you with every piece of paper we have——

Mr. BILIRAKIS. So that we know that your implementation plan is something, with all of its flexibility, is something that in fact will take place or can take place, rather than be subject to attack after attack after attack in the future. Now I think that's really simple.

Ms. BROWNER. If I might suggest, if it would be helpful to you— I don't know if this is the precise nature of the request that is pending, but if it would be helpful for you, to you and the other members, for us to cite the sections of the Clean Air Act which we rely on in the implementation plan, we will certainly do that. We will look forward to doing that.

Mr. BILIRAKIS. We have asked for, and I'm going to yield to Mr. Dingell in a moment, and maybe he'll follow up on this; I don't know; it's up to him, his 5 minutes.

But we've asked for your written legal opinions from your EPA lawyers and from the Department of Justice supporting the implementation plan.

Ms. BROWNER. We've given you 17,400 pages of documents; we'll give you anything else you want.

Mr. BILIRAKIS. But nothing regarding what I have just referred to.

Ms. BROWNER. I've said to you, I will give you a Clean Air Act section-by-section response—I don't know what else you want. We're more than happy to show you which sections of the Clean Air Act we rely on——

Mr. BILIRAKIS. Is this from the Justice Department and from your—and legal opinions from your lawyers?

Ms. BROWNER. Justice——

Mr. BILIRAKIS. Is this what you're referring to? You say, "We will give you"—who is "we?"

Ms. BROWNER. The EPA, and we're more than happy to do that in discussions with the Justice Department; they have been part and parcel of our discussions in crafting the implementation plan, and we're more than happy to do that if you would like us to do so.

Mr. BILIRAKIS. Can you give us a date?

Ms. BROWNER. We would. Part of the section-by-section is, I think, part of the response to the pending request. We will go ahead and get that to you in the timeframe we've agreed to, and then we will supplement that, based on this hearing and this request.

Mr. BILIRAKIS. I——

Ms. BROWNER. We'll do the whole thing in 2 weeks. You're going to get a piece of it sooner, was my point——

Mr. BILIRAKIS. We're going to get a piece of it by October 6 and the rest of it by—within another week?

Ms. BROWNER. That's fine. That's good. That's fine. Thank you. [The information provided is printed following the hearing.]

Mr. BILIRAKIS. All right, I will then yield to—it's convenient now for me to yield to you, sir.

Mr. DINGELL. I've heard a great deal from you this morning, but is there a piece of paper down there, either from your lawyers or the Department of Justice, which satisfies the requirements of you being properly informed of how this is going to work?

Ms. BROWNER. I met with our lawyers day-in and day-out.

Mr. DINGELL. I asked for a piece of paper; I didn't ask if you met with lawyers.

Ms. BROWNER. I asked——

Mr. DINGELL. Please answer the question.

Ms. BROWNER. There are pieces of paper; we've provided them. We'll provide any additional ones the committee may want.

Mr. DINGELL. So there is paper?

Ms. BROWNER. Mr. Dingell, we have provided 17——

Mr. DINGELL. From your lawyers and Department of Justice lawyers? Let's——

Ms. BROWNER. I don't know what the Department of Justice has. I can't——

Mr. DINGELL. Let us—let us focus on the question of opinions of lawyers, yours and the Department of Justice. Are there such kinds of written opinions on this matter there in EPA's files?

Ms. BROWNER. We'll give you whatever is in our files.

Mr. DINGELL. Yes or no?

Ms. BROWNER. I don't know the answer to that, with all due respect.

Mr. DINGELL. You have been meeting with these lawyers, and you have—have they presented you with written opinions on these matters?

Ms. BROWNER. You don't have a meeting and craft a plan and demand a written opinion; you work in an iterative process to get the best possible result.

Mr. DINGELL. Are you telling us that you have never seen a written opinion from your lawyers or from the Department of Justice attorneys on these matters?

Ms. BROWNER. The Department—the EPA lawyers were part and parcel of crafting the implementation plan——

Mr. DINGELL. Would you just answer this question, please, yes or no?

Ms. BROWNER. Do I have any paper? You have all the paper I have——

Mr. DINGELL. You have seen or you have not seen opinions from your lawyers or Department of Justice lawyers on these matters?

Ms. BROWNER. I have met with my lawyers——

Mr. DINGELL. Have you seen——

Ms. BROWNER. [continuing] frequently, and they have given me their opinion, and I have that.

Mr. DINGELL. [continuing] a piece of paper?

Ms. BROWNER. I don't know what is in writing. We're more than happy to give you whatever we have.

Mr. DINGELL. All right. Mr. Chairman——

Ms. BROWNER. We've given you much, and we'll give you more.

Mr. DINGELL. Now, Ms. Browner, I understand your testimony and the implementation plan EPA has put forward. EPA plans to dub some ozone attainment areas as transitional, not attainment areas. Would you give me a citation at this time of the specific sections of the act which set forth EPA's authority to set up transitional non-attainment areas? Please cite the section.

Ms. BROWNER. Yes, if I might ask the general counsel to read you those cites——

Mr. DINGELL. Please cite the section.

Mr. CANNON. Section 172(a)(1)——

Mr. DINGELL. All right.

Mr. CANNON. [continuing] of the Clean Air Act provides for the creation of——

Mr. DINGELL. All right, let us address—let us, because my time is limited, let us then address that section. This is a general reference, is it not, to creating classifications?

Ms. BROWNER. The section is referred to as non-attainment plan provisions classifications, and it gives us the authority to create classifications.

Mr. DINGELL. Does—please cite the language in that section which gives you and EPA authority to relax specific statutory requirements for non-attainment areas set out in sections 172 and 173 of the act. Please cite the specific authority which you are given in that section which you have cited to relax all or any of the requirements for non-attainment areas set out in sections 172 and 173. I await your quote.

Ms. BROWNER. We're more than happy to read——

Mr. DINGELL. I await your quote now.

Ms. BROWNER. I'll read it to you.

Mr. DINGELL. Please tell me where that language is in the section that you have just referred to.

Ms. BROWNER. We've all—I might just——

Mr. DINGELL. Please cite the language.

Ms. BROWNER. "On or after the date the Administrator promulgates the designation of an area as a non-attainment area, pursuant to section 74(s)(7)(d) of this title with respect to any national ambient air quality standard or any revised standard, including a revision of any standard in effect on November 15, 1990, the Administrator may classify that area for the purpose of applying an attainment date, pursuant to paragraph (2) and for other purposes. "In determining the appropriate classification"——

Mr. DINGELL. I'm not asking you to read——

Ms. BROWNER. That's what we relied on.

Mr. DINGELL. I'm asking you to give me the specific language which affords you the authority to waive requirements with regard to non-attainment areas set out in the statute.

Ms. BROWNER. I understood the question to be: Where's the authority to create a transitional classification. That is the section of the Clean Air Act that we rely on, and I was reading you the section.

Mr. DINGELL. That is the section upon which you rely. Please cite the specific language that gives you the authority to waive the clear requirements of the statute.

Ms. BROWNER. Sir, Congress clearly intended for the Administrator to have the ability to deal with classifications. There's an entire section of the statute, and that is what we reference.

Mr. DINGELL. Am I to infer, from your inability to respond to my question——

Ms. BROWNER. I responded to the question.

Mr. DINGELL. [continuing] by citing the specific language, that you cannot site the language, and that there is no language affording you the authority to waive those requirements of law?

The record at this time——

Ms. BROWNER. We read——

Mr. DINGELL. [continuing] will show that you have no authority. I await your citing——

Ms. BROWNER. With all due respect, we disagree. I have read you, and I will finish reading, the section of the statute that we rely on.

Mr. BARTON. Mr. Chairman?

Ms. BROWNER. We have responded in writing——

Mr. DINGELL. I could read the section of the statute as well as you've read it.

Mr. BARTON. Mr. Chairman?

Mr. DINGELL. I'm asking you to read the specific sections which afford you the authority to waive——

Mr. BILIRAKIS. Would you please hold——

Ms. BROWNER. We do not waive requirements. I mean, Mr. Dingell, with all due respect——

Mr. BILIRAKIS. Please hold a moment. The gentleman wants attention.

Mr. BARTON. The gentleman's time has expired. I would ask unanimous consent, since he's a former chairman, and was chairman when the Clean Air Act Amendments was passed in 1990, that he be given an additional 5 minutes.

Mr. BILIRAKIS. Is there objection? No objection. The gentleman is given an additional 5 minutes.

Mr. DINGELL. I want to thank my friend. I will wait for you to submit for the record, since you cannot at this time submit——

Ms. BROWNER. I'll do it right now.

Mr. DINGELL. [continuing] in response to the written question your authority to waive provisions of the law with regard to non-attainment.

Now my next question——

Ms. BROWNER. May I respond?

Mr. DINGELL. We will wait to permit you to submit the written response to the question, and I hope you will do it by the setting of the sun.

Ms. BROWNER. We have provided a letter which we'll make available to all the committee members today.

[The information provided is printed following the hearing.]

Mr. DINGELL. Let us address the next question, if you please. This is not a minor point. As I read the statute, it is very specific about the requirements that are imposed on EPA. These include "attainment demonstrations, new source review permitting requirements for new and modifying sources, including offsets and the lowest achievable emissions rate. That's LAER control technology. Reasonable further progress requirements, and reasonable available control requirements for major sources." Now isn't it fair to say that these are all the guts of pollution control efforts?

Ms. BROWNER. They are a piece.

Mr. DINGELL. They are——

Ms. BROWNER. A piece.

Mr. DINGELL. They are——

Ms. BROWNER. They are——

Mr. DINGELL. The Congress regarded them as being very important.

Ms. BROWNER. We agree——

Mr. DINGELL. Okay, now——

Ms. BROWNER. [continuing] but they are not the sole means by which we reduce pollution. They are a very important piece.

Mr. DINGELL. If, Ms. Browner, you try to relax the specific non-attainment requirements of section 172 and 173, won't EPA and the States or the source pay suits brought by citizens' groups under the citizens' suits provisions?

Ms. BROWNER. We will—first of all, I think it's important to understand that——

Mr. DINGELL. Let's just answer the question yes or no——

Ms. BROWNER. We don't agree with the question.

Mr. DINGELL. The question is, if you try to relax——

Ms. BROWNER. We don't agree with the premise of the question. I can explain why.

Mr. DINGELL. The specific question, which I think is answerable to by yes or no, is: If you try to relax the specific non-attainment requirements of sections 172 and 173, won't EPA or the States or the source pay suits brought by citizens' groups under the citizens' suits provision?

Ms. BROWNER. With all due respect, these are complex issues; they do not lend themselves to a yes-or-no answer. I am more than happy to explain.

Mr. DINGELL. So you—so you can't tell us whether citizens's suits will be brought if you seek to waive or to lighten the non-attainment requirements of the section——

Ms. BROWNER. If I might point out that——

Mr. DINGELL. You either know or you don't know.

Ms. BROWNER. This is not a yes-or-no question. There is a logic which we find inherent in the Clean Air Act which we subscribe to, which is common sense, which is about pollution reduction, and you can't simply look at one piece, with all due respect; you have to look at the package as a whole and how it goes about achieving cleaner air.

Mr. DINGELL. Well, you're telling me that citizens' suits will not then be filed?

Ms. BROWNER. Obviously, as you well know, whether or not something is filed is not something we have control over. How we defend against it is something we do have control over. We don't have control over what someone may do, as you well know.

Mr. DINGELL. Let us—since you can't answer this question, let us now——

Ms. BROWNER. Mr. Dingell, I'm trying to answer the question. They are not yes-or-no answers.

Mr. DINGELL. Please do not interrupt me. Let us now proceed to address the next question. Let me first express my admiration for the dexterity and the creativity of your implementation plan, but, as you know, the Clean Air Act allows lawsuits to be filed by private citizens and organizations, and these suits can compel EPA or the States to take enforcement actions. Is that not true? Yes or no? Or you don't know?

Ms. BROWNER. Nothing in the implementation plan creates a new cause of action. It does not in any way expand the citizens' suits, cause of action suits——

Mr. DINGELL. A citizen may sue whenever the citizen assumes the law is not being properly implemented. Is that not true?

Ms. BROWNER. No, that is not true. There is standing for when a citizen may file a suit. It is not——

Mr. DINGELL. Such a suit, as a matter of fact, is pending at this moment in southern California; is it not? Such a suit is now pending in southern California; is it not?

Ms. BROWNER. Not that we're aware of. They do not have appeared to have served us.

Mr. DINGELL. You're not aware of the suit?

Ms. BROWNER. No. We're more than happy——

Mr. DINGELL. Mr. Chairman, I think——

Ms. BROWNER. Not on this subject matter, no.

Mr. DINGELL. I think my time has expired. I thank you.

Mr. BILIRAKIS. It has. Thank you. Thank you, Mr. Chairman.

Mr. BILIRAKIS. Mr. Barton?

Mr. BARTON. Thank you, Mr. Chairman.

This is the 16th hearing on this issue that I've chaired or co-chaired in the last two Congresses. So there are some Congressmen that know a little bit about it, and I'm one of them. I'm not here to debate your judgment, but I think it's fair to say that these proposed standards are based on a judgment call. There's no clear bright line that says you have to do, but you have the authority to do it, and you've chosen to use that authority to exercise a judgment. That's decision's been made. The question before the Congress is whether we're going to accept that judgment or we're going to pass legislation that delays or changes it.

You are very gracious to come here. You have repeatedly, since the President adopted your position, talked about flexibility in the implementation, and I think Chairman Dingell, former Chairman Dingell, had some questions about how much flexibility the act really gives the EPA, but that's a question that he pursued.

I want to get real specific. I want to talk about a specific official in the Dallas-Ft. Worth region of EPA and the flexibility that that gentleman appears to have adopted. Congressman Frost, who I believe is a Democrat, and I believe he's chairman of the Democratic National Campaign Committee—so I believe he's somebody that would tend to be supportive of your position and the President's position—asked local businesses to meet with EPA and State officials to kind of get a data base on these proposed regulations. And a Regional EPA Air Quality Director, Dr. Allen Davis, attended that. And here's what this official said:

He said, "There are 33 metropolitan areas nationwide that have been in moderate non-attainment; 28 of those have cleaned up their air; 28 had a can-do approach," Dr. Davis said. "In Dallas-Ft. Worth, there's something wrong."

Now he didn't point out that in most of those 28 areas God cleaned up the air. The areas didn't do anything. Changing weather patterns and population growth helped them meet the standard.

He went on to say that if Texas officials turn in a weak anti-pollution plan for the metropolitan area, it will be disapproved. He also went on to say that political grandstanding of the worst sort has caused a cave-in—or something to that effect. I'm not sure that that's a direct quote. I think the direct quote was, "Political grandstanding and playing to the lowest common denominator have killed the more stringent emissions program." Now the State of Texas has turned in over 50 plans or action requirements to EPA in the last 2 to 3 years. They've all been approved. In the Dallas-Ft. Worth area, they've turned in approximately a dozen. They've all been approved.

The source of most of the pollution for ozone in the Dallas-Ft. Worth area is not heavy industrial pollution that can be controlled with central controls. It's cars and trucks, people driving back and forth to work. We meet the requirements, the tailpipe emissions requirements. So we've got a situation where 60 to 70 percent of the ozone is occurring because of mobile source pollution.

Now, first of all, do you know this Dr. Allen Davis?

Ms. BROWNER. I know who he is.

Mr. BARTON. Okay. Do you associate yourself with those remarks?

Ms. BROWNER. This issue of comments that may have been made was recently brought to my attention. I want to be very, very clear. First of all, if anybody was offended, I apologize on behalf of myself and the agency. Second, I do not associate myself with comments that suggest an unwillingness to work in partnership. As you well know, Mr. Chairman, we work very closely with your State on a number of very difficult issues, most recently the self-audit provisions; we found common ground. We moved forward with a State law that was important to you and many others in your State. And, again, I apologize if these kind of comments were made. They are not in keeping with what I would say is one of the fundamental operating principles of how we do our work, which is partnership with State, local governments, and industry.

Mr. BARTON. Well, here's the issue, Madam Administrator: We're not—I personally understand you've got the right to make some of these decisions, and the EPA has the requirement under the law to enforce our air quality standards. But I don't think this flexible approach filters down to the Dallas-Ft. Worth area, the State of Texas area. And if we're going to go ahead and do this—and I hope we don't; I hope we pass a law that delays it or changes it, but if we do it, we need flexibility. And it really irritates me and other members of the Texas delegation on a bipartisan basis——

Ms. BROWNER. I agree.

Mr. BARTON. [continuing] when the people responsible for enforcing the standards don't show any flexibility. Now I could talk to you about dirt. There's an EPA official in Dallas, Texas that is requiring homebuilders to spend $1,000 per lot to keep dirt from washing into the street when it rains in Texas in the spring. He's classified dirt as a hazardous material: D-I-R-T, dirt. And it's the only region in the country that apparently that interpretation is being given.

So I don't buy at the grassroots level this flexibility approach.

Mr. COBURN [presiding]. I think the gentleman's time has expired.

Mr. BARTON. My time has expired.

Ms. BROWNER. Is this public?

Mr. BARTON. If you'd care to respond to that——

Ms. BROWNER. If we might, the gentleman who you mentioned has decided to leave the agency, and will be leaving on Friday.

Mr. BARTON. That's a start.

Mr. COBURN. The gentleman from Ohio——

Ms. BROWNER. If I might, Mr. Chairman, in the opening to the question Mr. Barton suggested that the decision which the President announced, and I ultimately signed, was one not based on science, and with all due respect——

Mr. BARTON. I didn't say it wasn't based on science; I said it was a judgment call, and there is no bright line in the scientific evidence that says you have to do that. That's what I said.

Ms. BROWNER. And I would suggest to you that the science is compelling, the public health science, in terms of the effects on individual Americans——

Mr. COBURN. The gentleman's time has expired.

Mr. BARTON. We can debate that, you and I, man to woman, woman to man, Administrator to Congressman, in any forum in the country, with appropriate notification and preparation, and I believe I can more than hold my own in that debate.

Ms. BROWNER. I'll meet you anywhere.

Mr. COBURN. Taking control once again of the time, the members should know that we're going to continue and vote through this. We have people floating in and out.

And the gentleman from Ohio is recognized for 5 minutes.

Ms. BROWNER. Thank you.

Mr. BROWN. Thank you, Mr. Chairman.

Thank you again for being here, Administrator Browner, in another pleasant day in your life, in front of this committee.

Ms. BROWNER. Well, after how my morning began, this is better.

Mr. BROWN. Last night I had to watch the Indians blow a 6–1 lead against the Yankees. So that was pretty hard, too, but not quite on this level.

I would like to run through the implementation process and get some comments. Between 1998 to 2000, roughly 1,500 monitors will be put in place; correct?

Ms. BROWNER. We're now talking about particulate matter only; correct.

Mr. BROWN. Right. This is an undertaking that you've promised, you've committed to fund.

Ms. BROWNER. Yes.

Mr. BROWN. I guess much of that money is in the VA-HUD appropriations bill.

Ms. BROWNER. It has been resolved as of last night. We have an agreement on our appropriations, as I understand it.

Mr. BROWN. Will you be able to meet the—obviously, setting the funding aside for a moment, will you be able to meet this commitment without specific legislation from this Congress?

Ms. BROWNER. We have what we need in our appropriations bill; that is correct. Normally, what happens on the monitors, it's a cost-share with the States, and we worked with our appropriators to assume the full cost of the monitors. So we'll pay for that out of our budget.

Mr. BROWN. It was roughly $50 million, correct?

Ms. BROWNER. I think that's correct for all of them. The monitors are expensive pieces of equipment. They're very sophisticated pieces of equipment.

Mr. BROWN. But you can—beyond that, do this without specific legislation other than the funding.

Ms. BROWNER. Yes, we can. We can assume the full cost of the monitors with the appropriation and the language we agreed to with the appropriators. The amount we will spend actually, just to set the record straight, in 1998 is $35 million, and then there will be an additional expenditure in 1999 to complete the setup.

Mr. BROWN. Let me shift—the President's implementation memo states that EPA will issue the necessary rules and guidance by the end of next year, by the end of 1998. Will these rules provide States and industry the certainty they need that EPA will not change its mind regarding implementation plans for the new standards?

Ms. BROWNER. Precisely. I think that's a very, very good point to recognize—that it's not simply a memo from the President, which in and of itself is remarkable, but there is an entire regulatory rulemaking process with notice and comment, and the potential for litigation. Whatever rights people have in that process are preserved, and it goes into The Federal Register. So it's not simply a question of so-and-so said this on such-and-such a date. It is all subject to the procedures and gives people an opportunity for participation and certainty.

Mr. BROWN. And I understand all these dates and deadlines are consistent with the Clean Air Act?

Ms. BROWNER. Yes. Absolutely.

Mr. BROWN. Respond, if you would, to some other questions that were asked about citizen lawsuits. I have heard, as a public supporter of these standards, and as a recipient, if you will, from no small amount of criticism from industry in my district, that there will be citizen lawsuits concerning the implementation schedule, and that the EPA, Administrator Browner really might even welcome these lawsuits so she can implement this faster than she wants to——

Ms. BROWNER. No.

Mr. BROWN. What stops that from happening? Why don't those citizen lawsuits precipitate, as did, I believe, the Lung Association's lawsuit this summer.

Ms. BROWNER. What happened with the Lung Association is that EPA in past years, dating back, you know, 5, 10, 15 years ago, had ignored a provision in the Clean Air Act. In this case, the implementation plan in writing is based on all of the timeframes laid out in the Clean Air Act. It says that a State shall get "X" number of years to collect data for PM, 3 years. So there's no citizen suit that's available to anybody, because it's all within the law. The law clearly gives these timeframes, and we take advantage of what Congress very sensibly did in the 1990 amendments.

As I tried to point out to Mr. Dingell, we have no control over someone filing one of these. What we do have control over are the arguments we will make against them and the motions to dismiss, and we are quite comfortable that those kinds of suits that say, "3 years written in the statute should be read as a year," are not going to be successful. We are very comfortable in making that statement. And we will be the ones defending against those.

Mr. BROWN. What difference would it make if Congress provided you with legislation codifying the implementation schedule? Why would that be a problem?

Ms. BROWNER. It's in there. You don't need it. It's in there. All of what we have put out is in the law. We take it from the law. I mean, if the law says 3 years of data collection, I don't know what statute it is Congress would pass. It says 3 years. That's what the implementation plan is. I mean, I can go through each section, but it's all in the statute right now.

Mr. BROWN. My time has expired. Thank you.

Ms. BROWNER. May I—are you all in the middle of a vote? How long is the vote going to last?

Mr. BURR [presiding]. The Chair would recognize the gentleman for——

Mr. SAWYER. Mr. Chairman, I really appreciate this. I am going to be unable to return, and I just want leave in the record to submit questions in writing for a response as a part of this hearing. I appreciate the opportunity——

Mr. BURR. Without objection.

Mr. SAWYER. Thank you.

Mr. BURR. At this time we would adjourn for——

Ms. BROWNER. Five minutes?

Mr. BURR. [continuing] 5 minutes.

Ms. BROWNER. Thank you.

[Brief recess.]

Mr. BILIRAKIS. Let's have order, please.

Mr. Burr, to inquire.

Mr. BURR. Thank you, Mr. Chairman.

Ms. Browner, welcome. Welcome.

Ms. BROWNER. I was trying to understand if the Yankees or the Indians won. I apologize.

Mr. BURR. From the South, we never root for the Yankees regardless of what it is.

Ms. BROWNER. Excuse me—also being from the South, did you say we always root for the Yankees?

Mr. BURR. We always root against the Yankees.

Ms. BROWNER. That's what I thought. Okay. Well, Florida has a team this year.

Mr. BURR. I'll be very brief this morning because I think that you've certainly recovered some ground already treaded on many times, and I'll work with you to conclude that, through your work at EPA with Justice, that there are documents—there is some substantiation to make you believe, and to convince us, that you have within the statute of the Clean Air Act the ability to set this flexible implementation.

Let me ask you, with that confidence level that you have, would you be willing to instruct individuals at the EPA to work with this committee to codify into law in some way that implementation strategy, so that for your legal, for this committee, and for the localities across the country in their planning purposes of going into this implementation strategy, that we could do it with confidence of knowing that that plan would be followed?

Ms. BROWNER. We are very, very confident that within all of the portions of the law we have the authority, and we have used it to design an appropriate implementation schedule. Obviously, we are more than happy to meet with any member who has any thoughts on how to ensure that. We do not think it is necessary. We have worked with this committee and other committees to rewrite other environmental laws, where we have found them to be inflexible. This is not a law we find to be inflexible. I mean, I think it's perhaps one of the most brilliant pieces of environmental law crafting in its ability to adjust and to allow us to adjust to where we now find ourselves.

Mr. BURR. As I have said, I'll accept your confidence level on your ability within the statute to pursue this route, but also recognize the hesitancy that still exists within this committee, within the business community, within localities across this country, and would, therefore, ask you again: Would you, as the Administrator, endorse and instruct EPA individuals to work with this committee or with a member to codify into some type of legislative language this implementation strategy, so that there is no uncertain area in its implementation?

Ms. BROWNER. We do not believe legislation is necessary at this point in time, but we would certainly be more than happy to sit down with any member, with any committee, to look at proposals and to provide our thoughts on them; absolutely.

Mr. BURR. Allowing me some latitude to interpret that as a no——

Ms. BROWNER. No, we're more than happy to meet with you.

Mr. BURR. There's a difference, quite frankly, Ms. Browner, between meeting and supporting an effort, and I think——

Ms. BROWNER. I never say "never," because I

Mr. BURR. But—but——

Ms. BROWNER. We've thought about this—if we could write any bill we wanted, and if—I know this is not how the system works, but if the system would then say, "and it will not be changed"— would there be a proposal we would put on the table to this body? And I will tell you, in all honesty, at this point in time we cannot give you what that would be. It simply isn't necessary.

Now I'm not saying to you that someone else may not have a better idea. You know, we're not the keeper of ideas; you are.

Mr. BURR. Ms. Browner, you also cannot sit here and testify that you don't expect—as a matter of fact, I believe you would say you probably anticipate—a lawsuit over this flexible implementation?

Ms. BROWNER. With all due respect, I think we have something on the order of 600 lawsuits pending——

Mr. BURR. But you do expect a lawsuit on this implementation or you would not have sought the Department of——

Ms. BROWNER. The question is, do we expect——

Mr. BURR. You would not have sought the——

Ms. BROWNER. No.

Mr. BURR. [continuing] the Department of Justice's——

Ms. BROWNER. No, that's not accurate. You cannot simply take two points and connect them up in that way. That is not why we talked to the Department of Justice. We talked to the Department of Justice morning, noon, and night about how to correctly interpret provisions. We bounced ideas back and forth. It is not fair to say that we were afraid of being sued. You know, the question isn't, "are you afraid of being sued?" when you're in my position. The question is, "can you win the lawsuit?"

Mr. BURR. And I did not suggest that you were afraid; I suggested——

Ms. BROWNER. I wasn't——

Mr. BURR. [continuing] that you knew you would. Has anybody suggested to you or told you that they would sue you if you went with a flexible——

Ms. BROWNER. We've already been sued——

Mr. BILIRAKIS. The gentleman's time has expired.

Ms. BROWNER. We've already been sued, and I'm sure we will be sued many, many times more. The question is, can we win the lawsuit? Do we have the legal authority to secure a motion to dismiss?

Mr. BURR. Mr. Chairman, for the purposes of Mr. Klink and myself, let me just ask you to comment, since it might not get back around to me, Ms. Browner—do you interpret that restructuring of our electricity industry would, in fact, encourage more environmentally friendly generation facilities?

Ms. BROWNER. I think it would depend on the nature of the legislation Congress passes.

Mr. BURR. Thank you.

Mr. BILIRAKIS. I thank the gentleman.

Mr.——

Ms. BROWNER. It certainly could. It certainly could, and it should.

Mr. BILIRAKIS. Mr. Klink?

Mr. KLINK. Administrator Browner, let me just state: We have some very serious concerns about your implementation plan as it regards to citizens' lawsuit provisions, if you haven't picked that up from the members already, including Mr. Dingell.

The Department of Justice said to staff yesterday that they could not pass on the legality of your implementation plan because it is too indefinite. They already are raising the red flag to us. So I want you to understand that we've given this a tremendous amount of consideration.

Let me get on to——

Ms. BROWNER. Well, with all due respect, we did——

Mr. KLINK. That's not in the form of a question, Ms. Browner. I'm telling you what we're being told, and I would like to try to get to some questions here, so we could—if you want to respond to that later on in writing, that would be fine.

Ms. Browner, have you stated many times in the past, including at your testimony before the Agriculture Committee here in the House, that you do not think agriculture will be affected by the new standards?

Ms. BROWNER. That's correct.

Mr. KLINK. You also assured the construction industry that you don't think they're going to be affected; is that correct?

Ms. BROWNER. Are you asking—I just want to clarify something—you're asking about particulate matter?

Mr. KLINK. Yes, particulate matter.

Ms. BROWNER. Okay, I wanted to make sure we had the record straight here. We don't see any reason why construction activities that generate dust should be affected. This is not about dust.

Mr. KLINK. I understand.

Ms. BROWNER. This is not about dirt.

Mr. KLINK. I understand. In June, I went to Pittsburgh and met with the steelworkers' union management to discuss these new air quality standards. At that meeting I was informed by some of their staff people that had talked to EPA that they had been assured by the EPA that the steelworkers and the steel industry also were not the targets of the standards, but, rather, the steelworkers told us they were told that utilities and agriculture would be expected to share the responsibility of cleaning up the air.

Ms. BROWNER. I never had a meeting of that sort, and I am more than happy to provide to this committee a letter of April 30 sent to the United Steelworkers of America, a Mr. Michael J. White, which explains what precisely will occur under a standard, if a standard were to be changed. I would like, if—with your leave—to read the operative paragraph.

Mr. KLINK. Well, I'd rather not read it now, but——

Ms. BROWNER. But the important thing is that in here it explains, in response to their question—they asked, would closure of facilities occur over the next year, and the response is no.

Mr. KLINK. All right, regardless of that——

Ms. BROWNER. We'll submit this for the record.

Mr. KLINK. Okay. I would ask unanimous consent the letter be submitted for the record. It would be helpful to us.

Ms. BROWNER. Yes, thank you.

Mr. BARTON. Without objection.

Mr. KLINK. Thank you.

[The information referred to follows:]

UNITED STATES ENVIRONMENTAL PROTECTION AGENCY
WASHINGTON, D.C. 20460

APR 30 1997

OFFICE OF
AIR AND RADIATION

Michael J. Wright, Director
Health, Safety and Environment Department
United Steelworkers of America
Five Gateway Center
Pittsburgh, PA 15222

Dear Mr. Wright:

The Administrator has asked me to respond to your letter of April 17, 1997, asking for information on the process and timetable for implementing revised National Ambient Air Quality Standards for ozone and particulate matter. Specifically, your letter asks for EPA's response to concerns raised by some companies that the standards could cause the closure of some facilities over the next year.

I can assure you that if revised air quality standards for ozone and particulate matter are promulgated this summer, no industry facility will be required to comply with new pollution control responsibilities stemming from those standards in the coming year. In the first several years after revised standards are issued, EPA and the States will be engaged in intensive work to develop appropriate control measures at the federal, regional, and local level to attain those standards as cost-effectively as possible. Given planning and lead time requirements, it will be at least two years before any new requirements stemming from revised standards would affect new sources, and several years more for existing sources. The initial deadline to attain these standards will be approximately seven years after promulgation for ozone, and probably longer for particulate matter. Where warranted, the Act allows this initial deadline to be extended for up to seven additional years.

As you know, EPA is working with a subcommittee of our Clean Air Act Advisory Committee focused on implementation of ozone and particulate matter and regional haze requirements, of which you are a member. The subcommittee is helping EPA to develop innovative and cost-effective methods of meeting revised standards. EPA is also working with the 37 eastern-most States in the Ozone Transport Assessment Group to reduce long-distance transport of NOx emissions from large sources, such as power plants, which are contributing to unhealthy levels of ozone in downwind States. Reducing regional transport of pollutants from these large sources will be a key factor in implementing revised standards for both ozone and particulate matter.

As an additional benefit of the hazardous air pollutant standard established under the 1990 Clean Air Act, the steel industry's coke ovens are already making substantial reductions in

2

emissions that contribute to ozone and particulate matter. The coke oven standards were the product of a successful regulatory negotiation that involved both the industry and the union. That agreement is a model for achieving cleaner air consistent with the economic well-being of our basic industries.

In sum, there is no basis for concern that revised air quality standards for ozone and particulate matter could result in new requirements for steel industry sources in the next year, let alone in their closure.

Sincerely,

Mary D. Nichols
Assistant Administrator
for Air and Radiation

AX-9704649:DavidDoniger:6101

Controlled Correspondence For
OFFICE OF AIR AND RADIATION

CONTROL NO : AX-9704649
ALT NO:

EXT. DUE DATE:
ORIGINAL DUE DATE: 05/08/97
CORR. DATE: 04/17/97
REC. DATE: 04/22/97
CLOSED DATE:

OAR Due date 5/5/97

STATUS: PENDING

FROM: WRIGHT MICHAEL J

ORGANIZATION: UNITED STEELWORKERS OF AMERICA

SALUTATION: DEAR MR. WRIGHT

CONSTITUENT:

TO: ADMINISTRATOR

TO ORG: EPA

SUBJECT: REQ-EPA TIMETABLE AND PROCESS FOR IMPLEMENTATION OF PROPOSED NATIONAL AMBIENT AIR QUALITY STANDARDS FOR OZONE AND PARTICULATE MATTER

SIGNATURE: DIRECT REPLY

CC'S: ADMINISTRATOR
DEPUTY ADMINISTRATOR

ASSIGNED: OAQPS *David Doniger*

AX INSTRUCTIONS: DIRECT REPLY. SEND COPY OF REPLY TO OEX.

AX ADDTN'L INST:

OAR INSTRUCTIONS: FOR DIRECT REPLY. PLEASE SEND COPY OF RESPONSE TO OAR.

OAR ADDTN'L INST:

OAR COMMENTS:

	Assigned	Date Assigned	Code/Status	Date Completed by Assignee	Date Returned to AX :
Lead	OAQPS	04/25/97	ACTION	-	-
	OAQPS	04/25/97	ACTION	-	-

♻ EPA

OFFICE OF THE EXECUTIVE SECRETARIAT
CONTROL SLIP

DUE DATE:	
ORIG. DUE DATE:	05/08/97

CONTROL NO: AX-9704649

STATUS: PENDING

CORRES. DATE: 04/17/97
RECEIVED DATE: 04/22/97
ASSIGNED DATE: 04/24/97
CLOSED DATE:

FROM: WRIGHT MICHAEL J

ORG: UNITED STEELWORKERS OF AMERICA

SALUTATION: DEAR MR. WRIGHT

CONSTITUENT:

TO: ADMINISTRATOR
TO ORG: EPA

SUBJECT: REQ-EPA TIMETABLE AND PROCESS FOR IMPLEMENTATION OF PROPOSED NATIONAL AMBIENT AIR QUALITY STANDARDS FOR OZONE AND PARTICULATE MATTER

ASSIGNED: AIR & RADIATION
COMMENTS:

SIGNATURE: DIRECT REPLY

INSTs: DIRECT REPLY. SEND COPY OF REPLY TO OEX.

ADDTN'L INST:

CC's: ADMINISTRATOR
DEPUTY ADMINISTRATOR

IMS: SHARITA SHAW
IMT: GERALDINE BROWN

	Assigned	Date Assigned	Code/Status	Date Completed by Assignee	Date Returned to OEX:
Lead	OAR	04/24/97	ACTION	-	- -

United Steelworkers of America

AFL-CIO/CLC

Five Gateway Center
Pittsburgh, PA 15222

(412) 562-2400 • FAX (412) 562-2484

April 17, 1997

Carol M. Browner
Administrator
U. S. Environmental Protection Agency
401 M Street, S. W., Room 1101
Washington, DC 20460

Dear Ms. Browner:

no record in OEX

In March, the United Steelworkers of America filed comments supporting EPA's proposed National Ambient Air Quality Standards for Ozone and Particulate Matter. Some companies have charged that the standards will cause the closure of some facilities over the next year. We believe their motive is to pressure the union into opposing the standards, or to provide an excuse for management decisions that have nothing to do with EPA.

For our part, the union has argued that any new requirements for individual companies will not become effective for many years, during which time EPA, industry, environmental organizations, and unions can work together to implement the standards in a way that protects -- and perhaps even expands -- employment.

However, we could use your help in making that case. We would greatly appreciate a letter from your office describing the process and timetable for implementing the standards, and explaining why there is no immediate threat to any facility. It would be helpful if the letter also stated EPA's commitment to finding environmental solutions which preserve the economic vitality of our basic industries, and good jobs for American workers.

Thank you for your help.

Sincerely yours,

Michael J. Wright
Director
Health, Safety and
Environment Department

MW/np

Mr. KLINK. Well, let me just say this, Ms. Browner: They, when we met with them—and I'm not exactly sure the date that we met with them—in their discussions with us, they seemed to be very sure that they had your assurance.

Ms. BROWNER. They did not meet with me.

Mr. KLINK. Well, with you—I'm saying the generic "you," the EPA. They based their endorsement of these standards on that assurance in their discussions with us.

My problem is this, and that is that we don't understand how you can promise or predict what States are going to do. How can you say to agriculture, for example, that the State implementation plan won't require agriculture to be impacted——

Ms. BROWNER. Because——

Mr. KLINK. Let me please finish my question. I only have a certain amount of time.

From the record, we noted that EPA seems to think that tilling alone is 7 percent of the $PM_{2.5}$ problem, and considering the fact that there's a lot more to agriculture than just tilling, agriculture I think could be responsible for quite a big chunk of $PM_{2.5}$. You've left out other activities, like burning, ginning, milling, pesticides, not to mention other combustion activities. Considering this, they may look more like a target—more are a target to the States putting their implementation plan together than they do to the EPA. And if I remember your testimony to us back in May, when someone suggested that it was the EPA that would be saying, "You can't drive your car certain days. You can't use a lawnmower. You can't use charcoal grills," you said, "Wait. We're not saying that. The EPA is not saying that. It is the States that come up with their implementation plans."

So how can you assure any of these industries that the States, in putting their implementation plan together, won't lean on any of these sources of $PM_{2.5}$? How could you give assurance to Mr. Brown, as he said in his opening statement, that a State or a county wouldn't lean on the Amish to say, "We don't want you burning these wood-burning stoves. We want you to look at something else."? And I would give you time to respond to that.

Ms. BROWNER. Two reasons: One, we will be providing to the States guidance in how to craft those plans. Reason No. 2 is I trust the Governors. It is the Governor who sends us this plan, and I have yet to find a Governor out there who's going to say, "It's the farmer's dirt and I'm going to regulate the farmer's dirt." They're simply not going to do it. It doesn't make any sense under any scenario, political or substantive scenario. It just doesn't make any sense.

In the case of ozone——

Mr. KLINK. I see the red light. So let me just tell you this: You may—you trust the Governors, but I'm going to tell you something right now. The States have just—many of the States have just finished their State implementation plan, which, in essence, you seem to be telling them to throw out——

Ms. BROWNER. No.

Mr. KLINK. [continuing] the window, because we're now going to all new—a whole new set of regulations. We think, Administrator Browner, by promulgating those regulations, that you are going to open yourself up to citizens' lawsuits, and nothing that you've stated here today has yet convinced us that we're not going to see this as a lawyer's relief act. We think the more common-sense approach is, under the same timetable, to authorize the money for you to be able to build 1,500 $PM_{2.5}$ monitors, to deplore them, and to do the science. And we would have looked forward to having those discussions with you before we got to this point. Unfortunately, we weren't able to do that.

And I would yield back whatever——

Mr. BARTON. The gentleman's time has expired. Would the Madam Administrator like to make a brief reply to that before we go——

Ms. BROWNER. Well, in response to ozone, I would just remind the members of the committee, it is a State-drafted, a State-presented solution that we are adopting. It is not EPA-driven. It was crafted by 37 States over a 2-year period.

With respect to wood stoves, there is an on-the-book policy, if you will, that wood stoves that are the sole source of heating or cooking, as is the case with the Amish, are simply exempted; that you don't deny someone their sole source of heating or cooking in this case. It wouldn't make any sense. So we're more than happy to share that with the committee members who may have similar questions.

Mr. BARTON. Thank you.

The Chair would recognize Dr. Coburn of Oklahoma for 5 minutes.

Mr. COBURN. Thank you, Mr. Chairman.

Madam Secretary, you said something that I thoroughly agree with. You said our clean air regulations are not protective enough for our children. I agree with that.

Ms. BROWNER. I'm sorry, what was——

Mr. COBURN. You said our clean air regulations are not protective enough for our children. I agree with that. I want you to know that I agree with that.

And my objections through this whole process is where I think the scientific model has been tweaked and we have not maintained the purity with it. So please take in light my questions that I ask.

If, in fact, in Oklahoma, where lots of people supplement their heating to their homes with wood stoves, tons in my district, if it's not the sole source, what we're going to be saying—your implication is from what you just said is—the State may be forced in an implementation—maybe—not by the EPA, but maybe through the State, forced to regulate that and not allow that in the future—possibly.

Ms. BROWNER. I know a little bit about your Governor. I find it highly unlikely. You probably know more about him. If you think that's what he would do—I can't imagine.

Mr. COBURN. Well, you know, it's not quite as simple as saying what a Governor will or will not do. In the State of Oklahoma——

Ms. BROWNER. They'll sign the plan——

Mr. COBURN. [continuing] the legislature has much more power than the Governor, and in fact if their moneys are put at risk in terms of non-attainment, then a lot of things are going to happen that the Governor has no control over.

The point I was trying to make with the question is, if it's not sole source, we may not, in fact, be successful in limiting the impact, the very real impact, on many people whose family income is under $20,000 a year from utilizing a way to heat their homes.

Ms. BROWNER. Mr. Coburn, we are not aware, based on current data, that you will have any non-attainment areas for the new standard. If you have data—we get this data from your State——

Mr. COBURN. Yes.

Ms. BROWNER. If you have data, we would welcome it. Right now our modeling, on what your State has given us shows you don't have a problem. Now if you want to introduce evidence that suggests you do have a problem, we are more than happy to accept it. But right now, based on what we have——

Mr. COBURN. I'll probably take that statement and the assurance from the Secretary that we're not going to have a problem. My information is——

Ms. BROWNER. I'm saying, based on what your State has given us——

Mr. COBURN. We have six counties that are in non-attainment under your standards right now, the proposed standards. And I'd like to have a chart put up, because we happen to reside in region 6 of EPA, and this is 1994 EPA data on 2.5 particulate matter.

The point I want to make with this is the tallest one is region 6. There's six members on this committee in region 6. The only thing—the point I want to make with this chart, and assume that it's accurate—you can disagree with whether or not you think this chart is accurate. Just for——

Ms. BROWNER. Can I ask you if this is the chart we saw previously? I believe this is about direct emissions of 2.5, and not secondary emissions. Our concern is the secondary. Is that the chart we've seen once before? I can't see that far. I apologize. I'm just getting old.

Mr. COBURN. This is—I'll give you a chart. You can have a copy.

Ms. BROWNER. Okay, but have I seen this chart before? This is direct——

Mr. COBURN. Yes, you have——

Ms. BROWNER. This is direct emissions, not secondary; correct?

Mr. COBURN. This is 2.5 emissions.

Ms. BROWNER. Direct emissions——

Mr. COBURN. Right.

Ms. BROWNER. Not the secondary emissions. Okay. You do understand the distinction?

Mr. COBURN. Yes, I——

Ms. BROWNER. I know you understand it. But this chart only addresses the direct emissions or what you refer to as primary and not secondary. It is not what we would base an attainment designation on, just so you understand that.

Mr. COBURN. I understand that, and I'm applying this to region 6, not Oklahoma, just to region 6.

Ms. BROWNER. This is true for region 6 also.

Mr. COBURN. I understand that.

Ms. BROWNER. It's true of the entire country.

Mr. COBURN. But, nevertheless, the point is that in the direct particulate matter that's out there, less than 4 percent is available for us to change. The vast majority of it is background. The vast majority of this is natural occurrence of 2.5 direct—direct—and I admit that that's direct. So given that data and the limited ability for us to change the direct component, which will have an impact on our SIP, and it will impact the six counties which we've been told under our——

Ms. BROWNER. Before you were asking about ozone, and I responded to that; now you're asking about $PM_{2.5}$?

Mr. COBURN. I'm talking $PM_{2.5}$.

Ms. BROWNER. We agree that, based on modeling the State has given us, there may be four counties—again—may be. Your Governor gets 3 years of data collection which has not occurred, but there may be four counties, based on current modeling, with a 2.5 problem—we don't know——

Mr. COBURN. They all happen to be in my district. So, you know, I'm significantly concerned because we're an underdeveloped district and we're trying to recruit economic development to raise the standards for the people that I represent. So "may" is not a good enough answer for me right now——

Ms. BROWNER. Well——

Mr. COBURN. [continuing] and it's not a good enough answer for the people in my district, when in fact you can't say whether we will or won't.

Ms. BROWNER. Well, but that's the thing that——

Mr. COBURN. The question that I'm wanting to get to you——

Ms. BROWNER. But you don't want me to say right now, because the truth of the matter is your Governor has certain rights and you want me to honor those. If I didn't honor those, you would be absolutely right to criticize me. The law says your Governor gets 3 years of data collection. Don't you want me to do that?

Mr. COBURN. Yes, and the question——

Ms. BROWNER. Don't you want me to honor that?

Mr. COBURN. [continuing] I want you to answer in that regard is, if the data collection across this country on 2.5 particulate matter, direct and indirect, does not support all the scientific data that EPA has made the assumptions for this—if it does not, will it be the EPA's position that we should change these recommendations?

Ms. BROWNER. We have already said, and I will say here again, that before any State submits their final plans, we will complete another science—now I know you——

Mr. BARTON. Now the gentleman's time has expired.

Mr. COBURN. May I just get her to answer this one question, Mr. Chairman? If, in fact, the data doesn't support the assumptions that EPA has made to make these new regulations, will you in fact change the regulations?

Ms. BROWNER. Mr. Coburn, if a 5-year science review shows that the standard that we have adopted for 2.5 does not adequately protect the public's health, it will absolutely be adjusted. That's what the Clean Air Act says on the books today. It's in the law. It's been there. President Carter, President Bush retained it. We will absolutely, positively adhere to it. The President himself has said so. Nothing in that has changed—nothing.

Mr. COBURN. Mr. Chairman, I'd like to have this chart entered into the record, if I may.

Mr. BARTON. Without objection.
[The information referred to follows:]

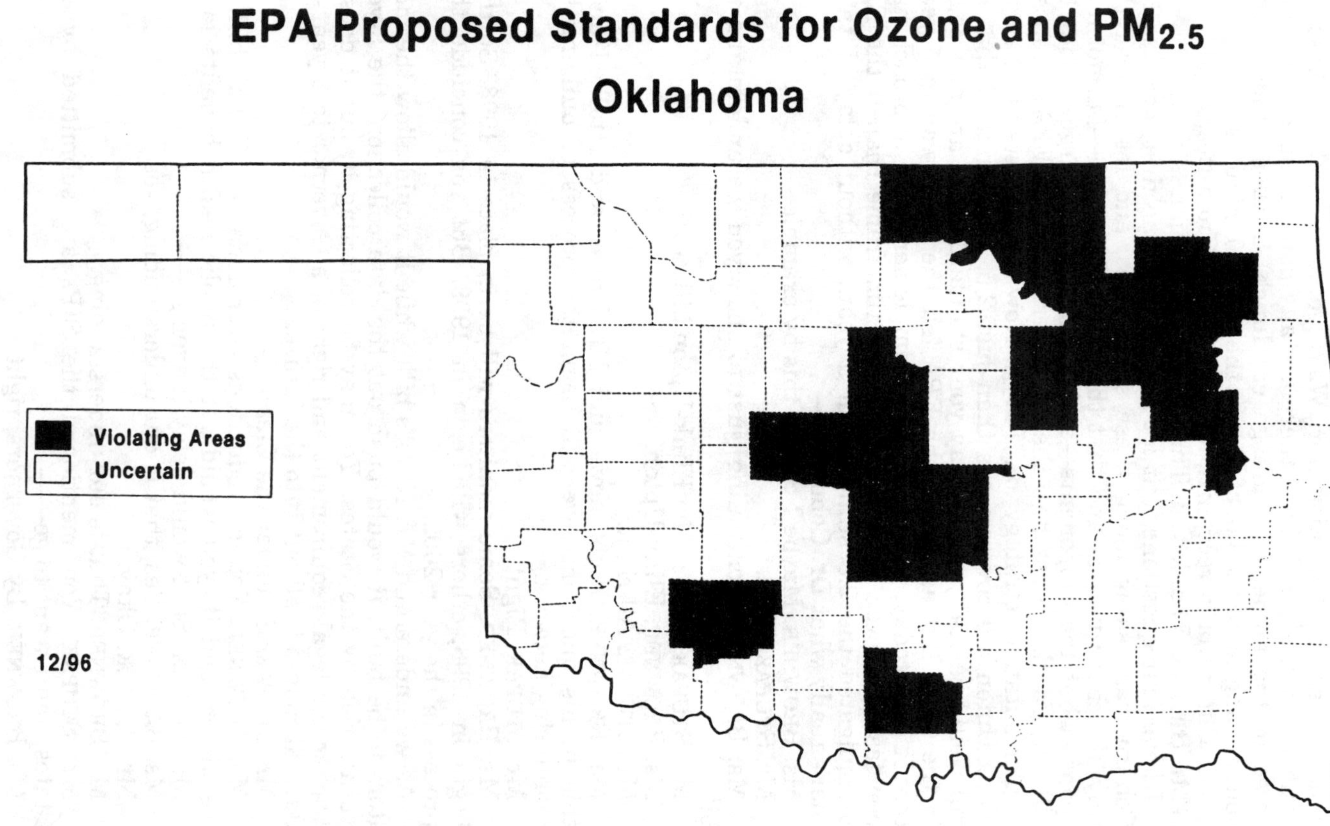

Mr. BARTON. The Chair wants to make an announcement. There's a pending vote. We're going to continue the hearing to try to give as many members opportunity to ask questions. The Chair would also announce that the second—there are two additional panels after the Administrator. We're going to combine those two panels into one panel. So if there are any panel members on panel three in the audience, don't go eat lunch. Stick around because you're going to be up, hopefully, in the next 30 minutes.

The Chair would recognize Mr. Stupak for 5 minutes.

Mr. STUPAK. Thank you. Thanks, Mr. Chairman.

I have four questions, but let me ask this one first, because Mr. Coburn was asking along these same lines, and let me ask this way, if I can: I understand that the Klink-Upton bill—I'm not a co-sponsor, but the proponents—and I don't want to argue about the bill, but I'd like to ask how the bill differs from the EPA's proposed implementation strategy on $PM_{2.5}$. Now I understand that the Klink-Upton bill covers more than just 2.5. I understand that the bill would basically wipe away your standards that are now found in the CFR. But assuming the EPA felt the same way in 5 years that it does today, would the EPA's implementation approach happen faster, slower, or under the same timeframe than if the EPA promulgated the 2.5 standard in 5 years without delay? I think that's really what Dr. Coburn——

Ms. BROWNER. Maybe I can do this by example——

Mr. STUPAK. Sure.

Ms. BROWNER. [continuing] just to show you we understand the bill.

Mr. STUPAK. This is the Klink-Upton bill now?

Ms. BROWNER. Klink-Upton.

Mr. STUPAK. All right.

Ms. BROWNER. Right now—and Mr. Brown and I talked about this in his questions—we will begin the process of building the monitoring network.

Mr. STUPAK. Right.

Ms. BROWNER. Some monitors will be in place in 1998. So they begin in 1998; others will begin in 1999. But you conclude that process in the year 2002.

As we understand Mr. Klink's bill, while it would allow the monitors to be built, it would push out the data collection, the 3-year verification by the States. You may be collecting it, but it doesn't trigger the legal requirement and right of a Governor to 3 years of data. It moves it all out into the future.

Mr. STUPAK. So under the Clean——

Ms. BROWNER. You get reductions later under Mr. Klink's bill. As we understand it, you would get the public health benefits later.

Mr. STUPAK. So it would delay it even if——

Ms. BROWNER. Yes, that is our understanding, yes.

Mr. STUPAK. Okay.

Ms. BROWNER. That is our understanding.

Mr. STUPAK. You mentioned the SIPs are submitted by the States, and I want to go——

Ms. BROWNER. By Governors, right.

Mr. STUPAK. Right, Okay. And SIPs are submitted by the States, and you said earlier you trust the Governors to do what's right. Can you overrule the Governors once they submit a SIP?

Ms. BROWNER. Just to back up for a second, the way the process works is we issue guidance, instructions, if you will——

Mr. STUPAK. Correct.

Ms. BROWNER. [continuing] to the States about how to do it. It's a performance standard. We can't say to a State, no, you should have picked that utility versus a different utility. It is up to the State.

Mr. STUPAK. Yes. I'm more concerned about——

Ms. BROWNER. It's a performance standard.

Mr. STUPAK. [continuing] the Governors because all my counties are in attainment except Bensey County, because we've got the air transport problem——

Ms. BROWNER. Right, that's how yours will be solved, right.

Mr. STUPAK. Right. And I understand it will be solved, but can the Governors then say, well, let's make this SIP area, and could the State of Michigan force controls on counties in my district even though the county's in attainment?

Ms. BROWNER. No.

Mr. STUPAK. Well, can they grab it? Can they grab part of my clean counties to offset a county that——

Ms. BROWNER. The way the implementation plan is structured, if your Governor—and we have every reason to think he will—signs up to the regional program, the cap-and-trade program, which he was part of creating, then that's it.

Mr. STUPAK. But can he expand that region to include part of my area—that's what I'm saying—part of my district?

Ms. BROWNER. No. I'm not sure I understand your question. If I understand it right, the answer is no. I may be misunderstanding something.

You're saying, could your Governor, even though a cap-and-trade utilities program gets your counties clean air, turn around and say, "I'm going to call that a non-attainment area."?

Mr. STUPAK. No, but he has to offset a region, right, of the State of Michigan? Doesn't he break the SIP plans down into regions?

Ms. BROWNER. Right, but, as I understand, in Michigan, you're going to get on a statewide basis—and this is true for most of the Midwest and Northeast—you're going to get the reductions you need across the State through the utility cap-and-trade program.

Mr. STUPAK. Let's say the utilities can't get it. So you've got to spread out the solution, if you will, over a larger area. Can you not go to a clean county and include it into a regional SIP plan to offset it——

Ms. BROWNER. Maybe an easier way to answer the question is, if your counties are not contributing to the problem——

Mr. STUPAK. They're not.

Ms. BROWNER. Right, which we agree with you on. That's our understanding of the situation. Then there would be no basis to expand the boundary or the requirements on your counties.

Mr. STUPAK. Right. Let me ask one more question. And you mentioned this New Jersey utility.

Ms. BROWNER. Yes.

Mr. BARTON. You don't have to, Madam Administrator, but you could.

Ms. BROWNER. I don't think a Governor could, quite frankly.

Mr. BARTON. Well, I think——

Ms. BROWNER. There's a common-sense answer to this: Why would a Governor do it? If an area is not contributing to the——

Mr. BARTON. Because it brings the overall region into compliance by expanding the region; that's why you'd do it. That's why they're thinking about doing it in the Dallas-Ft. Worth area.

The gentleman's time has expired.

Ms. BROWNER. But it's a contributing source—it either is or it isn't; that's a fact that you determine based on air quality data. That is an objective fact; it is not subjective.

Mr. STUPAK. You have better trust in our Governors than we do, but let me go to this last question.

Ms. BROWNER. I trust that they will use common sense.

Mr. BARTON. We have one more Congressman that has to go vote. So make this a brief question.

Mr. STUPAK. I understand that the proposal that the EPA is supporting would require NO_x reductions from midwestern electrical utilities of 85 percent——

Ms. BROWNER. Correct.

Mr. STUPAK. [continuing] over the 1990 levels or 35 percent over the reductions mandated by phase II of the acid rain program. Another issue this committee was slated to address is electrical deregulation. Although I understand you're not an expert in electrical deregulation, have you or the administration looked at the interaction between the reductions and the electrical deregulation? And I think your New Jersey one, you started to mention that, and——

Ms. BROWNER. Right. That is what the PSE&G study did. They looked at deregulation and cleaner air. As I said earlier, what they found is that you can have both the savings of deregulation and use some of that savings to achieve the cleaner air, and yet still pass onto the consumer a savings.

The administration has a process that we are actively involved in on restructuring. It is being led by the Department of Energy. In fact, I was in a conversation about it this morning. And the goal, obviously, is to ensure that you incorporate into a package ultimately brought to Congress—and, again, I am not the expert on this—but you ensure that you allow for the pollution reductions and the cost-effectiveness; don't do any damage to that——

Mr. BARTON. I am going to have to insist on regular order because——

Mr. STUPAK. Thank you, Mr. Chairman, and could I ask that that New Jersey utility study be included——

Ms. BROWNER. We'll submit that for the record.

Mr. BARTON. Without objection.

Mr. STUPAK. Thank you.

Mr. BARTON. Congressman Ganske, for 5 minutes.

Mr. GANSKE. Thank you, Mr. Chairman. I'm just going to make a few general remarks because we're about ready to vote.

I think the questions so far have pointed out a couple important points, and one is that it appears that the EPA does not have statutory authority to do the type of "flexible," changes in the implementation. And, quite frankly, I share your concerns about that degree of flexibility. I think that it's important for regulators to have some flexibility, but when we're talking about important levels and changes and differences from one area to the other in the country, I see a huge potential for political abuse.

We had an example of that with the Base Closing Commission problem during the Presidential election. And I don't want to see the clean air standards used in that way. So that's a major concern that I have.

I'm very interested in hearing testimony from the next panel on the problems with potential lawsuits arising this, quote, "flexibility," and not following a standard set of regulations, and so, with those caveats, I will yield back my time and appreciate the——

Ms. BROWNER. May I respond?

Mr. GANSKE. I'm sorry, but I have to go for a vote. Thank you, Ms. Browner, for coming today.

Ms. BROWNER. I would simply say to you that we believe the flexibility is within the law.

Mr. BARTON. The Chair would ask unanimous consent that former Chairman Dingell have one more round of questions while we're waiting for two members of the majority who haven't asked their first round to come back from the vote. Is there objection? Hearing none, so ordered.

Mr. DINGELL. Mr. Chairman, I thank you for that splendid unanimous consent request.

Mr. Chairman, I would insert into the—I would ask unanimous consent to insert into the record an article appearing in The Los Angeles times in which it says, "Three environmental groups notified Southland Smoke Board and the Wilson administration that they will file suits charging that the government agencies have failed to adopt two dozen anti-smog measures they committed to implementing 3 years ago." It goes on to say——

Mr. BARTON. Without objection.

Mr. DINGELL. It goes on to say as follows: "Under Federal law, the public has power to sue to enforce clean air plans after giving State and local agencies 60 days' notice, and similar lawsuits have forced smog agencies and the Federal Government to act."

[The article follows:]

SMOG PANEL GETS 60-DAY WARNING

Pollution: Environmentalists say they will sue because agencies have failed to implement key provisions as promised three years ago.

BY MARLA CONE

TIMES ENVIRONMENTAL WRITER

Three major environmental groups notified the Southland's smog board and the Wilson administration Thursday that they will file suit charging that the government agencies have failed to adopt two dozen anti-smog measures they committed to implementing three years ago.

The lawsuit, to be filed in August, seeks to force the South Coast Air Quality Management District and the California Air Resources Board to reinvigorate efforts to clean the Los Angeles region's air, which have slowed in recent years.

"If the AQMD and ARB won't listen to reason and stop their backsliding, they will have to listen to the courts," said Gail Ruderman Feuer, an attorney for the Natural Resources Defense Council, which is teaming up with the Coalition for Clean Air and Communities for a Better Environment.

Under federal law, the public has the power to sue to enforce clean air plans after giving the state and local agencies 60 days notice, and similar lawsuits have forced the smog agencies and federal government to act.

Among the rules at issue, one with broad potential impact would require scrapping about 75,000 high-polluting cars per year in the Los Angeles Basin. In 1994, the state Air Resources Board included that measure in its smog-fighting plan at the last minute to appease oil companies.

The companies, led by Texaco, vowed to help the Wilson administration persuade the Legislature to budget millions of dollars in annual funds to buy and destroy the older cars. A funding plan has been approved by the Legislature but it has not gone into effect. Two other state rules measures would increase the use of lower-emission diesel trucks and the scrapping of older trucks.

California's 1994 smog plan—which outlines more than 100 rules—establishes how the Southland will meet a national deadline to achieve healthful levels of smog by 2010. The plan was the first in California to be approved by the federal government as meeting the demands of the Clean Air Act.

The environmental groups charge that the AQMD has violated the federal law by failing to implement 21 regulations as promised. The AQMD has delayed or abandoned many of the rules, while others were implemented but do not require cuts in pollution as large as intended.

AQMD officials say the environmental groups named measures that are no longer considered politically feasible or cost-effective. A revised smog plan, which shelves about two dozen anti-smog proposals, was adopted in November and awaits approval from the U.S. Environmental Protection Agency.

"We don't think there's been a rollback," said AQMD spokesman Tom Eichhorn. "The AQMD has maintained its momentum toward clean air rule adoption. We feel we're ahead of the ARB [state air board] and EPA in doing our fair share."

Joe Irvin, an Air Resources Board spokesman, said the agency has made progress on many smog proposals—including an agreement with the diesel industry and federal government to cut fumes from heavy-duty trucks and buses in half by 2004.

"Clearly we're aware that there are deadlines to be met and we're confident we will pursue vigorously what it takes to meet our...obligations on time," Irvin said.

But he said, "in 1994 we made it clear that some of the measures may not work. Not all expectations are going to hit pay dirt, but we will find ways to meet our targets as appropriate."

Feuer said that if the rules in the 1994 smog plan were followed, about 200 fewer tons of smog-forming emissions would pollute the air.

In their suit, the groups will ask the EPA, which oversees the smog plans, to ensure the local and state boards comply.

The AQMD board last Friday deadlocked on whether to oust its executive officer, James Lents. Lents, who has managed the agency for over 10 years, must leave by July 31 unless his contract is extended by then.

Environmentalists say Lents' imminent ouster signals that the AQMD board will ease its attack on smog, but several board members say they are committed to controlling smog but looking for a leader more responsive to cities, counties, small businesses and state legislators.

Mr. DINGELL. Now, Ms. Browner, have you received any guidance or legal opinion from the Department of Justice with respect to the filing of citizen suits on the subject of your implementation plan?

Ms. BROWNER. I had conversations with our lawyers in crafting the implementation plan specifically on citizen suits. We can ask them if they had similar conversations with the Justice Department.

Mr. DINGELL. Now did—have you or your staff, or anyone else at EPA, had any communications, written or oral, with the Department of Justice about this matter?

Ms. BROWNER. As I said, I had conversations with our general counsel's office very specifically on citizens' suits, and I will certainly ask if we could respond for the record as to the nature of their conversations with the Justice Department on this matter.

Mr. DINGELL. Submit to us any written opinions or correspondence within or to—within EPA or between EPA and to the Department of Justice on this matter.

Ms. BROWNER. If there were oral conversations in the meetings on this precise matter, we will note that similarly, certainly.

[The information provided is printed following the hearing.]

Mr. DINGELL. We have received reports that you have stated that the Department of Justice informed you that the earliest any citizen suit could be brought involving these standards would be in 3 years. Is that correct?

Ms. BROWNER. I did not personally have conversations with the Department of Justice. I had conversations with numerous lawyers at EPA, in the Office of General Counsel. Whether or not the Justice Department said that, I would have to ask.

Mr. DINGELL. Are there any written opinions from EPA attorneys or from the Department of Justice on this matter?

Ms. BROWNER. We will check. I had numerous, as I said, meetings with our attorneys on this issue—numerous meetings in which this issue came up, as did many other issues.

Mr. DINGELL. Would you please submit such written opinions or correspondence for purpose of the record?

Ms. BROWNER. We'll also characterize the meetings where the issue was discussed for the record.

[The information provided is printed following the hearing.]

Mr. DINGELL. There are also reports widely circulated that you have stated that environmental organizations have assured you that they will not challenge the implementation plan in court. Is that statement true?

Ms. BROWNER. I would never presume to know what any environmental organization might do vis-a-vis the EPA.

Mr. DINGELL. Have you had discussions with environmental organizations regarding whether or not they will sue under the citizens' suits provision?

Ms. BROWNER. We're just asking a question as to whether or not an individual environmental group has a standing under citizen suits provisions. I think they may represent a citizen in a citizen suits provision.

Mr. DINGELL. Have you had any discussions—I don't want to belabor that question at this time. I simply want to ask, have you had any discussions——

Ms. BROWNER. Oh, I frequently say to environmental groups: Don't sue me——

Mr. DINGELL. [continuing] on this matter?

Ms. BROWNER. I frequently say to environmental groups, and I am sure I have said it on this and any other number of matters: Why do you guys constantly drag us into court? I mean, that is something I have been saying for 4½ years. We don't like having 600 lawsuits.

Mr. DINGELL. Have you discussed with them whether or not they would sue with regard to the implementation of these new regulations, if not done forthwith?

Ms. BROWNER. Jon Cannon, my general counsel, informs me that some of his attorneys may have actually asked groups whether or not this was something they intended to sue on. We can respond for the record on the nature of those conversations.

Mr. DINGELL. Would you please identify each and every one of those people for the record?

Ms. BROWNER. We'll do our best, certainly.

[The information provided is printed following the hearing.]

Mr. DINGELL. Would you submit any correspondence which you might have received from any person either within EPA, any environmental group——

Ms. BROWNER. Certainly.

Mr. DINGELL. [continuing] or any other governmental agency on that matter to this committee?

Ms. BROWNER. Certainly.

[The information provided is printed following the hearing.]

Mr. BURR [presiding]. If the gentleman would yield for 1 second for a clarification from Ms. Browner—I asked what I thought was an identical question earlier. Had anyone expressed to you that they would sue over the implementation, and I understood your answer to be no at that time.

Ms. BROWNER. That's correct. No one has said to me they will sue over the implementation.

Mr. BURR. Did we understand that your counsel has had conversations with individuals who have expressed that they would sue?

Ms. BROWNER. No, no. He can answer it. No. I understood the question to be slightly different, Mr. Chairman. I understood Mr. Dingell's question to be, did we, the counsel and myself, go to an environmental group and say, "Do you intend to sue over this?" or "Will you please not sue over this?" But it doesn't matter; at the end of the day, the truth of the matter is that I routinely say to environmental groups—I implore them not to sue me. They frequently turn around and sue me.

Mr. BURR. I thank the gentleman for yielding and would yield——

Mr. DINGELL. I thank my friend. I note my time is up, and I will yield back. There are a lot of questions——

Mr. BURR. I thank the gentleman from Michigan. The Chair would recognize Mr. Greenwood for 5 minutes.

Mr. GREENWOOD. Thank you, Mr. Chairman. Welcome.

I just want to summarize a little bit what we've heard so far and make sure that we're on the same page. You have indicated, I think, a number of times this morning that these control strategies would not go into effect for 5 years.

Ms. BROWNER. I don't think I've used the word "five years" this morning.

Mr. GREENWOOD. That's why we're——

Ms. BROWNER. Well, if it would be helpful to the committee, I can walk through the timeframe. Now I don't know if that's how you want to use their time. We have made timelines available for the better part, I don't know, 3 or 4 months now. They haven't changed. They flow from the implementation document put forward by the President. And we're more than happy to submit them for the record.

Mr. GREENWOOD. You have, I think—the reference that you've used is, until the next scientific review is completed——

Ms. BROWNER. Correct, for PM. I apologize. I wasn't sure what you were asking about. That is correct, and that is expressed in the time language we have made public.2.5

Mr. GREENWOOD. Well, what's the date?

Ms. BROWNER. The——

Mr. GREENWOOD. The date?

Ms. BROWNER. Do you want me to walk you through it, because there's several dates?

Mr. GREENWOOD. Well, not laboriously, just sort of get to the date.

Ms. BROWNER. Okay. EPA would complete a 5-year scientific review—and we have made this a public comment through The Federal Register, I believe—in the year 2002. Similarly, the States would complete the collection of monitoring data in the year 2002. Now what happens after a State completes the collection of monitoring data is between the years 2002 and 2005, under the statute—this is all based on schedules in the existing law—the designations occur. So my statement that we would complete a 5-year review prior to designation flows from the timelines provided in the existing law.

Mr. GREENWOOD. Okay. Most of what we've been talking about this morning has to do with the question of: Does the statute really enable you to offer the flexibility that you have repeated said you want to offer, and, clearly, there's questions about that.

Ms. BROWNER. Uh-hum.

Mr. GREENWOOD. One of the nice things about being a lawmaker is when you get to the point in the discussion where you start the question, "Is what we want to do consistent with the law," you don't necessarily have to stop there. You have the option of changing the law to make it certain that it's consistent.

Now Mr. Burr asked you a question with regard to the notion of codifying the implementation plan, and I think your response was something like you didn't reject the notion out of hand; you didn't think it was necessary, and you then went on to say something like, to be very honest with you, we have sat around and tried to think what would—if we could write the perfect law, I know it doesn't work that way, what would we do to make sure—and I think your point was to make sure that you have the standards and you also have the implementation plan. But we haven't, I think you said something to the extent that, we haven't been able out how to do that.

Ms. BROWNER. At this point in time we have no recommendation.

Mr. GREENWOOD. Okay. Could you share with me some of the impasses that you may have encountered in thinking through that scenario of developing a law, a bill that we could pass?

Ms. BROWNER. Oh, I'll give you——

Mr. GREENWOOD. What are some of the difficulties in doing that; that is, a law that would give you the flexibility you need, not diminish your flexibility, but assure us in a way that we've not been assured yet that the implementation plan will not collapse as a result of a lawsuit?

Ms. BROWNER. Well, I'll give you one easy example and build on the example we were just using in terms of the timeframe. We were very clear—I was very clear—that it would be important to complete a second science review before you—and as I think the Presidential memo makes very clear—before you require the States, for example, to submit their State implementation plans.

And I wanted to make absolutely sure that we had the authority to do that, to complete the science review and not require a State implementation plan until we had done so. We looked at the statutes, and are more than happy to provide to you all of the provisions that we rely on in making that judgment that on the face of the statute it in fact is very doable; that the statute is very clear—you get 2 years to do this; you get 3 years to do that, and then this happens. It lays out timeframes.

So then the question would be: What if someone filed—and these are the kind of conversations I had on numerous occasions with our lawyers—what if someone filed a citizen suit? What if a year into this someone said, "No, wait, we don't think they should get 3 years to collect data."? Do we have the ability to defend against that? Absolutely. The statute says a Governor gets 3 years. It's hard to be more explicit than that. So I don't know what I could offer you as a statutory revision that is any clearer than the current statute.

And I think we are all well aware that any time you change anything, that gives rise to its own litigation. These are provisions that have in many instances already been used and to some degree perhaps already litigated. We know how they've been interpreted; we know how to implement them.

Mr. GREENWOOD. Mr. Chairman, I'd ask unanimous consent for 1 more minute.

Mr. BURR. Without objection.

Mr. GREENWOOD. Thank you, and thank the members.

Ms. BROWNER. I'm sorry——

Mr. GREENWOOD. Suppose that we passed a statute that referenced your implementation plan and made it clear in statute that managing the Clean Air Act with regard to these standards is consistent with the implementation plan and cited it, cited that document; what would be the difficulty with that? I mean, that would certainly answer the question. It would make moot the question, all of the questions, about whether or not the implementation plan is consistent with statute, because we would say so in law.

Ms. BROWNER. With all due respect, it is simply not necessary, and moreover——

Mr. GREENWOOD. Let's assume it's not necessary. Let's assume that it's not necessary, but it would make us feel good. Would it do any damage?

Ms. BROWNER. Well——

Mr. GREENWOOD. Would it harm anything?

Ms. BROWNER. I would ask you to just consider one fact. As I've explained, you have the Presidential memo; you have——

Mr. GREENWOOD. I have limited time. Would it do any harm to do what I have just described?

Ms. BROWNER. It is important to the process that the rulemaking, the regulatory phase of this, the stakeholder meetings, all move forward. That is an incredibly important next step, and I will be honest with you; just hearing what you're saying—I would be concerned that you could unintentionally do damage to that process and the best use of that rulemaking process.

Look, no one would ever say never in our business—in your business and in my business. If in the same way that we did on drinking water, we go into the nitty-gritty of this and 3 years down the road we're through the rulemaking process, and we can all see that here's a piece that, quite frankly, is not going to work well—I can't tell you any of those pieces right now—of course, we'll be back here to say, look, we all worked together—States, local government, industries—and here's something no one could have dreamed of on the front end. But you'll get the best decisionmaking by allowing the next steps of the process to unfold, as they have previously.

Mr. BURR. The gentleman's time has expired, and, Ms. Browner, I don't think it's the last time we'll ask you to answer that.

At this time the Chair would recognize Mr. Upton for 5 minutes.

Mr. UPTON. Thank you, Mr. Chairman.

Ms. Browner, again, I want to welcome you to the hearing. I want to walk you again through as the new standards impact southwestern Michigan.

I represent six counties; all of them meet the old standard or the present standard at 120 parts per billion——

Ms. BROWNER. For 1 hour.

Mr. UPTON. For 1 hour.

Ms. BROWNER. Right, okay.

Mr. UPTON. And for sure, they'd make it for 8 hours, too.

Ms. BROWNER. That's true.

Mr. UPTON. Under this new regulation, all of my counties would not be in attainment. That's purely because of the transient air coming across Lake Michigan—Gary, Milwaukee, and Chicago—and there are some pretty good studies to show, in fact, that about 95 percent of the ozone is from those sources. It's my understanding that by putting my counties in a transition area or designating them as a transition area, there would be relief, if Michigan went along with the OTAG group; is that correct? Or went along with the——

Ms. BROWNER. Just to make sure we have the facts straight here, we believe there are potentially six counties in your district that——

Mr. UPTON. Five of the six would be hit. St. Joe County I think is not because it's more than——

Ms. BROWNER. That's not one of the counties we have; right.

Mr. UPTON. Yes, but Kalamazoo County, which was not on the original list, we're told ozone levels have exceeded 80 parts per billion at least 34 times already this year. So at whatever point——

Ms. BROWNER. Oh, no, we would know if someone had 34 exceedances this year, I assure you. Los Angeles hasn't had 34 exceedances this year.

Mr. UPTON. Well, that's what my State tells me.

Ms. BROWNER. We got the counties from data from your State, and I'm more than happy to read them to you or show them to you afterwards, but——

Mr. UPTON. Well, I want to go over the general——

Ms. BROWNER. [continuing] it doesn't matter. The general point—as I understand the situation in your part of the State, and we will go back and check this, and if there's any need to clarify, we will do so is—you're right; it is largely a transport problem. While you, obviously, generate some amount of the pollution in those counties, a significant amount of it is coming from somewhere else.

Mr. UPTON. That's right, from other States, in fact.

Ms. BROWNER. From other States. From other places. If—and this is true for all of the States that are part of OTAG—if they participate in the program that they crafted—and we have every reason to think they will because it's their program—and the counties that meet the 1-hour, but might not be able to meet the 8-hour standard, can meet that through the regional controls, then they would be eligible for the transitional classification.

Mr. UPTON. That's right, but if Michigan does not go along and participate in the OTAG, then they're sunk.

Ms. BROWNER. Well, no——

Mr. UPTON. The counties.

Ms. BROWNER. Michigan, under the existing standard, under the existing requirements, has to deal with the pollution problems. What's being offered them, quite frankly, in OTAG is a much more sensible way to deal with the transport issue.

Mr. UPTON. Well, it's my understanding that by participating in the OTAG we would have to have our utilities reduce their emissions by some 85 percent. I think that's one of the underlying principles.

Ms. BROWNER. The cap-and-trade program is designed to focus on utilities across the States; that is correct.

Mr. UPTON. And I don't know where they got the 85 percent. I know in my——

Ms. BROWNER. The States came up with that.

Mr. UPTON. Well, I know in my district we have two nuclear plants, and their emissions are negligible.

Ms. BROWNER. They probably won't be part of this then.

Mr. UPTON. And I know for—I think for every——

Ms. BROWNER. I don't believe they are envisioned to be part of this. It's largely coal-fired plants.

Mr. UPTON. And, again, for Michigan, I think we've been burning low-sulphur coal that is shipped from Montana versus Kentucky and West Virginia for a good number of years as well. So I think it would be very hard—as I understand it, as I've talked to our DEQ folks, it would be very, very difficult to come up with a plan to reduce the emissions for our utilities by 85 percent, which is probably one of the reasons that they're not anxious to sign onto the OTAG.

Ms. BROWNER. As you point out, you are receiving pollution from other places. The pollution your utilities generate is impacting someone else. That's why a regional approach makes so much sense.

Mr. UPTON. I don't know that it's ours; it might be Ohio's or Illinois'. I don't know that it's——

Ms. BROWNER. Well, your pollution is going somewhere. It's not simply disappearing. It's going somewhere. I mean, that's just a fact. At some point, if people would like, we could demonstrate the computer models that were developed with the States. They're actually kind of interesting to watch how the pollution moves. You can begin to see where the pollution goes. It may no be impacting your citizens; it's impacting somebody.

Mr. UPTON. Well, let's say that somehow——

Mr. BARTON. This will have to be your last question, Mr. Upton. You can have one more question——

Mr. UPTON. I can put it all in about a 10-minute question.

Ms. BROWNER. Obviously, your State has the right to do what they ultimately want to do under the SIP. What the 37 States all focused on was, what is the most cost-effective way to deal with NO_X reductions?

Mr. UPTON. Well, the question that I have is, if—who is—what evidence do you have that these transitional—if everything works out okay—let's say that we get the transition; let's say that somehow this thing works. What assurances do you have or not have from the Justice Department or others to try and have these things stand up in court, when someone goes to the courthouse door, once these things have been stamped——

Ms. BROWNER. We take our authority from provisions on the books today, and we are very comfortable that within the existing Clean Air Act we have the authorities. For example, to go back to Mr. Greenwood's example, if someone were to file some lawsuit saying, "Get this done in 1 year," we would respond with the law, which says a Governor gets 3 years. That's the law, pure and simple. That's what Congress did.

Mr. BARTON. Madam Administrator, Madam, that's simply not true. It's——

Ms. BROWNER. We'll read you the section——

Mr. BARTON. It gives the EPA—the EPA has the discretion to set rules and regulations, but nowhere in the act does it say 3 years. You have, by rule, I'm told, promulgated 3 years, but that is not in the statute.

Ms. BROWNER. Well, I'm more than happy to read you where the years are. I can read you the sections. It's in two different places. In one place a year is provided; in a second place 2 years are provided.

Mr. BARTON. Well, cite the specific authority.

Mr. UPTON. Reclaiming my time just for one nanosecond, it's the——

Ms. BROWNER. Sorry.

Mr. BARTON. We'll give you more than one nanosecond.

Mr. UPTON. You're either in or you're out, is the way that the statute reads.

Ms. BROWNER. No.

Mr. UPTON. There's no transition area provision that allows you to have the flexibility to do these——

Ms. BROWNER. I think the specific question I was being asked to give a citation on was the data collection by Governors, and I would refer you to section 7407(d)(1)(a) and (b). Okay? And (d)(1)(a)—I'm reading it—right here it says 1 year. Then I flip the page and in (b) it says 2 years. Now 2 plus 1——

Mr. BARTON. Read them. Read the provisions, because staff doesn't agree with your interpretation.

Ms. BROWNER. Do you want me to read it?

Mr. BARTON. That's what I asked you to do.

Read it out loud.

Ms. BROWNER. I'm trying to figure out how not to read you 3 pages. I mean, I'm trying to——

Mr. BARTON. I know you're trying to be responsive.

Ms. BROWNER. Okay. "Designations. Generally, submission by Governors of initial designations following promulgation of newer revised standards. By such status, the Administrator may reasonably be required, but not later than 1 year after promulgation of"—we'll give the clerk the cite—"on the new or revised National Ambient Air Quality Standard for any pollutant under section 7409 of this title, the Governor of each State shall, and at any other time the Governor of any State deems appropriate, the Governor may submit to the Administrator a list of all areas," blah, blah, blah, blah, blah.

I know, those were well-crafted, thought-out words, but, "Upon promulgation or revision of a National Ambient Air Quality standard, the Administrator shall promulgate the designation of all areas or portions thereof submitted under paragraph (a) as expeditiously as practicable, but in no case laster than 2 years from the date of promulgation of the new or revised National Ambient Air Quality standard. Such period may be extended for up to 1 year in the event the Administrator has insufficient information to promulgate the designation." That adds up to 3 years.

Mr. BARTON. But, Madam Administrator, we're not trying to be argumentative just to prove that Congress can be argumentative, but the things that you just cited gives you the discretion and some timetables when you have to act, but nowhere does it say you give the Governor—that the Governors are given 3 years. Now that's the way I read it.

Mr. KLINK. Will the chairman yield?

Mr. BARTON. I would yield on that point.

Mr. KLINK. In fact, at the end of what she was reading it says, "Such period may be extended for up to 1 year in the event the Administrator has insufficient information to promulgate the designations." That's the whole question. That's our discussion, is whether or not they have insufficient information, and that's the reason that we wanted to move forward by doing the studies first, by—in 1984 we wanted to authorize the expenditure of $75 million a year to build the equipment first, to set up the monitors, and to collect the data for 3 years. So I, again, think that——

Ms. BROWNER. Can we cite a different section on——

Mr. KLINK. I'm sorry, I'd ask for regular order, Mr.——

Mr. BARTON. Well, the regular order is Mr. Green's asking the questions. So we're a little beyond regular order.

Mr. UPTON. Mr. Chairman, I haven't yielded back the balance of my time.

Mr. BARTON. I would yield to the distinguished former chairman.

Mr. DINGELL. I thank the chairman.

I would observe that, first of all, we're discussing whether the Administrator has the authority to give 3 years.

Mr. BARTON. I would say——

Mr. DINGELL. She has not cited an authority to give 3 years. She has cited authority to give up to a year where she has insufficient information. I am driven, reluctantly, to the assumption that she's telling us that she has insufficient authority—rather, insufficient information to proper act. This is one of the theses of the hearing in which we're now engaged, and that is that she lacks adequate information upon which she may properly proceed. But she's not cited the authority for giving 3 years.

Mr. BARTON. I agree with that.

The Chair would now recognize Mr. Green for the last 5 minutes of questions of the very patient Administrator.

Mr. GREEN. Thank you, Mr. Chairman, and, again, I apologize to our witness for, again, having votes and not being here for other responses to the questions. And so if it's repetitive, if you'll forgive me, but also I'd like to just say, on a personal note, I'm glad to see my friend, Mr. Barton, in the chair. The last person who was in the chair said he didn't like Texans. So I'm glad to have a friendly Texan in the chair.

But, Madam Administrator, I think you may have answered this before, at least from what I've picked up, but I want to walk you through for the reason I feel like that I—and in my opening statement said I co-sponsored 1984, was that we actually needed a statutory change to allow the EPA to give the flexibility for the implementation rules, because the reason we're here is because of a lawsuit that EPA had to respond. And so by the—does the EPA have the authority to carry out the implementation plan and conjure up—because, again, as a lawyer, there will be a lawsuit, if someone is not—whether it be a citizen, whether it be a group, or whatever, and I'm concerned that 4 or 5 years down the road, when you give us that kind of time, whether the Governors have the authority or the EPA Administrators have the authority, that there will not be the ability to deal with that, and we'll be back where we are, unless we see a statutory change to give you that——

Ms. BROWNER. Let me just say—maybe it's obvious, but from my perspective, the statutory change would also be litigated. What I'm suggesting to you, with all due respect, is that in the body of the existing law the authorities exist to defend against any lawsuits of the sort that you raise. Look, if we get 3 or 4 years down the road and one of these lawsuits, which we don't think are going to be successful or some judge does some truly amazing thing, and that certainly has happened——

Mr. GREEN. It's happened a lot——

Ms. BROWNER. [continuing] where will we be? Right here asking you to work with us to resolve it. But right now there is nobody in this body that could begin to say with absolute certainty what might happen in any of that litigation. I can tell you that we are very comfortable that we have the authority to defend against those lawsuits, but I will not, obviously, speak to what an individual judge may do. But I will give you my word that if a judge does something beyond what any of us could have ever dreamed of, the first place we will be is back here. If you write language now, we will merely litigate that language. That's just the nature of the beast. It's the nature of the system. Why not stay the course? Many of these provisions have been litigated. We understand them. We know what the courts will say about them.

Mr. GREEN. Well, I just feel more comfortable with litigating a statute than your response to a statute or interpretation—Would the gentleman yield?

Ms. BROWNER. But that's what we'll be litigating. We won't be raising my interpretation. We will be taking the section of the statute and presenting it to the court, and obviously when you do that, you present the historical use of that section. That's the way the system works.

Mr. GREEN. And I think the reason my colleague, my two colleagues from both Michigan and Pennsylvania have the bill was to give you that assurance that you could have that implementation schedule that you've said—you testified to, and we've heard—let me get to another question before I run out of time.

In the areas of non-attainment, where emission reductions are needed beyond what is achievable on current technologies, again, in relationship to the district I represent and the city of Houston, the EPA—I have heard it once established that emissions could be reduced for an average cost of $10,000 per ton of emission. Are you familiar with that figure, and is it correct?

Ms. BROWNER. No, I'm not familiar with it.

Mr. GREEN. Okay.

Ms. BROWNER. If you have a document, I'm more than happy to look at it.

Mr. GREEN. Okay.

Ms. BROWNER. I don't know what document you're referring to. I'm more than happy to look at it.

Mr. GREEN. Okay. So you're not familiar with the $10,000-per-ton estimate?

Ms. BROWNER. Not particularly, no.

Mr. GREEN. Okay. Okay.

Ms. BROWNER. It may exist; I don't know.

Mr. GREEN. We will submit a question——

Ms. BROWNER. Okay.

Mr. GREEN. [continuing] and give the background, and ask for a response.

We're going to hear later in the panel from the National Council of State Legislatures, and I've read the testimony, and the concern of what the requirement will be on the State legislatures. For example, can you describe what financial and administrative burdens may fall on State and local governments if the implementation schedule from the EPA is successfully challenged in court? Again, I know we're looking into the future and crystal balls, but will that raise the specter of the 50 States having additional resources that they may not have either now or in the future?

Ms. BROWNER. I'm not sure I understand the question. As far as I am aware, most of the States either have or are on their way to having passed the necessary State laws, because they're not unique to this standard. They are clean air laws passed by the States, so the can assume day-to-day operation of the program.

Mr. GREEN. Well, and I know——

Ms. BROWNER. And they've done that in their funding, and we give them money in our budget every year. Beyond this monitoring issue, there is money we give to the States; you all give to the States.

Mr. GREEN. Yes, but there's also maybe under current—you know, before the July interpretation or the July regulations, the States have set a level where they're going, and now we have additional standards, and, again, depending on the lawsuits on the implementation schedule is challenged—again, it's—I know we're reading into the future, but——

Mr. BARTON. The gentleman's time has expired.

Mr. GREEN. [continuing] that's my concern.

Mr. BARTON. Let there be a response. The gentleman has been very gracious to wait his turn. So we want to give him adequate opportunity.

Ms. BROWNER. The States do a lot of the day-to-day work, and I think we all agree that that's the best way to manage these programs. Many of them have tremendous expertise. I'm not really sure I understand your question still. In terms of what will they have to do? Obviously, the Clean Air Act envisions that they'll write a plan.

Mr. GREEN. If the Clean Air Act revisions stand, but also with the implementation schedules, do you foresee what financial and administrative burdens may fall upon the State and local governments in——

Ms. BROWNER. It's what they have to do already, I think, except for the monitoring, which we're paying for. So if they have some specific instances, we'd be more than happy to sit down and talk to them.

Mr. GREEN. Well, Mr. Chairman, I know our next panel—and we have a member from the Council of State Legislatures; they may want to add that to their testimony in response to it. Okay?

Mr. BARTON. The gentleman's time has expired. The Chair would recognize, for the last questions of this witness, Mr. Pallone of New Jersey for 5 minutes.

Mr. PALLONE. Thank you, Mr. Chairman, and I want to apologize, Ms. Browner; if this has already been answered, you know, just tell me, because I haven't been here for the last hour or so.

But in your statement, not only in your oral statement, but also in your written presentation, you had talked about the cap-and-trade program for NO_X, and I just wanted, if you could, to just elaborate a little more on the controls that utility sources play in EPA's plans for achieving attainment of these new ozone standards, specifically the cap-and-trade program for NO_X. And I notice you mention in your written statement that regional emissions cap-and-trade systems similar to what you're proposing now exist for the acid rain program. And I just wanted—maybe you could tell us a little bit about the success of that and how you're going to try to implement that in—you know, with NO_X and with the utilities now.

Ms. BROWNER. Cap-and-trade programs were developed I think back under President Bush, when they were put into place for the first time, and they have proven to be incredibly cost-effective. And, simply, the way they work is we don't worry about which individual facility achieves the reductions as long as an average across all facilities within a region is achieved, and that allows a business decision in part to be made. If one facility is older, it might be above the average. They could buy credits from the newer facilities that's below. It allows for the public health protections to be met in the most cost-efficient manner.

Acid rain is a great example. When the program was originally proposed, industry did some studies and they suggested it was going to cost as much as $1,000 per ton of reduction. EPA did a study that said $600, and I think one of the members mentioned earlier today that on the Chicago Board of Trade, today a credit sells for $78 per ton, demonstrating just how cost-effective these programs can be. They have enjoyed tremendous industry support, and we're looking, actually, in other areas beyond just this area; we think they're so helpful.

Mr. PALLONE. The other thing is, how do the costs of controls on utility sources compare with obtaining the necessary emissions reductions from other sources? In other words, you know, I guess the impression I get is that you feel that you get a bigger bang for the buck, so to speak, by doing utilities before other sources, and if you could just comment on that?

Ms. BROWNER. The studies have looked at—and, again, these tend to end up being sort of worst-case scenarios; that for utilities it will run approximately $1,000 to $1,500 a ton. If you looked to other sources, if you move beyond the regional utilities approach, you are talking of as much as $10,000 per ton of pollution reduction. So it's significantly less costly.

And, again, remember, when you factor in a cap-and-trade program, that may well have the benefit of further reducing the cost to the utilities.

Mr. PALLONE. And then, last, you did——

Ms. BROWNER. There are credits already being sold. I should just explain this. Some States are already running a cap-and-trade program for NO_X. I believe New Jersey and Massachusetts have similar type programs in place. Massachusetts, the last time I checked, was trading at approximately $150.

Mr. PALLONE. And you——

Ms. BROWNER. I mean, it's not exact—it's not identical, but that gives you an idea.

Mr. PALLONE. Okay. You also made reference to another New Jersey study, I think, in your opening——

Ms. BROWNER. Yes.

Mr. PALLONE. I think it was PSE&G—I'm not sure—where you said that the costs of doing the controls are a fraction of the profit utilities will realize during deregulation. How is that figured? What was the basis of that?

Ms. BROWNER. Well, probably the easiest thing to do is show you the report. It was done by a utility, and what it shows you in this chart is the savings they would expect to realize through deregulation, and the cost of the pollution reductions, relative to those savings. And as they point out, it is 7 to 1. In other words, for every $7 saved through deregulation, they would anticipate a dollar spent for pollution reduction, which leaves you with $6, if I did my math right, in savings. And we have agreed to submit this to the committee for the record. Again, this was done by a utility who went out and actually looked at what, as they described it, a regional program will result in.

[The study is retained in subcommittee files.]

Mr. PALLONE. Okay, thank you. Thank you, Mr. Chairman.

Mr. BARTON. Does the gentleman from New Jersey yield back?

Mr. PALLONE. Yes.

Mr. BARTON. That concludes the first round and the only round of questions. The Chair is going to ask unanimous consent that Mr. Burr have an additional round to ask some questions, during which time he will yield to any other member still here who may have 1 or 2 follow-up questions.

Ms. BROWNER. How much time?

Mr. BARTON. Is there objection? Mr. Burr is recognized.

Mr. BURR. All right. I thank the chairman. I thank the members.

Ms. BROWNER. Five minutes?

Ms. Browner?

Mr. BARTON. Ten minutes.

Mr. BURR. If the EPA—let's make an assumption. If the EPA is sued and loses on the implementation plan that you currently have, do you have a contingency plan, and if so, can you share it with this committee?

Ms. BROWNER. The way in which a party would take access to a court would be around particular steps in the process, not the implementation plan at large. So, for example, if we say you get "X" number of years for a Governor to run the monitors, someone would say no. I'm just playing out your example.

Mr. BURR. I'm going to ask you to be very brief, if you would.

Ms. BROWNER. Well, we are very clear of our legal authority to grant that time, and we would defend against the lawsuit. As I said earlier, if some weird thing were to happen with a judge and they were to say X, Y, or Z, at that point we would obviously come back to Congress and say, despite all of our agreements that it was 3 years, someone has said something else; fix that provision; clarify that provision.

Mr. BURR. I'm going to ask, and then yield to Mr. Dingell, that if there is written contingency plans for certain sections that you anticipate that exist at EPA, that you share them with this committee, and I would yield——

Ms. BROWNER. Can I respond?

Mr. BURR. Yes, ma'am.

Ms. BROWNER. Obviously, there wouldn't be because I can't begin to tell you what some judge may do. I can tell you—and this is what you have every right to demand of me—what is the legal authority, and what are the sections of the Clean Air Act upon which we premise the implementation plan. And what I can also tell you is that if a judge were to do something outside the face of the statute, the first place we would be—as we have done in other instances—we have come back here and said, "Could you make a correction? Everyone read it this way, but we found the one person in the United States who read it a different way."

Mr. BURR. I will take this opportunity to yield to the gentleman from Michigan, Mr. Dingell, to let you sell him on the fact that, his original line of questioning, that he got an answer to, because I don't think he did.

Mr. DINGELL. I thank the gentleman.

Mr. Chairman, I ask unanimous consent to insert in the record three documents, the first of which is a letter to the Honorable Dennis J. Kucinich, House of Representatives, Washington, DC. This relates to—and it's from Mary Nichols, in which she says, "The Administrator has asked me to respond to your letter of April

23, 1997 in which you requested information from the Environmental Protection Agency concerning a number of assertions made by Governor George Vonivich of Ohio. Specifically, Governor Vonivich raised concerns about the impact of these standards on a Ford automobile manufacturing plant in Lorraine, Ohio, and a Ford casting plant in Brook Park, Ohio."

And the letter goes on, then, to say, "As to impacts of efforts to meet these new standards specifically on Lorraine in Cuyahoga County, where the two facilities Governor Vonivich referred to are located, these counties would likely not have to meet the proposed new ozone or PM standards, had these standards been in place during the 1993-1995 period. However, based on current data, it is likely nothing other than continued implementation of the 1990 Clean Air Act will be necessary in order for Lorraine County to meet the proposed new standard.

It goes on to say, "In essence, it is not likely that either county would require additional local controls in order to meet the new ozone standard."

The next is a letter——

Mr. BARTON. Mr. Chairman, we will take those under advisement. The majority staff would like to look at the documents, and once we've looked at them, we'll almost certainly accept them.

Mr. DINGELL. You'll find these most helpful.

Mr. BARTON. Okay.

[The information referred to follows:]

U.S. House of Representatives
Committee on Commerce
Room 2125, Rayburn House Office Building
Washington, DC 20515–6115

June 11, 1997

The Honorable Carol M. Browner
Administrator
U.S. Environmental Protection Agency
401 M Street, S.W.
Washington, D.C. 20460

Dear Administrator Browner:

In recent weeks you have contacted Members of Congress as well as state, county and local officials to offer assurances as to how your agency proposes to handle the disruptions and costs inevitably associated with EPA's proposed new standards for ozone and fine particles. For instance, we have reviewed a letter to Representative Kucinich in which Assistant Administrator Mary Nichols essentially asserts that an automobile manufacturing plant and an automobile casting plant in Rep. Kucinich's district would not have to put on additional controls even if the plants are located in counties found to be violating any new ozone standard. We also understand that you have assured certain mayors and other local officials, as part of your efforts to bolster support for EPA's proposed standards, that EPA would not require jurisdictions to take any additional steps even if the area were found to violate the new ozone standard.

We applaud your apparent recognition of the problems EPA's proposals will create. But we are concerned with these assertions both because we question EPA's legal authority to suspend Clean Air Act requirements and because of the appearance that EPA is seemingly trying to win support for its proposals through extra-legal promises of leniency for selected areas during implementation of the standards.

Your assertions raise three fundamental questions: (1) who has received assurances of leniency, (2) what were those assurances, and (3) how much value those assurances should be given, in the face of contradictory statutory provisions and the expansive citizen suit rights found in the Clean Air Act?

These questions are set out in detail in an attachment to this letter. We would appreciate your response to these questions by June 20, 1997.

Sincerely,

JOHN D. DINGELL
RANKING MEMBER
COMMITTEE ON COMMERCE

RON KLINK
RANKING MEMBER
SUBCOMMITTEE ON OVERSIGHT
AND INVESTIGATIONS

Attachment

QUESTIONS

: Please describe in detail all instances where EPA has provided assurances that specific localities or sources will not have to take certain actions to reduce emissions or meet certain specific Clean Air Act nonattainment requirements despite the area's projected violation of EPA's proposed air quality standards for ozone or particulate matter.

: As you are aware, Title I of the Clean Air Act Amendments sets out the steps that EPA and States must take once a new national ambient air quality standard is established Pursuant to section 107, EPA is to promulgate area designations based on the new standard "as expeditiously as practicable, but in no case later than 2 years from the date of promulgation of the new or revised national ambient air quality standard." Once an area is designated nonattainment, are there any requirements that apply as a matter of law with the designation? For instance, major sources in nonattainment areas must seek special construction permits before they can build new emissions units or modify existing facilities. This preconstruction review program requires sources to install stringent controls and secure offsetting emissions reductions and it causes substantial delays in projects. When would these construction permit requirements apply to sources in an area designated nonattainment? Similarly, many sources in nonattainment areas must secure an operating permit as required by Title V of the Clean Air Act Amendments of 1990. When would the requirement for operating permits apply to sources in a nonattainment area?

3 Within three years after promulgation of a national ambient air quality standard, States must submit implementation plans which must include numerous planning and control requirements as well as enforceable schedules and timetable for sources to comply with the requirements. These measures must include, pursuant to section 172(c), the requirement that sources in the nonattainment area retrofit all emissions units with reasonably available control technology. How does EPA plan to ensure that the two plants in Representative Kucinich's district will not have to retrofit reasonably available controls?

4 Please list all other Clean Air Act nonattainment requirements that will apply to nonattainment areas and indicate which ones EPA will be enforcing and which EPA will not be enforcing after promulgation of EPA's proposed ozone and fine particle standard For each of the requirements EPA will not be enforcing, please provide an explanation of the rationale and legality of EPA's decision.

5 States that fail to take actions required under the nonattainment provisions of the Clean Air Act face sanctions of increased offsets and loss of highway funds The Act also requires EPA to step in and directly order controls, issue permits and take other actions to enforce nonattainment requirements when a State fails to act in a timely manner These sanctions and federal implementation provisions are written as mandated actions, once a State defaults on its duties. Can EPA assure States that these sanctions and federal implementation provisions will not apply to any nonattainment requirements that EPA hopes to "suspend" during implementation of the new ozone and fine particle standards?

6 Section 304 of the Clean Air Act provides for citizen enforcement of most requirements established under the Act. What assurances can EPA provide that a citizen will not force EPA and States to implement fully all of the nonattainment requirements, notwithstanding any EPA decision to suspend certain requirements? What assurances can EPA provide that citizens will not be able to sue sources directly for violations of suspended Clean Air Act requirements?

UNITED STATES ENVIRONMENTAL PROTECTION AGENCY
WASHINGTON. D.C. 20460

''''' ' 6

Honorable Dennis J. Kucinich
House of Representatives
Washington, DC 20515

Dear Congressman Kucinich:

The Administrator has asked me to respond to your letter of April 23, 1997 in which you requested information from the Environmental Protection Agency (EPA) concerning a number of assertions that were made by Governor George Voinovich of Ohio. Those assertions focused on the Governor's view of the impact of proposed revisions to national air quality standards for ozone and particulate matter (PM). Specifically, Governor Voinovich raised concerns about the impact of these standards on a Ford automobile manufacturing plant in Lorain, Ohio and a Ford casting plant in Brookpark, Ohio. He expressed concern that the standards would provide no significant health protection and have a devastating impact on these facilities and other small businesses in Ohio.

As you know, EPA has proposed new standards for ozone and for particulate matter based on multi-year scientific reviews. As an orderly means of incorporating the best, most recent science into the air quality standards, the Clean Air Act requires such reviews to be completed every five years for each of six major air pollutants. Analyses conducted during these reviews indicated that serious adverse health effects were occurring in children, the elderly, and others, both at and below the level of existing air quality standards. This information led us to propose a tightening of the ozone standard, and to propose setting a new particulate matter standard which focuses on very small particles.

The proposed PM standard is estimated to achieve each year 15,000 fewer premature deaths; 9,000 fewer hospital admissions, and many fewer emergency room visits; 60,000 fewer cases of symptoms associated with chronic bronchitis; 250,000 fewer incidences of respiratory symptoms in children, such as aggravated coughing and difficult or painful breathing; and 250,000 fewer incidences of aggravated asthma. The proposed ozone standard is estimated to achieve each year 1.0 million fewer incidences of significant decreases (15% to over 20%) in children's lung functions (such as difficulty in breathing or shortness of breath); 200,000 fewer incidences of moderate to severe respiratory symptoms in children, such as aggravated coughing and difficult or painful breathing; 500 fewer hospital admissions; and 2,000 fewer emergency room visits for individuals with asthma and reduced risks of premature aging of the lungs. In short, we believe these protections to be significant improvements in public health. We are now reviewing the substantial public record on these proposals, and will make final decisions by July 19, 1997.

Regarding implementation efforts which would follow from any new standards, as you may know, the first several years following the setting of a new standard involve monitoring, designations and planning activities. These activities are important to the ultimate attainment of an air quality standard, but do not require new controls or expenditures on the part of pollution sources. In practice, additional pollution controls on existing sources, if necessary to meet a new standard, would not be required until at least six years from the date of setting a new ozone standard, and probably longer for a new fine particle standard. The setting of a standard based on public health information is the essential step, because it provides local communities with a clear measurement of local air quality, i.e., is it healthy or is it not. But the setting of the standard itself does not trigger immediate pollution controls.

As to the impacts of efforts to meet new standards, specifically on Lorain and Cuyahoga Counties where the two facilities Governor Voinovich referred to are located, these counties likely would not have met the proposed new ozone or PM standards had these standards been in place during 1993 to 1995 period, the most recent three year period for which we have complete data. However, based on current data, it is likely that nothing other than continued implementation of the 1990 Clean Air Act (CAA) will be necessary in order for Lorain County to meet the proposed new ozone standard. In addition, based on current data, we believe that continued implementation of the Act, plus the application of a regional control strategy in the Eastern United States (focused on reducing large emissions from power plants, large industrial sources and new automobiles), will allow Cuyahoga County to meet the proposed ozone standard as well. In essence, it is not likely that either county would require additional local controls in order to meet the new ozone standard.

In fact, of the 335 counties nationwide projected to potentially not meet the proposed ozone standard, EPA estimates that 226 (67%) of these counties would come into attainment without additional local controls, with continued progress under the 1990 Clean Air Act and a application of a regional control strategy. Such combined strategies will go a long way towards meeting any revised ozone standard in the most cost-effective manner.

Regarding particulate matter, as you may know estimates of nonattainment with the new PM 2.5 standards rest on projections from PM10 data, not on comparisons to actual fine particle air quality monitoring, because very little long term monitoring of these small particles has been done. Once actual fine particle monitors are in place gathering real data, it may be that counties now projected as potentially non-attainment will be found to not have a fine particulate matter problem. In addition, for those counties in nonattainment, significant reductions in $PM_{2.5}$ levels will be achieved through the full implementation of the acid rain provisions of the Clean Air Act and the regional control strategy for NO_x discussed above. In many cases, these strategies will reduce or eliminate the need for local controls.

Concerning our efforts to help manage the ultimate costs of meeting any new standards, as you may know, we are conducting economic analyses to go along with new standards. The information about potential economic costs and benefits developed through these analyses will be helpful in informing the debate about potential costs and benefits, and in pinpointing potential high cost areas of implementation that can be avoided.

In addition, EPA has established an advisory committee comprised of representatives from industries including small businesses, State and local air pollution agencies, local governments, environmental groups, and others. The purpose of the advisory committee is to make recommendations to EPA regarding the development of new, cost-effective strategies for implementing any new ozone or PM standard. An important part of the advisory committee's work has been the consideration of regional scale strategies to help reduce the transport of these pollutants of concern. In developing these new strategies, EPA will take into account the emissions reductions to be achieved when the 1990 Clean Air Act is fully implemented, and assess costs associated with any proposed strategies.

In addition to the advisory committee, EPA, the Small Business Administration, and the Office of Management and Budget within the Office of the President, are holding a series of meetings with representatives from small businesses, small governments and other small organizations. The Administration is very interested in making sure that small entities have a special opportunity to express their views on these proposed standards and will maintain an ongoing dialogue with their representatives throughout the implementation development process. We are committed to working with States, industry, environmental groups and the public through the process of achieving any new standards. Ultimately we believe that setting standards based on public health, and working collectively to meet these standards in cost-effective ways, will allow us to continue to clean up our air, and maintain a health economy.

I appreciate this opportunity to be of service and trust that this information will be helpful to you.

Sincerely yours,
ORIGINAL SIGNED BY

MARY D. NICHOLS
Mary D. Nichols
Assistant Administrator
for Air and Radiation

UNITED STATES ENVIRONMENTAL PROTECTION AGENCY
WASHINGTON, D C 20460

AUG 1 1997

OFFICE OF
AIR AND RADIATION

Honorable John D. Dingell
Ranking Member
Committee on Commerce
U.S. House of Representatives
Washington, D.C. 20515-6115

Dear Congressman Dingell:

This is in response to your letter of June 11, 1997, to Administrator Browner in which you ask a number of questions related to implementation of the new ozone and particulate matter air quality standards. As you know, on July 16, 1997, Administrator Browner signed the final rulemaking packages for the new national ambient air quality standards for ozone and particulate matter. On that same date, President Clinton issued a memorandum to the Administrator, providing direction for implementation of the standards. All of these documents were published in the Federal Register on July 18, 1997.

The revised standards are critical for protecting public health, in particular the health of the elderly, the infirm, and children. Although we are prohibited from considering costs in setting these standards, consistent with the President's memorandum, EPA is considering the economic impact of the standards in the course of developing an implementation strategy. EPA believes that the Clean Air Act provides sufficient flexibility to allow cost-effective implementation of the standard, while achieving significant public health and environmental benefits. Below, I set forth the general outline of the Administration's implementation strategy for the new air quality standards.

For the past two years, 37 eastern states have worked collectively, as the Ozone Transport Assessment Group (OTAG), to assess the regional transport of ozone pollution across state boundaries. In late June, OTAG agreed upon a set of recommendations for reducing regional emissions of nitrogen oxides (NO_x). The OTAG States concluded that regional NO_x reductions, particularly if they are achieved through a regional cap-and-trade system, are extremely cost-effective. After an independent analysis of the recommendations and considering the reductions necessary for both the 1-hour and 8-hour ozone standards, EPA will issue this September a proposed rule calling on States to reduce significantly NO_x emissions.   After an opportunity for public comment, including additional input from the affected States, EPA will issue a final rule by September 1998. A regional NO_x strategy will provide significant benefits not only for areas needing to meet the new 8-hour ozone standard but also those areas not attaining the 1-hour ozone standard. Moreover, EPA anticipates that nearly all areas within the OTAG States meeting the 1-hour standard, but not meeting the 8-hour standard, will be able to attain the 8-hour standard as a result of the implementation of regional NO_x controls. Because EPA believes that many of these new areas will be able to attain the new 8-hour standard with no or minimal additional local controls, EPA plans to exercise its authority under the Act to minimize the need for States to adopt additional control requirements on local sources in such areas.

With respect to the new particulate matter standards, it will be five years before EPA has sufficient monitoring information to determine the attainment status of areas for the new PM-2.5 standard. As a result, EPA will initially designate areas as "unclassifiable" for fine particulate matter. During this time, EPA in conjunction with the States will be establishing a nationwide comprehensive monitoring network. In addition, States will be developing emission inventories and ensuring that their existing plans provide adequate authority to regulate fine particulate matter. Although the Act does not mandate local controls for areas until such time as an area is designated nonattainment, public health benefits will be achieved by continued reductions from the acid rain program and other possible national or regional control programs. In addition, States may voluntarily move more quickly to adopt and implement local controls.

In summary, over the next 18 months, EPA will issue the rules and guidances to implement the strategy identified above. The strategy will be centered on five basic principles:

Continue today's progress toward cleaner air by keeping current ozone standards and plans in place until they are achieved, respecting agreements already reached by communities and businesses and not disrupting progress currently being made

Provide new tools for areas to address regional sources of pollution, most importantly a trading plan -- based upon recent recommendations developed by the OTAG States -- for emissions from utilities that will address violations far downwind and that will provide the most cost-effective emission reductions

Do not subject areas that will achieve early reductions through these approaches to potentially burdensome enhanced planning and pollution reductions requirements for traditional non-attainment areas

Initiate a new round of review of the particulate matter science to be completed within 5 years, before areas are designated as not attaining the new standards and before any pollution controls would be required

Cap air pollution control costs for industries, consistent with the responsibility of the States to provide for attainment of the standards, by allowing them to pay into a State Clean Air Investment Fund -- the proceeds of which would be used to find more cost-effective pollution reductions where controls would be too costly

The standards will not require local controls until 2004 for ozone and 2005 for fine particulate matter, with no compliance determination until 2007 and 2008, respectively, and with possible extensions thereafter. The Administration's targeted, yet flexible, approach will ensure that the updated standards can be achieved without economic or lifestyle disruption and with a minimal impact upon small businesses.

I appreciate this opportunity to more fully explain the basis for the Administration's strategy for implementation of the revised ozone and particulate matter standards. The enclosed attachment responds directly to the questions presented in your letter.
I also have enclosed a copy of the Statement issued by the President on July 16, 1997.

Sincerely,

Mary D. Nichols
Assistant Administrator for
Air and Radiation

ATTACHMENT

:   Please describe in detail all instances where EPA has provided assurances that specific
localities or sources will not have to take certain actions to reduce emissions or meet certain
specific Clean Air Act nonattainment requirements despite the area's projected violation of
EPA's proposed air quality standards for ozone or particulate matter.

RESPONSE: Prior to promulgation of the new or revised ozone and particulate matter
standards, inquiries were made to the Agency regarding the potential impact of the standards.
However, the responses to these inquiries typically do not address the ramifications for specific
localities or facilities. Examples of this type of response are attached. We have identified two
letters that address specific localities or specific sources. The first, a letter to Representative
Dennis Kucinich, which you specifically identified in your request, addresses two sources
located in Representative Kucinich's district. In the letter, EPA does not provide any
assurances that such sources will not be required to comply with Clean Air Act requirements.
Rather, EPA simply states that a regional control strategy will likely be sufficient for such area
to attain the new ozone standard. The second letter, responding to an inquiry from
Representative Sherrod Brown, addresses the possible impacts of the proposed ozone and
particulate matter on the Amish community in Representative Brown's district. Again, EPA
does not provide any assurances that this community will not need to comply with Clean Air
Act requirements. The letter explains EPA's existing policy with respect to wood burning
stoves and fireplaces. In addition, it states that regional, not local, controls will be the
solution for particulate matter problems caused by regional transport. A copy of this letter is
attached.

On July 16, 1997, President Clinton issued a memorandum to the Agency setting forth
a strategy for implementing the new standards. The Agency has conducted preliminary
analyses concerning the potential implications of that implementation strategy and shared with
interested parties the Agency's analyses for various areas. These analyses are based on
currently available data and are not definitive determinations for any area.[1] EPA employees at
all levels within the Agency have continued to respond orally to inquiries regarding the
potential application of the implementation strategy in particular circumstances.

2.   As you are aware, Title I of the Clean Air Act Amendments establishes the steps that EPA
and States must take once a new national ambient air quality standard is established. Pursuant
to section 107, EPA is to promulgate area designations based on the new standard "as
expeditiously as practicable, but in no case later than 2 years from the date of promulgation of
the new or revised national ambient air quality standard." Once an area is designated
nonattainment, are there any requirements that apply as a matter of law with the designation?
For instance, major sources in nonattainment areas must seek special construction permits
before they can build new emissions units or modify existing facilities. This preconstruction
review program requires sources to install stringent controls and secure offsetting emissions
reductions, and it causes substantial delays in projects. When would these construction permit
requirements apply to sources in an area designated nonattainment? Similarly, many sources
in nonattainment areas must secure an operating permit as required by Title V of the Clean Air
Act Amendments of 1990. When would the requirement for operating permits apply to
sources in a nonattainment area?

RESPONSE: EPA must designate areas two years after promulgation of a new or revised
national ambient air quality standard (NAAQS), but may take an additional year if "the
Administrator has insufficient information to promulgate the designations." CAA §
107(d)(1)(B). For those areas designated nonattainment, the general planning provisions of
subpart 1 of part D of title I of the CAA apply.[2] In particular, section 172(b) requires EPA to
establish a deadline "extending no later than 3 years from the date of designation" for purposes
of State submission of plans "meeting the applicable requirements of [section 172(c)] and
section 110(a)(2)." In addition, the conformity requirement in section 176 applies to areas
designated nonattainment. Subpart 1 is phrased in general terms and provides ample room for
flexibility in implementation. EPA has historically developed policy and guidance documents

to further delineate these and other planning requirements for states. However, for certain types of submissions, such as conformity and new source review (NSR), EPA has also issued regulations.

As part of the Administration's approach to implementing the new 8-hour standard, EPA will create a "transitional" classification. Section 172(a) provides EPA with the discretion to create classifications for areas designated nonattainment. Classifications may be created for the purpose of establishing attainment dates or for "other purposes." In developing such a classification scheme, EPA has the authority to interpret and apply the provisions of subpart 1, including new source review and transportation conformity, in a way that recognizes the particular circumstances of areas. Areas that have attained the 1-hour standard but that have air quality that violates the 8-hour standard will be eligible for this classification.

The Clean Air Act provides two basic regulatory schemes for new sources. In general, "nonattainment" new source review applies to new or modified sources constructing in a nonattainment area (see, CAA § 173); the prevention of significant deterioration (PSD) program applies to sources constructing in any area designated attainment or unclassifiable for any pollutant (see, CAA § 161). Hence, once EPA promulgates a new or revised NAAQS, the PSD program typically would apply to all areas prior to designation. In addition, for areas designated attainment or unclassifiable, PSD would continue to be the applicable new source requirement.

For areas newly designated nonattainment, the Clean Air Act does not specify whether nonattainment NSR applies upon designation or upon Part D SIP submission and approval.[3] Because the statute does not speak directly to the new source requirement that applies during the period between designation and plan submission, EPA has authority to determine what policy would be appropriate in newly designated nonattainment areas. EPA will establish a transition policy by December 1998.

In addition to determining the appropriate new source requirement for the period between designation and plan submission, EPA also must consider the appropriate new source program that will need to be included as part of an attainment plan. For those new areas that will attain the standard with no or minimal additional local controls, EPA believes that it has flexibility to establish a new source program that is different from the traditional program. A control technology requirement and a system for achieving offsets will be part of such a new source program. Within the next 18 months, EPA will finalize a new source rule that will define the applicable requirements for these areas.

Operating permits under Title V are required for all major stationary sources. The general definition of a major source for purposes of criteria pollutants such as particulate matter and ozone, which definition applies in attainment areas and many nonattainment areas, is a source that emits or has the potential to emit 100 tons or more per year of any pollutant. The general definition would apply for purposes of areas subject to the subpart 1 nonattainment area requirements, which will apply to new areas designated nonattainment for the 8-hour ozone standard and the fine particulate matter standard. Therefore, since the major source definition is identical for attainment areas and for areas newly designated nonattainment for the new ozone or particulate matter NAAQS, States will not be required to change procedures and requirements for permitting under title V of the Act due to the new designations.

3. Within three years after promulgation of a national ambient air quality standard, States must submit implementation plans which must include numerous planning and control requirements as well as enforceable schedules and timetables for sources to comply with the requirements. These measures must include, pursuant to section 172(c), the requirement that sources in the nonattainment area retrofit all emissions units with reasonably available control technology. How does EPA plan to ensure that the two plants in Representative Kucinich's district will not have to retrofit reasonably available controls?

RESPONSE: The planning requirements of section 172(c) are linked to a designation of nonattainment, rather than to promulgation of the NAAQS.[4] As stated above, EPA is required to establish a timeframe for submission of the applicable requirements under section 172(c); this time frame may not exceed three years from designation of an area as nonattainment. See, section 172(b).

The requirement for States to promulgate reasonably available control measures (RACM), including reasonably available control technology (RACT), is provided in section 172(c)(1) and, therefore, would not be due until as late as three years following a nonattainment designation. Historically, for purposes of applying RACT to sources that emit volatile organic compounds (VOCs), an ozone precursor, EPA has interpreted the RACT requirement to apply independent of a State's ability to demonstrate that an area will attain the ozone standard. Conversely, for particulate matter and other pollutants, EPA has interpreted the RACM/RACT requirement to mean that only those measures that are needed to demonstrate attainment by the applicable attainment date are "reasonably available" and thus required. EPA believes it has the authority to extend the interpretation it has taken for particulate matter for determining whether additional stationary source controls are "reasonably available" for purposes of obtaining emission reductions for ozone precursors. In addition, EPA believes that a trading program may be sufficient to meet the RACM/RACT requirement. The letter to Representative Kucinich conforms with these interpretations.

4. Please list all other Clean Air Act nonattainment requirements that will apply to nonattainment areas and indicate which ones EPA will be enforcing and which EPA will not be enforcing after promulgation of EPA's proposed ozone and fine particle standard. For each of the requirements EPA will not be enforcing, please provide an explanation of the rationale and legality of EPA's decision.

RESPONSE: As previously stated, section 172(c) provides the list of applicable requirements for areas designated nonattainment for an 8-hour ozone standard or for a PM-fine standard. These include RACM/RACT, a demonstration of attainment, reasonable further progress toward attainment, an initial (and periodic updated) emissions inventory, permits for new and modified major stationary sources and contingency measures.[5] In addition, the conformity requirements under section 176 would also apply to such areas.

EPA has not determined that any of these requirements would not be applicable for purposes of an area designated nonattainment. However, the language in section 172(c) is, for the most part, written broadly and provides EPA with flexibility in determining what each of these requirements means for any particular area designated as nonattainment. As provided in response to Question 1, in the context of creating classifications under section 172(a), including a "transitional" classification, EPA has the authority to interpret or apply the requirements under section 172(c) in a manner that recognizes the circumstances of areas within each classification. EPA believes that it has comparable flexibility under the conformity requirements of section 176(c). Finally, EPA is also exploring whether construing the nonattainment new source review requirement in light of de minimis principles would help provide a common sense approach to implementation of the new or revised standards. EPA will continue to develop implementation policies and rules to further explain how these provisions might be interpreted. As provided in the President's memorandum, EPA will issue final rules revising its transportation conformity and new source review rules by September 1998.

5. States that fail to take actions required under the nonattainment provisions of the Clean Air Act face sanctions of increased offsets and loss of highway funds. The Act also requires EPA to step in and directly order controls, issue permits and take other actions to enforce nonattainment requirements when a State fails to act in a timely manner. These sanctions and federal implementation provisions are written as mandated actions, once a State defaults on its duties. Can EPA assure States that these sanctions and federal implementation provisions will not apply to any nonattainment requirements that EPA hopes to "suspend" during implementation of the new ozone and fine particle standard?

RESPONSE: Since EPA does not intend to suspend CAA requirements for any area. States will not face sanctions or federal plans for "suspended" requirements. However, as you have noted, the Clean Air Act establishes two repercussions for areas that fail to plan, develop a defective plan, or do not follow through on implementing an approved plan; these repercussions will be available if areas do not respond to EPA's call for regional NO_x reductions or do not adopt, submit or implement other required planning measures. First, such areas may be subject to the highway funding and the offset sanction.

6. Section 304 of the Clean Air Act provides for citizens enforcement of most requirements established under the Act. What assurances can EPA provide that a citizen will not force EPA and states to implement fully all of the nonattainment requirements, notwithstanding any EPA decision to suspend certain requirements? What assurances can EPA provide that citizens will not be able to sue sources directly for violations of suspended Clean Air requirements?

RESPONSE: Section 304 of the Clean Air Act provides for citizen enforcement against state and federal regulators and against individual sources for violation of CAA requirements. With respect to EPA, those actions would solely consist of claims that EPA failed to act (or act in a timely manner) where required to do so under the Act. In addition, section 307 provides citizens with the opportunity to challenge in the court of appeals alleged errors regarding final agency action.

In general, EPA cannot prevent citizens from seeking redress in the courts. As more fully described above, EPA is not providing guarantees that applicable requirements will not be enforced; rather, EPA will issue policies and rules that more fully explain how the applicable requirements of the Act will be interpreted or applied to areas based on their classification and in a manner that recognizes the circumstances of areas within each classification. EPA and the Department of Justice will vigorously defend these actions in court if challenged. With respect to interpretations of the Act regarding implementation, EPA believes that citizens would primarily seek redress in the courts of appeal against EPA rather than challenging States and sources. In any case, the relevant implementation provisions of title I of the Clean Air Act generally do not regulate sources directly. Thus, citizens would likely be unsuccessful in a suit directly against sources based on alleged inadequacies in federal or state planning efforts.

---

[1] These analyses are largely based on monitoring information from calendar years 1993-95. In accordance with section 107(d)(1), EPA will determine whether areas have air quality meeting the new standards within 2-3 years. Such designations will be based on data from 1996-98 or 1997-1999. It will be these designations that will trigger the planning requirements that are the basis for most inquiries.

[2] Subparts 2-5 set forth standard- or pollutant-specific requirements that may also apply to certain nonattainment areas. For example, subpart 4 would apply for purposes of a revised PM_{10} standard, but does not apply for purposes of the new $PM_{2.5}$ standard. Subpart 2, which sets forth detailed requirements for a 0.12 ppm 1-hour ozone NAAQS, does not apply for purposes of the new 8-hour ozone NAAQS.

[3] Areas designated nonattainment for the 1-hour standard will continue to apply their current new source programs. However, EPA is committed to re-examining the NSR program applicable to current nonattainment areas to determine whether it is appropriate to revise the program so that new sources are treated more equitably across all areas.

[4] The SIP requirements due no later than 3 years following promulgation of a new standard are the planning requirements provided in section 110(a)(1) and (2). This planning obligation applies to all areas, regardless of the area's designation.

[5] Areas designated nonattainment for the current standards are subject to similar requirements as provided in subparts 2 (for ozone) and 4 (for PM-10) of part D of title I of the Clean Air Act.

Mr. DINGELL. These measures—and then I sent a similar letter down inquiring about this, on which I have a response to me from EPA, in which it says, "Historically, for purposes of applying RACT to sources that emit volatile organic compounds and ozone precursors, EPA has interpreted RACT's requirement to apply independent of the State's ability to demonstrate the area will attain the ozone standard"—diametrically opposed to what was told to my other colleague.

"Conversely, for particulate matter, other pollutants, EPA has interpreted that the RACT requirement to mean that only those"——

Mr. BROWN. Mr. Chairman, I ask for another 3 minutes.

Mr. BARTON. Well, I think I must have set the clock wrong. That was not 10 minutes.

Mr. BROWN. Another 5 minutes, okay.

Mr. BARTON. Yes, but we're going to get it——

Mr. DINGELL. I'm helping the gentleman from Ohio. He needs to know this because he's the one——

Mr. BROWN. That's why I asked for the additional 5 minutes, Mr. Chairman.

Mr. BARTON. I would urge the gentleman to do the help more expeditiously.

Mr. DINGELL. "In addition, EPA believes that a trading program may be sufficient to meet the RACT requirement." The letter to Representative Kucinich conforms to these interpretations.

I'm asking you to cite now, if you please, the specific authority for you to treat—to afford those reliefs to Cleveland, the specific statutory authority, if you please, now—my time is limited.

Ms. BROWNER. We'll get you the cite.

[The information provided is printed following the hearing.]

Mr. DINGELL. So——

Ms. BROWNER. We're looking it up.

Mr. DINGELL. Then we may come to the interpretation that EPA is relaxing in a transitional ozone non-attainment area as a requirement that major sources of pollution put on reasonably available controls; is that correct? That's the import of what has been said to me and to Representative——

Ms. BROWNER. I do not agree with your statement.

Mr. DINGELL. You don't agree?

Ms. BROWNER. I'd be happy to explain why.

Mr. DINGELL. So are you telling me, then, that——

Ms. BROWNER. I don't agree with your characterization.

Mr. DINGELL. Are you telling me that you are going to treat all areas of the country in exactly the same fashion on that matter?

Ms. BROWNER. Not all areas of the country have the same levels of pollution.

Mr. DINGELL. Well, who are those whom you will you afford this kind of preferential treatment, and that you will not require RACT and RACM performance from those sources?

Ms. BROWNER. Not all areas of the country have the same problem. A one-size-fits——

Mr. DINGELL. So you are going to then pick and choose, and you're going to afford some areas in the country special treatment and others not?

Ms. BROWNER. As Congress has asked me to do under——

Mr. DINGELL. Yes or no?

Ms. BROWNER. Mr. Dingell——

Mr. DINGELL. Yes or no?

Ms. BROWNER. It is not a question to which I can answer yes or no; it's a complicated issue which I am more than happy to answer.

Mr. DINGELL. You are greatly incapable of answering yes or no. What I'm saying is you are here assuring everyone that you are going to afford special treatment; you are going to permit people to evade the requirements of the citizen suit and engage in other special preferential treatments in a program that clearly cannot work.

I thank you.

Ms. BROWNER. That's simply not true. That is simply not a fair characterization of our implementation plan.

Mr. BURR. Reclaiming my time for 1 minute, I promised to recognize Mr. Coburn for one question, and then I'll certainly yield to anybody else.

Mr. COBURN. Ms. Browner, if you could explain to me, in the year 2002, you're going to do a scientific review——

Ms. BROWNER. You're talking about particulate matter?

Mr. COBURN. Yes, ma'am.

Ms. BROWNER. Uh-hum.

Mr. COBURN. And, also, the States will have collected all their data by then?

Ms. BROWNER. The final—let me just make sure I have this right—the monitoring data will be collected between 1998 and the year 2002; that is correct.

Mr. COBURN. And so we're going to have a concurrent review in 2002 by the EPA, and that's the same year we're going to finish collecting the data——

Ms. BROWNER. There's two—this is a—you know, this is——

Mr. COBURN. I haven't asked a question yet. Let me get to it, and then you can answer it.

Ms. BROWNER. You're using apples and oranges, with all due respect, in the question, and maybe we could have a better conversation if I could just clarify.

Mr. COBURN. I'll be happy to let you clarify.

Ms. BROWNER. The data that is collected by the States, as is true for ozone and other ambient air pollutants, is done to determine the quality of the air in a particular city, and whether or not steps will need to be taken to reduce that pollution.

Mr. COBURN. Madam Administrator, I understand that completely.

Ms. BROWNER. It is not what's used in the scientific——

Mr. COBURN. That's why I want to get to my question. My question——

Ms. BROWNER. It's not used for the scientific study.

Mr. COBURN. [continuing] has to do with your research plan, the research plan which you have as well, and my concern is, we are going to be collecting data on 2.5 particulate matters; in 50 collectors, I believe, we're going to be looking at the chemical complexes, the CAT ions. How, in fact, will your scientific review take into consideration those areas, those limited 50 areas, where we're going to collect the chemical data, and how will that play out in regard to the review and how you will implement the changes when, in fact, we may see significant differences of impact on particle size based on the CATion complex that's associated with them?

Ms. BROWNER. There are already PM monitors in place. There are PM monitors that have been in place that are collecting the data for the scientific studies. It's really apples and oranges. The monitoring network is what the Clean Air Act allows for in terms of Governors making decisions about areas and what steps must be taken to reduce the pollution. Although they may physically look the same, it is different than the monitors that are used in the scientific analysis.

Mr. COBURN. I understand that fully, but the fact is, is the science not there right now to tell us whether it's a chemical reaction or particulate size that's causing the problem——

Ms. BROWNER. That's where we have——

Mr. COBURN. If we're going to have the States doing it based only on particulate matter, aren't we giving up a whole lot?

Ms. BROWNER. These monitors will measure 2.5. They measure the size of the particle.

Mr. COBURN. Right.

Ms. BROWNER. That the States will be using——

Mr. COBURN. They're not going to be measuring the chemical component, except in 50 out of 1,500 monitors.

Ms. BROWNER. But that doesn't mean we don't have data on that.

Mr. COBURN. But that data——

Ms. BROWNER. We have data.

Mr. COBURN. [continuing] is different by chemical reaction in every area of the country.

Ms. BROWNER. This is an issue that has been the focus of a great deal of scientific discussion in the CASAC process.

Mr. COBURN. I understand—I guess my question to you is: Will you give us some assurance that if, in fact, you find limited data in these 50 areas, that in fact you will come back here and say, "Whoa. Time out. We need to be doing chemical—collection of the chemical aspects of these 2.5 particulates throughout the country," because in fact we're going to find that there is significant chemical interaction, not just particulate size, and I would like the assurance that we're going to change the approach, if in fact that's the nature. It may be that you can have a 5.0 micron if it has no significant CAT ion and not affect health.

Ms. BROWNER. That's what the Clean Air Act already says.

Mr. COBURN. All I'm asking——

Ms. BROWNER. It says, do a 5-year review——

Mr. COBURN. [continuing] for is the assurance that you're——

Ms. BROWNER. [continuing] and we're going to do it. We have promised to do the 5-year review, and in fact in the budget agreement that was reached for EPA I think last night, there is $45 million for a research program, and that will be incorporated into the statutory requirement of a 5-year review, and at the end of it a decision will be made by whomever is the Administrator of the EPA.

Mr. COBURN. But you'll have 1,450 where you have no data on chemical complexing whatsoever.

Mr. BARTON. The Chair—does Congressman Bilbray wish——

Ms. BROWNER. Actually, can I just clarify one point here? We are now talking to the States, I'm reminded, to identify places for over 300 monitors that would actually speciate, which is the issue that you're raising, the speciation. So it's 300 monitors that we're talking to the States about.

Mr. COBURN. That's new to your implementation plan?

Ms. BROWNER. We've been working with the States in the specifics of the monitors. We've now agreed to pick up the cost, as I explained, and Congress has very helpfully provided us with the money, so we are talking to the States.

Mr. COBURN. It is true that you all—that were fifty of them, though, is that—am I in error on that? Did you all not give us information that there were fifty?

Ms. BROWNER. I think that may have been what was originally envisioned and now we have been able, through the good graces of Congress and the money, to grow that.

Mr. COBURN. Okay.

Mr. BARTON. The Chair is going to recognize Mr. Bilbray for 5 minutes, in which he expects him to yield some of that time to Mr. Klink of Pennsylvania, so that this will actually be the last round, the last question period with Ms. Browner.

Mr. BILBRAY. Mr. Chairman, I'd yield that time at this time to Mr. Klink.

Mr. KLINK. I thank the gentlemen for yielding. I understand, Ms. Browner, when I was out of the room at one of the votes, that the response to one of the questioners, that you were asked something about H.R. 1984 and I'm at a disadvantage because I wasn't here, but you said that you felt this would put the monitoring—I'm talking about PM$_{2.5}$——

Ms. BROWNER. No, I don't believe that's what I said.

Mr. KLINK. Maybe you could clarify for me, so I could——

Ms. BROWNER. I'm trying to remember the question. It was Mr. Stupak's question. I believe the question was, would there be—I don't want to restate the question.

Mr. KLINK. Well, all right. Well, let me let me ask you this: Our intention with H.R. 1984 was to say, look, we don't want to promulgate first; we want to do the science first, which would really plays off of what Mr. Coburn was saying. Because if you've only got a limited number of monitors out there, what happens if—and we're all worried; what if you find out something about the chemical makeup of the PM$_{2.5}$ that tells you on the final day it isn't only size that counts; there are other things that are important?

So the question is, if we give you 2 years and we authorize the $300 million over a 4-year period, $75 million over a 4-year period for EPA to be able to buy and to—you seem to be shaking your head like you haven't read the act. The point is that what we are saying, we would be, under 1984, allowing the monitoring to go forward during 1997 to the year 2002. I don't understand where this is a problem for the EPA, and in fact we are going back and asking for additional funds for EPA to be able to do this. This seems, I would say, Ms. Browner, this seems to be a very reasonable position.

There were a lot of people here in Congress that wanted to really be much more adversarial toward the EPA; I'm sure you could imagine. We thought that this was a very reasonable step to take, and I don't understand why you would have a problem with this course of action.

Ms. BROWNER. Our concern—and you will undoubtedly correct me if we misunderstand your proposal—is that you would delay the public health protections that the public is entitled to under the current law; that you would be moving out certain actions and the date in which they would occur.

Mr. KLINK. No.

Ms. BROWNER. Well, with all due respect——

Mr. KLINK. With all due respect, that's not what we would do. And if we would have been able to sit and have these discussions with the administration, we think that we could have worked all these matters out. That's not the intent; it is in fact the intent not to delay one moment and not to roll back anything. It is our intent simply—so you understand—our intent is simply to build the monitors, to collect the data, and then to promulgate, and to have that happen over the next 5 years. According to the schedule that I've seen, your implementation schedule, you're not going to be doing anything anyhow, and the testimony that you gave before the Agriculture Committee, you said that it would be 2009 before——

Ms. BROWNER. Mr. Klink, can I ask a question?

Mr. KLINK. The bottom line—one moment, just one moment.

Ms. BROWNER. I'm sorry.

Mr. KLINK. The bottom line, perhaps the most important thing I can say about the implementation, is nobody is legally required to reduce air pollution until the year 2009. Those are your words. We would say—this was the Agriculture Committee, September 16. Now our point is this: If 15,000 premature deaths a year are occurring, we don't want to see—and I'm sure you and no one at the EPA wants to see—this drug out any farther than it needs to be. If 100,000 people, as you gave in your testimony today, are having decreased lung capacity due to these kinds of pollutants, we don't want it to drag out a day, a month, a year longer.

All we're saying is at the end of the day we don't want you to say, "Whoops, we've collected the data; we've had the 3 years to collect the data; we've built the monitors and now we found out that there is, indeed, some kind of a chemical reaction that is taking place within certain types of $PM_{2.5}$ that we didn't know about back in 1997. Before you promulgate, before there is the threat that a citizens' lawsuit would be filed that would say, "Why are you doing this in Lorraine and Cleveland and you're not doing it in Pittsburgh? Why are you doing this in Los Angeles and you're not doing this in Kansas City?"—we simply think that before you promulgate, let us in Congress, in a bipartisan fashion, give you the money and authorize the money, so that you would have enough money to build and deploy these monitors, collect the data, and then we'll promulgate. We don't want to put anything at all off. And if that's a problem for you, I'd like to understand why.$_{2.5}$

Mr. BILBRAY. I'm getting—if I may reclaim my time—Administrator, do you want to respond?

Mr. KLINK. Go ahead.

Mr. BILBRAY. I read so much of the implementation strategies and they are looking at stationary sources as being the major—basically will address some major stationary sources. We're talking about Midwest. My concern is that when we think in California, especially San Diego, that we can eliminate every stationary source in San Diego, and between mobile sources and transport from Los Angeles and Mexico, we will not be able to fulfill existing standards, let alone the standards that are being projected here.

My question is that in the plan you're proceeding that stationary sources need to basically go back to best available technology. Are you proposing now that best available technology now be mandated in those areas where mobile sources are a major problem?

Ms. BROWNER. Can I back up for 1 second?

Mr. BILBRAY. Yes.

Ms. BROWNER. San Diego has a plan. Everybody now has a plan with a deadline for meeting the ozone standard. No area is required to undertake additional activities until they have completed work, until their deadline. So, I mean, San Diego does has a plan. It's all been agreed to. Los Angeles has a plan for the steps they are going to be taking.

Mr. BILBRAY. Excuse me. The issue here though is that San Diego's plan is to reach attainment by the year 1999. We've got— 10 years later Los Angeles comes along. The trouble is, we're still confronting the issue that, what is going to be the administration's strategy about the gap of our ability to reach attainment after 1999? Are we going to then be required to mitigate for Los Angeles and Tiajuana air problems coming in, especially with the fact that the Federal Government is allowing Los Angeles to go 10 years longer than San Diego?

Ms. BROWNER. Let me answer the Mexico question first. No area, whether it be San Diego or any of the border communities, Texas, et cetera, will be—I don't know if you've used this word—but "penalized" for the Mexico pollution problem. I mean, we've crafted some very reasonable binational agreements, border agreements, for example, in El Paso and Juarez to deal with pollution problems. But if the premise of your question is, would San Diego be "held responsible" for Mexico's pollution, the answer is no.

Mr. BILBRAY. Administrator, after 10 years of working on this, 13 percent of our mobile source are commuters crossing across the border that we're being held accountable for right now. If you're saying now that 13 percent of our mobile source will be forgiven and that the act gives you the authority to forgive that 13 percent, I mean, I'd like to hear it. Now next month—let me just say that next month we are going to have a hearing on this issue, and I'd like to see the administration be brave enough to step forward and support the legislation to address this problem. But I have never heard you or never read anywhere in the law that you or anyone else has the authority to hold harmless 13 percent of the mobile source that's crossing across the border. No matter what your agreements are with Mexico, where does the act give you the ability to hold us harmless of that 13 percent of mobile source?

Ms. BROWNER. I understood the question to be the new standard in terms of the existing standard. You do have a plan and we support plan. We support your plan your State, your city developed.

Mr. BILBRAY. So, in other words, you feel that we're safe; we don't have any problems with somebody drawing up the fact that the new standard now is going to require us to now address those issues before 1999?

Ms. BROWNER. Well, first of all, as I said, areas that have a plan and are staying the course to implement their plan, don't have to take the additional steps until they have completed work.

Mr. BILBRAY. By 1999. But now after 1999, with Los Angeles 10 years behind us, still having Catalina impacting us, what is going to be our status while Los Angeles is in that lag period? Are we going to be again back on the burner, the fact that we have mitigate for our neighbor's pollution and for the lack of the Federal Government's action?

Mr. BARTON. The gentlemen's time has expired. Does the Administrator wish to answer that last question?

Ms. BROWNER. Why don't we answer it for the record? I mean, the California issues tend to be different than the rest of the country, and there is also provision in the Clean Air Act which we would like to call the members' attention to dealing with international border problems and this is——

Mr. BILBRAY. Mr. Chairman, I'd ask unanimous consent to enter into the record a letter that was sent to the Commissioner, actually Mary Nichols, and the responses by the EPA to these issues. California is sort of unique because we are 20 years ahead of the curve, and the rest of America should look at what we are doing in California, and have had to do in California, and get ready for the future.

Mr. BARTON. We will let the minority staff review that, and if there is no objection, we will put that in the written record.

[The information referred to follows:]

UNITED STATES ENVIRONMENTAL PROTECTION AGENCY
WASHINGTON, D.C. 20460

AUG 7 1997

**OFFICE OF
AIR AND RADIATION**

The Honorable Brian Bilbray
Committee on Commerce
U.S. House of Representatives
Washington, DC 20515

Dear Congressman Bilbray:

This letter is in response to the questions faxed to the U.S. Environmental Protection Agency dated June 12, 1997 regarding the proposed ozone and particulate matter national ambient air quality standards (NAAQS). These are follow-up questions from the May 15, 1997 hearing on the NAAQS. Questions 1-3 concern the proposed ozone standard; questions 4-7 concern the proposed $PM_{2.5}$ standard; and questions 8-12 concern implementation of the standards. The enclosure restates the questions and provides a detailed response.

I appreciate this opportunity to be of service and trust that this information will be helpful to you.

Sincerely yours,

Mary D. Nichols
Assistant Administrator
for Air and Radiation

Enclosures

ENCLOSURE

QUESTIONS ON THE PROPOSED OZONE STANDARD

1. Data has been presented during the comment period that indicates that background concentrations of ozone in some "pristine" areas, without contributions from upwind or transported sources, already approach the levels of the new proposed standard for ozone. What is the likelihood that urban areas, particularly well-regulated urban areas that have fully implemented and sophisticated emission control programs, can ever attain the currently proposed standards? If so, what would be a realistic time frame?

Response: Data on "background" ozone concentrations submitted during the comment period are consistent with information included in the Environmental Protection Agency's (EPA's) Criteria Document, and were thus considered in EPA's ozone criteria and standards review. The EPA's analysis of air quality values at remote sites produced an estimated range of daily 1-hour maximum ozone concentrations near sea level of 0.03 to 0.05 ppm during the ozone season. Because the distribution of ozone air quality values at remote sites is relatively flat, a reasonable estimate of the 8-hour daily maximum ozone background during the summer season is also 0.03 to 0.05 ppm. These analyses also show that 1- and 8-hour ozone background concentrations will typically be much lower than 0.05 ppm. These estimates of background levels of ozone were carefully assessed during the development of the ozone staff paper and were explicitly agreed upon by the Clean Air Science Advisory Committee (CASAC).

The EPA recognizes, however, that at some high altitude remote sites 1-hour ozone concentrations exceeding 0.07 ppm have been reported on peak days within the ozone season. Most of the highest reported values were associated with forest fires, while other high values may result from occasional atmospheric conditions that are conducive to intrusion of stratospheric ozone into the troposphere. Even considering these peak concentrations at selected remote sites, the 3-year average of the 4th-highest maximum ozone concentrations at these sites was well below the proposed 8-hour standard level of 0.08 ppm. However, in recognition of the occurrence of such natural or exceptional events that can contribute to atypically high ozone concentrations, EPA will prepare implementation guidance on the use of data affected by such events in determining whether the new ozone standards have been met.

An urban area, such as Los Angeles, that is well-regulated and that has implemented sophisticated controls may have to rely on technology advances and innovative market mechanisms to achieve additional emission reductions; it is obviously difficult to predict timing on newer technology. In some cases, controls that are technologically feasible, but more expensive than those currently being implemented, may have to be employed; timing on the implementation of these expensive measures is also difficult to predict.

The EPA, in consultation with State and local agencies, will be looking at all alternatives to achieve the NAAQS by the attainment dates specified in the Act.

2. In developing the form of the standard, how did EPA consider and evaluate the relative importance of seasonal variations, transport from one air basin to another, and daily and weekly variations in industrial and public activities that contribute the most to the ozone problem? How would these conclusions and assumptions apply to the San Diego air basin, in EPA's view?

Response: In developing the form of the proposed ozone standard, EPA considered the distributions of hourly and daily maximum 1- and 8-hour average ozone concentrations over a number of ozone seasons, based on monitored data in the Aerometric Information Retrieval System (AIRS) database. Thus, daily, weekly, and seasonal variations and the effects of long-range transport are reflected in these evaluations. These variations in air quality are influenced by changes in industrial and public activities. The EPA considered the public health implications and the relative stability of alternative standard forms, based on analyses of these ozone air quality distributions, in selecting the proposed form.

As San Diego prepares a State implementation plan (SIP) to address the revised NAAQS, it will have to account for the seasonal and daily variability of emissions, for ozone and its precursors transported from outside its boundaries, and for variations in industrial and public activities. This accounting will be done through the technical analyses required in the SIP development process in order to demonstrate attainment of the NAAQS. San Diego's ozone problem may be complicated by transported emissions from Mexico. The EPA will consider the impact of transport on international border areas, such as San Diego, in developing the implementation strategy for the revised ozone standard.

3. *If EPA proceeds with establishing a new goal or standard for ozone, under what schedule or timetable would EPA proceed to implement such a standard?*

Response: The Clean Air Act (Act) sets out time frames for implementing a new or revised standard. EPA published a final standard in July 1997; the Act schedule generally would be as follows:

Activity	Date
States submit nonattainment/attainment/unclassifiable recommended designations (1 year after NAAQS promulgation)	July 1998
EPA designates nonattainment/attainment/unclassifiable areas (3 years after date of promulgation of standard)	July 2000
States submit SIPs containing attainment demonstrations for nonattainment areas (3 years after designation of nonattainment areas)	July 2003
Attainment date (no later than 5 or 10 years from designation of nonattainment areas; two 1-year extensions may be allowed)	No later than July 2012

In accordance with a Presidential memorandum issued July 16, 1997, the EPA is developing a plan for States to meet the new standards affordably through common sense measures. Over the course of the next 18 months, EPA will issue guidance and rules that will provide States and local governments the flexibility they will need to meet the new, more protective health standards in a reasonable, cost-effective way.

-- EPA will create a new "transitional" classification that will be available for areas in which anticipated regional measures will provide the bulk of the needed ozone reductions. In addition, areas that are not subject to a regional control strategy may also be eligible for the "transitional" classification if they make an early submission demonstrating how the area will attain the revised standard. Areas that receive a "transitional" classification will need to do little or no additional local planning in order to attain the standard.

-- The EPA will work from a regional plan developed collectively by 37 States over the last 2 years to call on States to address the long-distance transport of ozone. The plan developed by the States focuses on major power plants (which offer the most cost-effective opportunities for reducing pollution) to reduce nitrogen oxide, a key ingredient of smog. The reductions called for by EPA should be enough to allow most of the newly non-attainment counties to be able to comply with the new standard.

-- By participating in the regional ozone reduction plan, counties that are not in compliance with the new standard will quickly achieve significant reductions in their ozone levels.

The strategy for implementing the revised standards is described in the Presidential memorandum, a copy of which is attached (Attachment 1).

QUESTIONS ON THE PROPOSED PM$_{2.5}$ STANDARD

4. *What are the relative emission contributions of stationary sources, mobile sources, area sources, and natural or indigenous sources of PM$_{2.5}$ and/or its precursors?*

Response: PM$_{2.5}$ is composed of a mixture of primary particles directly emitted into the air and secondary particles formed in the air from the chemical transformation of gaseous pollutants. In general, it is the secondary particles that dominate measured PM$_{2.5}$.

As measured at ambient air monitoring sites in the Eastern United States, contributions from stationary sources comprise about 60 percent of PM$_{2.5}$ (52 percent from boilers and 8 percent from other industrial stationary sources). Almost all of the contribution from boilers is in the form of secondary sulfate particles formed from gaseous SO$_2$ emissions. Area sources comprise about 26 percent (6 percent from soil entrained by vehicles, construction and agricultural operations, 20 percent from directly emitted particles from combustion area sources -- home heating, waste combustion and managed burning in forests, and about 1 percent from natural fires). Mobile sources comprise the remaining 14 percent (8 percent directly emitted from diesels, split equally between on and off road, 4 percent from gas-powered vehicles -- mostly as secondary nitrates and sulfates, and the remaining 2 percent directly emitted from ships, aircraft and rail). In the Western United States, the composition is more variable, with the sulfates being lower and the nitrate, mobile source and soil components each being somewhat higher.

It is noteworthy that the soil component is only about 5-15 percent, even in very arid and actively farmed areas. The main difference between PM$_{10}$ and PM$_{2.5}$ is that the soil component in PM$_{2.5}$ is much less than that found in PM$_{10}$. This is because soil particles are generally larger than the combustion-related and secondary particles and they dominate the size fraction between 2.5 and 10 μm

5. *Please describe EPA's view of the incremental public health benefits (beyond those which would occur with the current standards) which will derive from attainment of its proposed PM$_{2.5}$ standard.*

Response: The EPA has calculated a potential range of projected incremental benefits for partial implementation of the proposed PM$_{2.5}$ standards for its regulatory impact analysis. This analysis indicates that the incremental benefits of the proposed PM$_{2.5}$ standards -- over both the current PM$_{10}$ standards and full implementation of the 1990 Clean Air Act Amendments requirements -- would include annual health improvements ranging as high as tens of thousands of fewer premature deaths; thousands fewer respiratory-related hospital admissions; hundreds of thousands fewer incidences of aggravated asthma and respiratory symptoms; and tens of thousands of fewer cases of chronic bronchitis.

6. *What are the ambient concentrations, and seasonal variations in ambient concentrations, of PM$_{2.5}$? What does EPA consider a "natural" background ambient concentration of PM$_{2.5}$ in a dry, semi-arid region such as San Diego?*

Response: Currently, approximately 200 sites report PM$_{2.5}$ data to the EPA in the AIRS. Table 1 lists summary statistics of PM$_{2.5}$ concentrations from these sites. These data were collected from 1985 to 1996. The columns are labeled as follows:

> STATE - The two character postal code for the State in which the monitor is located,
> MSA - The metropolitan statistical area (MSA) in which the monitor is located,
> AIRS SITE-ID - The nine digit code for the site as given in AIRS,
> ANN. MEAN - The average of the annual means from all data from the site,
> MAX - The average daily maximum value in a year reported by the site,

SAMPLE SIZE - The average number of samples reported by the site in a year
YEARS - The number of years over which the statistics were averaged.

The average annual means for these sites range from 4 5 µg/m³ to 37.5 µg/m³, while the average annual maximums range from 13.5 µg/m³ to 142.7 µg m³ (there is an 8.0 µg/m³ in the table but this is from 16 samples for 1-year and probably does not represent the maximum value for that area very well).

Since current $PM_{2.5}$ data are somewhat limited. the seasonal variation of $PM_{2.5}$ can be estimated through the analysis of visibility data. On this basis. EPA's staff paper showed estimated seasonal $PM_{2.5}$ concentrations for the entire United States. which indicated that the summer season consistently had the highest levels. The eastern half of the country exhibited significantly lower levels (by about 20–40 percent) for all other seasons. whereas the western half of the country showed much less variability, the other seasons averaging only about 5-10 percent lower. As $PM_{2.5}$ data are collected at more sites, more detailed seasonal trends will be quantified.

Background concentrations as defined by the EPA in the particulate matter (PM) staff paper are those levels that would be expected to occur in the absence of manmade emissions of PM and precursor compounds. Background levels of PM vary by geographic location and season. The exact magnitude of the natural portion of PM for a given geographic location cannot be precisely determined because it is difficult to separate the long-range transport of manmade particles or precursors. The background levels reported for the Western United States range from one to four ug m³. The Interagency Monitoring of Protected Visual Environments (IMPROVE) network has monitoring sites located in National Parks and wilderness areas and are considered to measure background concentrations of $PM_{2.5}$. The closest IMPROVE network site to San Diego is the San Gorgonio wilderness area. This site reported annual means of 9 54. 10 06. 8.89. 7.45. 7.88. 7 62. and 7 37 µg m³ of $PM_{2.5}$ from 1989 to 1995 Because the site is close enough to the Los Angeles basin to be influenced by transport. actual background levels should be lower than these annual means. While this gives an unclear picture of the background levels in San Diego. it is the best information available.

7 *What is the likelihood of actually reaching attainment of the proposed $PM_{2.5}$ standard. and how long does EPA estimate it would take to reach attainment? Has EPA projected or estimated the likely costs of attaining this standard in the implementation process?*

<u>Response</u>: The EPA believes that many of the areas projected to violate the new $PM_{2.5}$ standard would be expected to come into compliance as a result of the SO_2 emission reductions already mandated under the Act's acid rain program. which will be fully implemented between 2000 and 2010. The regulatory impact analysis, which EPA sent to Congress on July 18, 1997, provided detailed analysis for the likelihood and timing of attainment as well as the projected costs of attaining the new $PM_{2.5}$ standards. A copy of the regulatory impact analysis is attached (Attachment 2).

QUESTIONS ON IMPLEMENTATION

8. *The EPA has repeatedly stated that under the Act. it cannot consider cost in establishing new standards. Subsection 109(a)(1)(C)(iv) of the Act requires that the independent scientific review committee advise EPA of any adverse public health. welfare. social. economic. or energy effects that may result from strategies for attainment of any revised ambient air quality standard This language would appear to require EPA to consider the effects associated with such strategies which must be implemented in meeting revised ambient air quality standards. Given this language. is it EPA's position that Congress did not intend for these effects. including economic effects. to be considered by the EPA in establishing revised ambient air quality standards? If so. why would Congress have required in the Act that the review committee is to advise EPA on these effects? Is it EPA's position that the independent committee's review is limited to health-*

related economic impacts, as opposed to consideration of the economic costs associated with the strategies required to achieve any revised standards?

<u>Response</u>: It is EPA's position that Congress did not intend for the effects enumerated in section 109(a)(1)(C)(iv), including economic effects, to be considered by EPA in establishing revised NAAQS. From the language and structure of section 109(d), it is clear that the independent scientific review committee's responsibility to advise on these factors is separate from its responsibility to review and recommend revision of air quality criteria and NAAQS, and that the advice pertains to the implementation of NAAQS rather than to setting them. The legislative history confirms this view, indicating that the advice was intended for the benefit of the States and Congress. (<u>See</u> H.R. Rep. No. 95-294, at 183 (1977)). Further, it is EPA's longstanding position that the independent committee's review is limited, with respect to NAAQS, to providing advice and recommendations on revisions of air quality criteria and the NAAQS themselves, and that the cost associated with attainment of NAAQS is to play no role in setting them.

9. The EPA is aware that as part of California's ongoing efforts to achieve attainment with the existing NAAQS, individual air districts have proposed control measures on consumer products such as barbecue lighter fluid and deodorants, and controls on commercial charbroilers as are used in restaurants. Does EPA envision other states ultimately turning to similar measures in their efforts to achieve attainment of any revised standards? Would EPA agree that states may be required to adopt control measures of significant cost and marginal air quality benefit in its efforts to achieve attainment with new standards?

<u>Response</u>: In his July 16, 1997 memo to the EPA Administrator on implementation of the revised NAAQS, President Clinton supported a plan developed by EPA in consultation with other Federal agencies. This plan was developed in recognition of the President's desire to maximize common sense, flexibility, and cost effectiveness. Consistent with the States' ultimate responsibility to attain the standards, the plan indicates that EPA will encourage the States to design strategies for attaining the PM and ozone standards that focus on getting low cost reductions and limiting the cost of control to under $10,000 per ton for all sources. Market-based strategies can be used to reduce compliance costs. The EPA will encourage the use of concepts such as a Clean Air Investment Fund, which would allow sources facing control costs higher than $10,000 a ton for any of these pollutants to pay a set annual amount per ton to fund cost-effective emissions reductions from non-traditional and small sources. Compliance strategies like this will likely lower the costs of attaining the standards through more efficient allocation, minimize the regulatory burden for small and large pollution sources, and serve to stimulate technology innovation as well.

10. In the majority of urban areas, mobile sources cause the overwhelming majority of ozone-forming emissions, as much as 90 percent of total emissions in some regions. Title II of the Act reserves to the Federal government the authority to regulate mobile source emissions. How does EPA envision States achieving attainment with any revised NAAQS, given this lack of direct control of a significant emissions source? Would EPA agree that any revised emission control strategy should focus on the sources which present the greatest challenge in terms of air quality?

<u>Response</u>: The EPA does not believe that mobile sources contribute as much of a proportion to the ozone problem as indicated in your letter. For example, in the emissions inventory compiled by the States in the Ozone Transport and Assessment Group (OTAG) region, on-road mobile sources account for about 30 percent of anthropogenic ozone precursor emissions. For the past 2 years, EPA has been working with OTAG, which was chartered by the Environmental Council of States for the purpose of evaluating ozone transport and recommending strategies for mitigating interstate pollution. The OTAG completed its work in June 1997 and forwarded recommendations to EPA. OTAG essentially recommended a NOx reduction strategy, aimed at reducing emissions from large stationary sources (i.e., utilities). Since the focus of ozone reduction is shifting from local to regional strategies, the 30 percent value noted above should be generally representative of the eastern ozone transport region. Moreover, as noted above in the

response to question 9, a number of control programs are expected to be issued and implemented at the national level. A number of mobile source-related controls -- generally related to exhaust emissions and fuels -- are among those contemplated. Therefore, States will be able to rely on reductions from those programs in their attainment demonstrations.

Under the Act, States have a great deal of discretion in their choice of emission control programs to use in achieving the NAAQS. In some cases, the Act preempts States from regulating fuel content or emissions from motor vehicles or nonroad equipment. The Act, however, does not preempt a number of control measures that States may undertake in these areas, and the Act also provides for waiving preemption in some cases. State inspection and maintenance programs and transportation control measures are just two examples of the kinds of programs a State can implement to reduce emissions from mobile sources.

11. *In EPA's February 10, 1997 response to Chairman Joe Barton, EPA contended that it has the authority to establish revised attainment dates if the proposed revised standards are adopted. The EPA's legal justification for this is that significant portions of the Act pertain specifically to the existing ambient air quality standards and therefore would no longer be valid if new standards are adopted. This position would have the practical effect of nullifying significant portions of the Act without Congressional action. In EPA's option, does the Act specifically give EPA the authority to modify or nullify any portion of the Act? Does EPA believe that its current interpretation of the Act allows EPA to in effect "override" Congressional intent and oversight?*

Response: In EPA's February 10, 1997 response to Chairman Joe Barton, the Agency addressed whether subparts 2 and 4 of part D of title I of the Act would be directly applicable for purposes of a new or revised ozone and fine particulate matter NAAQS. In short, the Agency reached the conclusion that these portions of the Act would not apply to areas for purposes of a revised or new ozone or fine PM NAAQS. However, EPA did not conclude that these provisions were in any sense nullified or modified.

The EPA initially determined that subpart 2 would not continue to apply directly to areas designated nonattainment for the 0.12 parts per million 1-hour standard after that standard was revoked and replaced with an 8-hour standard. The EPA now believes that subpart 2 will continue to apply to areas designated nonattainment for the 1-hour standard until such time as EPA determines each area has air quality meeting that standard. While EPA has now changed its position with respect to the continued applicability of subpart 2 to certain ozone nonattainment areas, it is critical to note that EPA has never taken the position that it was either modifying or nullifying these provisions in the Act. Rather, EPA initially took the position that subpart 2, by its own terms, is only applicable in the context of the 1-hour standard; subpart 2 does not apply to areas that have achieved the 1-hour standard or areas for which a 1-hour health-based standard no longer exists.[1]

The EPA believes that the statute does not specifically address the continued application of subpart 2 to areas that have not yet met the 0.12 ppm 1-hour standard when EPA revises the ozone standard. In its proposed interim implementation policy, EPA examined many statutory references and adopted the reading of the statute that subpart 2 applied only for purposes of the 1-hour standard and that this section would not directly apply to areas once a different standard was promulgated. However, in the past several months, EPA has continued to examine the issue and has concluded that an alternative interpretation is also well-supported by the statute and will provide additional policy assurances that areas will continue along the path toward achieving air quality consistent with the 1-hour standard by the dates set forth in section 181 of the Act, and will also provide benefits to ensure timely attainment of an 8-hour standard. More specifically, EPA now believes that subpart 2 continues to apply directly to areas that have been designated nonattainment for the 1-hour standard until such time as EPA determines that those areas have air quality meeting that standard.

Although subpart 2 will continue in effect for areas designated nonattainment for the 1-hour standard, for the reasons set forth in the letter to Chairman Barton, EPA believes that subpart 2 will not apply to areas meeting the 1-hour standard but newly designated nonattainment for the new 8-hour standard. These areas will be subject to the more general planning provisions of subpart 1. Subpart 1 provides the time frames that EPA may establish for areas to attain a new or revised NAAQS. Areas designated nonattainment for both the 1-hour standard and the 8-hour standard will primarily plan in accordance with subpart 2 until the 1-hour standard is met. However, such areas will need to provide for the incremental planning needed to move from attainment of the 1-hour standard to attainment of the 8-hour standard by the attainment date for the 8-hour standard. In most cases, EPA believes that the attainment date for the 8-hour standard will be several years after the continuing attainment date (under section 181(a)) for the 1-hour standard.

The EPA has retained the annual PM_{10} standard and the 24-hour PM_{10} standard in a revised form -- subpart 4 will continue to apply to areas that are designated nonattainment for PM_{10}. Therefore, areas that remain nonattainment for PM_{10} or that are newly designated nonattainment for PM_{10} will need to meet the planning requirements of subpart 4. However, as with an 8-hour ozone standard, nonattainment planning for a new fine PM standard will be subject to the requirements of subpart 1.

12. The EPA has a separate docket to address implementation issues related to the proposed new ambient air quality standards. Of great concern to states such as California, which have pre-existing stringent air quality control measures in place, is the issue of equity. The EPA should ensure that states such as California be given full credit for past emission reduction commitments. Therefore, if any baselines are established by EPA from which new emissions reductions are to be achieved, California should be given full credit and consideration for the substantial emission reductions already achieved as part of the state's efforts to meet its own air quality requirements. California industries and air districts have achieved significant emission reductions as part of their ongoing efforts to meet the requirements of the federal and California Clean Air Acts.

What is EPA's position on this issue, and does it intend to credit States (and California in particular) with the emissions reductions achieved to date? How does EPA intend to do so?

<u>Response</u>: It is EPA's intent that States be given all the credit toward reasonable further progress (RFP) for reductions possible under the Act. It should be noted that <u>any</u> reduction in appropriate emissions, regardless of when they occur, will serve to make progress toward attainment, which is the primary function of the RFP requirement. It should also be noted that the FACA Subcommittee, which is providing recommendations on implementation of the revised NAAQS, is considering how to interpret the RFP requirement for purposes of the revised standards. The RFP requirements of the Act appear in part D of title I. For the current ozone NAAQS, §182 in subpart 2 sets forth fairly firm requirements related to the baseline and those reductions for which credit can be taken. California and other States can continue to take credit for reductions under subpart 2 under current rules since subpart 2 will continue to apply until areas have attained the .12 ppm NAAQS.

For the PM_{10} NAAQS, §189 in subpart 4 lays out more general requirements concerning the baseline and credits for reduction. In subpart 1 ("Nonattainment Areas in General"), RFP is defined in §171(1), and the very general requirement that nonattainment SIPs provide for RFP appear in §172(c)(2). The EPA has not yet developed proposed procedures for dealing with the issue of credit for RFP, and intends to do so when it publishes all of the proposed guidance on implementation of the revised NAAQS.

Table 1. PM-2.5 data reported to AIRS from 1985 to 1996

Identification			Average statistics			
State	MSA	AIRS site ID	ann. mean	max	sample size	years
CA	NOT IN AN MSA	060250004	22.0	61.0	32.5	2
CA	NOT IN AN MSA	060250005	13.9	41.0	48.0	1
CA	NOT IN AN MSA	060251003	13.2	40.0	53.3	3
CA	NOT IN AN MSA	060270004	6.3	15.5	29.5	2
CA	NOT IN AN MSA	060271001	4.5	13.5	48.5	2
CA	NOT IN AN MSA	060271003	6.6	36.0	36.0	2
CA	NOT IN AN MSA	060310003	18.2	70.7	48.0	3
CA	NOT IN AN MSA	060310004	16.2	22.0	5.0	1
CA	NOT IN AN MSA	060311003	17.1	70.0	53.0	1
CA	NOT IN AN MSA	060510001	18.0	79.0	54.0	1
CO	NOT IN AN MSA	080070002	16.6	53.0	47.0	2
CO	NOT IN AN MSA	080450001	8.9	19.5	38.0	2
CO	NOT IN AN MSA	080510004	24.2	91.0	26.0	3
CO	NOT IN AN MSA	080670005	13.6	32.0	16.0	1
CO	NOT IN AN MSA	080670006	15.8	25.0	16.0	1
CO	NOT IN AN MSA	080671001	4.5	8.0	16.0	1
CO	NOT IN AN MSA	080970003	15.7	44.7	28.0	3
CO	NOT IN AN MSA	080970620	11.6	24.0	17.5	2
ME	NOT IN AN MSA	230050002	11.5	58.0	26.0	1
ME	NOT IN AN MSA	230070001	20.5	55.5	20.0	2
ME	NOT IN AN MSA	230090003	7.5	32.5	93.0	4
ND	NOT IN AN MSA	380570001	9.6	27.0	61.0	1
OR	NOT IN AN MSA	410170001	14.9	67.3	50.3	3
VT	NOT IN AN MSA	500230003	16.3	44.0	26.0	1
CA	BAKERSFIELD, CA	060290004	27.9	106.0	53.0	2
CA	BAKERSFIELD, CA	060290014	19.1	65.0	57.5	2
CA	BAKERSFIELD, CA	060290014	18.3	66.0	52.0	2
CA	BAKERSFIELD, CA	060292004	10.9	55.0	57.5	2
MD	BALTIMORE, MD	245100008	26.1	108.0	60.0	1
MD	BALTIMORE, MD	245100038	22.1	49.7	30.7	3
ME	BANGOR, ME	230190002	13.0	26.0	25.5	2
ND	BISMARCK, ND	380150003	9.7	23.0	42.5	2
MA	BOSTON, MA-NH	250250002	19.2	48.7	64.3	3
SC	CHARLESTON-NORTH CHARLESTON, SC	450190046	11.2	25.9	37.0	9
IL	CHICAGO IL	170310001	13.7	32.0	61.0	1
IL	CHICAGO, IL	170311019	13.9	28.0	57.0	1

IL	CHICAGO, IL	170311901	14 2	31 0	58.0	1
IL	CHICAGO, IL	170312001	14 2	28.0	60.0	1
OH	CLEVELAND-LORAIN-ELYRIA, OH	390350013	28 4	65 0	66.0	1
CO	COLORADO SPRINGS, CO	080410008	10 1	32.0	27.5	2
SC	COLUMBIA, SC	450790007	19 1	48 8	28.8	8
SC	COLUMBIA, SC	450790007	18 0	63 5	16.5	4
TX	CORPUS CHRISTI, TX	483550012	11 6	42.5	129 0	4
CO	DENVER, CO	080010001	15 4	47 3	29.7	3
CO	DENVER, CO	080310017	12 3	41 3	28 3	3
MI	DETROIT, MI	261630001	15 5	39 8	49 4	5
MI	DETROIT, MI	261630002	22 4	57 0	49 7	3
MI	DETROIT, MI	261630033	17 8	52 7	37 0	3
MI	DETROIT, MI	261632002	17 2	54 0	47 0	2
MN	DULUTH-SUPERIOR, MN-WI	271370032	10 4	32.3	46.3	3
TX	EL PASO, TX	481410037	14 3	55 8	126.8	4
OR	EUGENE-SPRINGFIELD, OR	410390013	13 6	39.0	35.5	2
TX	FORT WORTH-ARLINGTON, TX	484390060	12 2	32.0	122.0	2
CA	FRESNO, CA	060190005	24 6	108.5	58.5	2
CA	FRESNO, CA	060190008	17 8	60 5	60.0	2
CA	FRESNO, CA	060190008	18 0	54 5	60 0	2
CA	FRESNO, CA	060191002	15 4	87 0	44 0	1
CA	FRESNO, CA	060390001	13 6	40.5	52.5	2
CA	FRESNO, CA	060390002	26 6	116.0	58.0	1
TX	GALVESTON-TEXAS CITY, TX	481671002	13 2	37 5	89.5	2
NC	GREENSBORO—WINSTON-SALEM—HIGH POINT, NC	370670009	20 3	57 5	48.5	4
NC	GREENSBORO—WINSTON-SALEM—HIGH POINT, NC	370670020	20 1	49.6	43.4	5
TX	HOUSTON, TX	482010024	15 6	50 3	113.5	4
CA	LOS ANGELES-LONG BEACH, CA	060370002	27 0	76 3	48.8	4
CA	LOS ANGELES-LONG BEACH, CA	060374002	22.9	78 8	55 0	4
OR	MEDFORD-ASHLAND, OR	410291001	11 4	33 7	44 3	3
OR	MEDFORD-ASHLAND, OR	410293001	32.2	142.7	47 3	3
MN	MINNEAPOLIS-ST PAUL, MN-WI	271230047	13 0	35 0	49 0	2
CA	MODESTO, CA	060990002	14 5	49 0	57 5	2
CA	MODESTO, CA	060990003	18 2	75 0	47 0	1
TN	NASHVILLE, TN	470370006	23 6	61 0	45.0	2
LA	NEW ORLEANS, LA	220710010	15 6	37 4	86.6	5
NY	NEW YORK, NY	360610056	23 6	62.6	41.9	10
NY	NEW YORK, NY	360610069	27 7	62 3	38.9	8
NY	NEW YORK, NY	360610077	37 5	83 6	51 3	9
NY	NEW YORK, NY	360610079	21 1	57 0	44 5	2

NY	NEW YORK, NY	360610095	31 5	68.0	31.0	1
NJ	NEWARK, NJ	340130011	20 7	57 8	52.5	6
NJ	NEWARK, NJ	340390004	21 5	56 5	53 8	6
PA	PHILADELPHIA, PA-NJ	421010004	20 9	52 5	41.5	6
PA	PITTSBURGH, PA	420030021	24 8	67 0	40 0	4
PA	PITTSBURGH, PA	420030027	22 2	77 0	154.7	7
PA	PITTSBURGH, PA	420030064	24 1	77 0	109.0	1
ME	PORTLAND, ME	230050014	27 0	51.5	13.0	2
OR	PORTLAND-VANCOUVER, OR-WA	410050004	10.2	31.0	61.0	1
OR	PORTLAND-VANCOUVER, OR-WA	410510015	15 0	47.3	53.7	3
OR	PORTLAND-VANCOUVER, OR-WA	410510015	11 5	20.0	8.0	1
OR	PORTLAND-VANCOUVER, OR-WA	410510080	15.5	42.0	35.0	2
OR	PORTLAND-VANCOUVER, OR-WA	410510080	12.9	41 5	44.0	2
CA	RIVERSIDE-SAN BERNARDINO, CA	060658001	34.5	102.3	53.0	4
CA	RIVERSIDE-SAN BERNARDINO, CA	060710006	11.0	23.0	48.0	1
CA	RIVERSIDE-SAN BERNARDINO, CA	060710014	11 3	23.0	49.0	1
NY	ROCHESTER, NY	360556001	15 8	37 3	29 3	4
CA	SACRAMENTO, CA	060670010	14.2	47 5	75.5	2
MO	ST. LOUIS, MO-IL	291890001	14.8	41.0	36.7	3
MO	ST. LOUIS, MO-IL	291892003	15.1	43.2	48.7	9
MO	ST. LOUIS, MO-IL	291895001	15 6	46.0	101.7	6
MO	ST. LOUIS, MO-IL	291895001	12.8	37.7	44.1	7
MO	ST. LOUIS, MO-IL	291897001	16.0	30 3	14.7	3
MO	ST LOUIS, MO-IL	291897002	14 5	44 3	47.0	6
UT	SALT LAKE CITY-OGDEN, UT	490350003	29 3	89 7	40.3	3
UT	SALT LAKE CITY-OGDEN, UT	490351001	11 0	22.0	8.5	2
UT	SALT LAKE CITY-OGDEN, UT	490353001	20 6	74 0	41.0	2
TX	SAN ANTONIO, TX	480290036	11.8	43 3	142.5	4
CA	SAN JOSE, CA	060850004	15.1	55 7	48.0	3
CA	SAN LUIS OBISPO-ATASCADERO-PASO ROBLES,CA	060791005	16.5	25 0	13.0	1
MA	SPRINGFIELD, MA	250154002	10 5	41 7	103.3	3
OH	STEUBENVILLE-WEIRTON, OH-WV	390811001	25 7	96.0	51.0	1
OH	STEUBENVILLE-WEIRTON, OH-WV	390811002	18.9	53 0	46.0	1
OH	STEUBENVILLE-WEIRTON, OH-WV	390811012	12.6	29 5	32.5	2
OH	STEUBENVILLE-WEIRTON, OH-WV	390812003	18 9	50 0	42.0	1
CA	STOCKTON-LODI, CA	060771002	15 1	64 0	52.3	3
NY	SYRACUSE, NY	360671014	31 4	70 8	33 3	4
NY	SYRACUSE, NY	360671016	14 8	48 1	50 9	8
CA	VISALIA-TULARE-PORTERVILLE, CA	061072002	22.8	84 3	59.0	4

BRIAN P. BILBRAY
49TH DISTRICT, CALIFORNIA

COMMERCE COMMITTEE

SUBCOMMITTEE ON
HEALTH AND ENVIRONMENT

SUBCOMMITTEE ON
FINANCE AND HAZARDOUS
MATERIALS

SUBCOMMITTEE ON
OVERSIGHT AND
INVESTIGATIONS

Congress of the United States
House of Representatives
Washington, DC 20515

September 26, 1997

#2   WASHINGTON OFFICE:

1530 LONGWORTH HOUSE OFFICE BLDG.
WASHINGTON, DC 20515
(202) 225-2040
FAX (202) 225-2948

DISTRICT OFFICE:

1011 CAMINO DEL RIO SOUTH
SUITE 330
SAN DIEGO, CA 92108
(619) 291-1430
FAX (619) 291-8956

INTERNET:

E-mail:
brian.bilbray@mail.house.gov

World Wide Web:
http://www.house.gov/bilbray

Ms. Carol Browner, Administrator
U.S. Environmental Protection Agency
401 M Street SW
Washington, D.C. 20460

Dear Administrator Browner:

I am writing to continue our dialogue on the new National Ambient Air Quality Standards (NAAQS) which have been proposed by the Administration, which we have had the opportunity to develop over the last six months via correspondence and a number of hearings in the Commerce Committee and elsewhere on Capitol Hill.

As I have expressed to you publicly and privately, while I am not opposed, I am of a different mind about the directions you have proposed we take in the form of these NAAQS. While I believe that that proposed move from a one-hour to an eight-hour standard for monitoring ground level ozone will help to improve our ability to protect the public health, I nonetheless remain concerned that the proposed new standard for fine particulate matter (PM 2.5) may not be comprehensive enough to provide the highest levels of protection possible. Let me reiterate here that I believe that too little information at this time exists as to our understanding of the relationship between both the size and composition of fine particles, and how they affect the human respiratory system as a result. I would feel more comfortable implementing a new national standard for particulates if it reflected a more robust understanding of the science involved, such as the State of California's current comprehensive standard for larger particulates (PM 10).

Having said this, however, I do not intend to take action to block the new standards, and instead wish to work with you and my colleagues in the House and Senate to ensure that the implementation of these standards over time provides the best possible protection of the public health. Because the implementation process as outlined in the guidance provided by EPA will occur over a period of many years, however, I want to also ensure that the implementation process does not unintentionally interfere with existing attainment schedules and deadlines in San Diego and elsewhere. To this end, I would appreciate additional clarification from you as to how such a scenario can be avoided. As you are aware, the Health and Environment and Oversight Subcommittees of the Commerce Committee will hold a joint hearing next week to examine these issues; it may ultimately be useful for additional hearings on this topic to occur, as I can only speculate that other of my colleagues may have similar concerns.

I have consulted with the San Diego Air Pollution Control District, locally elected officials, and the regulated community to focus and narrow a series of questions to which I would like you to respond as soon as possible. As you know, in San Diego we are proud of the progress we have been able to make in reducing overall levels of air pollution, and in order to continue this line of progress I would appreciate having your answers to the following questions as soon as possible.

1   How does EPA intend to allow consideration of air pollution transport between air basins within the same state in determining an area's attainment status, and the severity of its nonattainment problem?

2   What analytical tools will EPA consider acceptable for analysis of air pollution transport? Specifically, will EPA require an actual analysis for <u>each</u> exceedance of the ozone standard, or will modeling of typical events be allowed to demonstrate that such an exceedance occurred due to transport?

3   How will EPA handle the shift from a one-hour analysis to an eight-hour analysis for ozone?

4   How will EPA address the issue of ozone transport which occurs during some, but not all, of the eight hours of an exceedance of the ozone standard? Specifically, how will EPA address transport events when the individual hours within an exceedance occur as a result of a mix of local and transported air pollutants? Will such a contribution be prorated, and if so, how?

5   How will EPA allocate funding for the data collection, purchase of the analytical tools needed to conduct the required analysis, and the staff time to perform the analysis?

6   How does EPA intend to analyze exceedances of the new PM 2 5 standard to account for the impact of transported air pollutants?

7   Will areas which are likely to meet the PM 2.5 standard be allowed to conduct minimal data collection demonstrate compliance?   A comprehensive track record for PM 10 is already in place. The unnecessary expenditure of limited resources can be avoided if an area is allowed to demonstrate that the standard has been met, by developing an appropriate ratio between PM 10 and PM 2.5, and applying it to historical PM 10 monitoring data.

8   If an area attains the new one-hour standard for ozone in 2001 (the second extension of a 1999 attainment deadline) and then submits a State Implementation Plan which meets all specified requirements, would it then be classfied by EPA as "transitional"?

9   Will areas which receive a "transitional" classification for ozone be designated Non-attainment Unclassifiable for the 8-hour standard? Please further define "Non-attainment Unclassifiable".

10   In what ways will planning requirements be reduced for areas which are designated as "transitional" for ozone?

Carol, I appreciate your attention to these questions, and believe that more detailed discussion of these issues will lead to both a better public understanding of this process, and an improved ability of local public health officials to work with the federal government to most effectively protect our air quality. Please don't hesitate to contact me directly if you would like to discuss these questions further, or if I or my staff can provide you with additional information. I look forward to your swift response so that we can get on with the substantive business of protecting the public health.

Sincerely,

Brian Bilbray
Member of Congress

UNITED STATES ENVIRONMENTAL PROTECTION AGENCY
WASHINGTON D C   20460

NOV 0 6 1997

NOV - 5 1997

OFFICE OF
AIR AND RADIATION

Honorable Brian Bilbray
House of Representatives
Washington, D.C. 20515

Dear Congressman Bilbray:

This is in response to your letter of September 26, 1997 to Administrator Carol Browner regarding the implementation of the new ozone and particulate matter (PM) standards. In your letter you reiterated your concerns about the lack of information on the size and composition of the fine particles and their effect on the human respiratory system. You also expressed a concern that the implementation of the new 8-hour national ambient air quality standard (NAAQS) for ozone will interfere with the existing attainment schedule and deadlines in San Diego and elsewhere. In addition, you asked ten specific questions on the implementation of the new standards

The Environmental Protection Agency (EPA) is committed to developing more information on the health effects of fine PM and on the composition of the fine particulates. In his July 16, 1997 memorandum to the Administrator of the EPA concerning implementation of the revised air quality standards for ozone and PM, President Clinton directed that within five years the next review on the PM standard is to be completed, including a review by the Clean Air Scientific Advisory Committee. This review schedule will permit the analysis of additional health studies on the impact of fine particles before the EPA designates areas as nonattainment for the $PM_{2.5}$ standard and States develop their implementation plans (SIPs) for attainment of the new $PM_{2.5}$ standard. In addition, within the next two years, EPA will deploy $PM_{2.5}$ samplers at 1500 sites across the nation. The PM samples collected at approximately 300 of these sites will be routinely analyzed to determine composition. These analyses will provide a robust data base for SIP development.

EPA does not believe that the implementation of the revised ozone standard will negatively impact the progress areas are currently making. Areas that are not attaining the old 1-hour standard will need to concentrate on attainment of that standard. The control measures which an area adopts to meet the 1-hour standard will also meet requirements applicable for attaining the new 8-hour standard. Once an area attains the old standard, EPA will revoke the standard for that area. This will allow the area to concentrate on developing the additional control measures necessary to meet the additional increment of reductions for the new standard. Therefore, implementation programs for the 1-hour standard can serve as a glide path toward attainment of the new standard. The answers to your specific questions are included in the enclosure to this letter.

I appreciate this opportunity to be of service and trust that this information will be helpful to you.

Sincerely yours,

Richard D. Wilson
Acting Assistant Administrator
for Air and Radiation

Questions on the Implementation of the New Standards

1. How does EPA intend to allow consideration of air pollution transport between air basins within the same State in determining an area's attainment status, and the severity of its nonattainment problem?

Answer: Nonattainment designations are based on the ambient concentrations of a pollutant in an area. The source of the pollutant is considered in determining the appropriate control strategy for an area and in developing the area's SIP. The portion of the ambient concentration attributable to transport and the reductions in the transport that will result from upwind control programs are modeled in the SIP development process. A state must ensure that the attainment demonstration and control strategies contained in its SIP address the emissions that are transported between air basins within the state.

2. What analytical tools will EPA consider acceptable for analysis of air pollution transport? Specifically, will EPA require an actual analysis for each exceedance of the ozone standard, or will modeling of typical events be allowed to demonstrate that such an exceedance occurred due to transport?

Answer: EPA's guidance for analyzing prospective strategies for meeting the new 8-hour ozone air quality standard is being developed over the next year. Much of the new guidance is likely to be similar to that which exists for the 1-hour standard. The existing guidance does not require an analysis of each occasion on which the level of the standard is exceeded at one or more monitoring sites. Instead, State or local agencies are asked to group observed exceedances into broad meteorological classifications. The most important factor in making the distinction among meteorological classifications is generally wind speed and direction. Representative days are chosen for further analysis from each of the meteorological classifications. One or more of the identified meteorological classifications could well be representative of cases with predominant transport. For such cases, regional emission controls or controls in upwind areas would likely be identified as needed to meet the air quality standard on such days. Air quality models are the principal means used to determine appropriate levels of regional or local controls necessary to meet air quality goals.

3. How will EPA handle the shift from a 1-hour analysis to an 8-hour analysis for ozone?

Answer: The same model is used for determining both the 1-hour and 8-hour ambient concentrations. The calculations of the 1-hour and 8-hour concentrations are done in a computer program that arrays the model output data in different formats. However, the planning process for the two standards may differ slightly. The EPA will work with the affected States to allow them to focus on the appropriate standard. For example, EPA will revoke the 1-hour standard for areas as they attain it so that such areas can turn their attention to attainment of the 8-hour standard (if they in fact violate that standard). Areas that do not attain the 1-hour standard will continue to develop and implement programs for attainment of that standard. Those control programs will also be helpful in the attainment of the 8-hour standard.

4. How will EPA address the issue of transport which occurs during some, but not all, of the 8-hour period of an exceedance of the ozone standard? Specifically, how will EPA address transport events when the individual hours within an exceedance occurs as a result of a mix of local and transported air pollutants? Will such a contribution be prorated, and if so, how?

Answer: With ozone, almost all ambient concentrations result from a combination of local emission sources and transported pollutants. The portion of the ambient concentration which is attributable to transport varies from one area to the next. In some cases, the transported pollutants are a significant portion of that concentration, while in other cases the transports represent only a small portion. When the transport crosses State boundaries and contributes significantly to nonattainment or an inability to maintain the standard in a downwind state, EPA has the authority to require the responsible state to require emission controls. For example, October 10, 1997, EPA announced a call for 22 States and the District of Columbia to revise their SIPs to address transport of nitrogen oxides. When the pollutants are transported between air basins within state boundaries, the State must ensure that the SIP addresses transport and design control strategies that will allow all areas of the State to attain the standard.

5. How will EPA allocate funding for data collection, purchase of analytical tools needed to conduct the required analysis, and staff time to perform the analysis?

Answer: The EPA estimate of costs for the establishment of a $PM_{2.5}$ network is based upon the following factors: network design, site preparation, instrument procurement and supplies, operation and maintenance, sample analysis and quality assurance, data management, and personnel. These costs were derived from a survey of selected State/local air pollution control agencies and updated to reflect revised assumptions specific for $PM_{2.5}$ sampling. Each State's portion of the 1500 $PM_{2.5}$ site network program has been determined from the requirements specified in the regulations. The costs, and associated hourly burdens, are prorated for each State agency based on their share of the national network.

The Administrator has committed to providing 100 percent of the funding needed to develop and implement the network. In support of this commitment, Congress has approved an additional $24,743,000 in the Agency's FY 1998 budget specifically for the $PM_{2.5}$ monitoring network and has directed the Agency to award these funds using the authority of section 103 of the Clean Air Act. Section 103 grants do not require State matching funds, which allows EPA to provide 100 percent of the funds for $PM_{2.5}$ monitoring. (Generally, EPA provides annual funding for State air pollution control agencies using the authority of section 105 of the Clean Air Act, which requires a 40 percent State match). These funds will enable EPA to fully fund the cost of purchasing and operating the monitors, as well as the associated analytical costs.

In addition to providing 100 percent funding for the $PM_{2.5}$ monitoring network, the Agency is in the process of implementing programs to offset some of the burdens that would inherently fall upon the individual States. For example, EPA is establishing a national contract that will provide for the purchase of the required $PM_{2.5}$ monitors. It is anticipated that the majority, if not all, of the States will utilize this mechanism to purchase their $PM_{2.5}$ equipment. We are also establishing a quality assurance program for the States, and we are expanding capabilities at several strategically located laboratories to assist the States in analyzing samples.

The remaining funds will be allocated to the EPA Regional offices for awarding to their respective States based upon the cost estimates described above. These funds will cover the costs associated with the operation and maintenance of the network.

6. How does EPA intend to analyze exceedances of the new $PM_{2.5}$ standard to account for the impact of transported air pollutants?

Answer: The EPA will require each State to install and operate a Regional Transport and a Regional Background monitoring site for $PM_{2.5}$. Data from these sites, in conjunction with data from other required monitoring locations, will be used to assess the contribution of nearby emission sources as compared to long-range transport of the pollutants. It is recognized that transport could be potentially more important for $PM_{2.5}$ than has previously been the case for PM_{10}. However, the influence of atmospheric transport cannot be adequately assessed at this time. With the installation of the extensive $PM_{2.5}$ network in 1998, the assessments of transport will become more quantitative.

7. Will areas which are likely to meet the $PM_{2.5}$ standard be allowed to conduct minimal data collection to demonstrate compliance? A comprehensive track record for PM-10 is already in place. The unnecessary expenditure of limited resources can be avoided if an area is allowed to demonstrate that the standard has been met, be developing an appropriate ratio between PM-10 and $PM_{2.5}$, and applying it to historical PM-10 monitoring data.

Answer: The EPA is developing guidance to allow reduced sampling frequency for those areas not likely to violate the $PM_{2.5}$ standard. This should reduce the monitoring burden in those areas. Designations under the $PM_{2.5}$ NAAQS will be determined only when 3-years of FRM (or equivalent method) data are available. As the Clean Air Scientific Advisory Committee recognized in its review, there is wide variability in $PM_{2.5}/PM_{10}$ ratios in time and location that precludes defining uniform ratios. Because of the substantial uncertainties associated with $PM_{2.5}$ projections using ratios, EPA believes that compliance with the $PM_{2.5}$ standard should be based on actual ambient $PM_{2.5}$ data. For these reasons, in the NAAQS rule, EPA provided that three years of monitoring data are needed to determine whether an area is meeting the standard.

8. If an area attains the new 1-hour standard for ozone in 2001 (the second extension of a 1999 attainment deadline) and then submits a SIP which meets all specified requirements, would it then be classified by EPA as "transitional"?

Answer: To qualify for the transitional classification, areas must be attaining the 1-hour standard at the time EPA designates areas for the 8-hour standard.

9. Will areas which receive a "transitional" classification for ozone be designated nonattainment unclassifiable for the 8-hour standard? Please further define "nonattainment unclassifiable."

Answer: Section 107(d)(1) of the Act requires EPA to designate areas as "nonattainment," "attainment," or "unclassifiable." In general, EPA does not use the term "unclassifiable" with respect to designated nonattainment areas. However, with respect to the 1-hour standard a designated nonattainment area was identified as "nonclassifiable" if it did not fit one of the five listed classifications in section 181(a) of the Act. With respect to the designation of areas for the new 8-hour standard, EPA has created the "transitional" classification for certain nonattainment areas. Thus, since these areas will already have a classification, EPA will not need to classify them as "nonclassifiable."

10. In what way will planning requirements be reduced for areas which are designated as "transitional" for ozone?

Answer: As provided in the implementation strategy, EPA will be revising its new source review and conformity rules to allow transitional areas to meet those requirements through minimal changes to their existing plans. For the areas that will attain the standard through the implementation of a regional nitrogen oxide strategy, EPA believes that new local controls will not be necessary. EPA anticipates such areas will be able to use regional-scale modeling and will not be required to conduct local modeling. Over the next 15 months, EPA will issue guidance interpreting the requirements applicable for these areas.

Mr. BARTON. Madam Administrator, we appreciate the courtesy of coming. There will be additional written questions for the record. There may also be a request in the very near future for you to attend again, simply because of the fact we have had so many members that couldn't participate today, and these are important issues; I think that you understand that. So——

Ms. BROWNER. I'm sorry, there may be a request for——

Mr. BARTON. We may, after consultation with Chairman Bliley, schedule another hearing. I'm not saying we will.

Ms. BROWNER. Fine. Whenever you need me to be here, we will certainly work to accommodate you.

Mr. BARTON. Thank you.

Ms. BROWNER. Thank you.

Mr. BARTON. You may be excused.

Ms. BROWNER. And we have two matters that you had agreed to insert in the record.

Mr. BARTON. Beg your pardon?

Ms. BROWNER. We had two documents which you had agreed to make a part of the record today.

Mr. BARTON. If you will provide those to the staff, they will be apart of the record.

Ms. BROWNER. Thank you.

[The information provided is printed following the hearing or retained in subcommittee files.]

Mr. BARTON. The Chair is going to announce that there may be a series of votes on suspensions on the floor, but, because of the time of the day, we are going to call the second two panels at this time and attempt to get your testimony into the record in between these votes. So as soon as Administrator Browner vacates the chair, we would ask the next two panels to come up in a combined fashion.

[Brief recess.]

Mr. BARTON. The committee will come to order.

Our next panel is a combined panel of elected officials and business representatives. We have the Honorable Pat McCrory, who is the mayor of the city of Charlotte. We have the Honorable William Walaska, who is a State senator from the State of Rhode Island. We have the Honorable Steve Morse, who is a State senator, the State of Minnesota. We have the Honorable Brian Frosh, who is a State senator from the great State of Maryland. We have Mr. Glen Heilman, who is vice president for Heilman Pavement Specialties, representing the NFIB, and we have Mr. Lee Ottenberg, who is the president and CEO of Ottenberg Bakery, and he is representing the Independent Bakers Association.

Gentlemen, it is the tradition of the Oversight Subcommittee to take all testimony under oath. Do any of you refuse to testify under oath? It is also the tradition, consistent with you rights under the Constitution, that you have the right to be advised by counsel during your testimony. Do any of you so wish to take advantage of that constitutional guarantee? Will you please stand, then, and raise your right hands?

[Witnesses sworn.]

Mr. BARTON. Be seated. We are going to start with the mayor of Charlotte. Your entire written statement is put into the record. We would ask that you summarize that, and we will recognize Mr. McCrory for 5 minutes.

STATEMENT OF HON. PAT McCRORY, REPRESENTING THE U.S. CONFERENCE OF MAYORS; ACCOMPANIED BY HON. WILLIAM WALASKA, STATE SENATOR, STATE OF RHODE ISLAND; HON. STEVEN MORSE, STATE SENATOR, STATE OF MINNESOTA; HON. BRIAN E. FROSH, STATE SENATOR, STATE OF MARYLAND; GLENN HEILMAN, REPRESENTING THE NATIONAL FEDERATION OF INDEPENDENT BUSINESS; DAVID G. HAWKINS, REPRESENTING THE NATURAL RESOURCES DEFENSE COUNCIL; AND LEE OTTENBERG, REPRESENTING THE INDEPENDENT BAKERS ASSOCIATION

Mr. McCrory. Thank you very much. I just want to say that I didn't know anything that would seem to last longer than the Monday night football game between the Panthers and the 49ers, but you have succeeded.

Mr. BARTON. Welcome to congressional hearings.

Mr. McCrory. Thank you very much. We do have comments for the record, and I would like to just give a brief summary to those comments. In fact, I came here this morning from Charlotte and the headline in The Charlotte Observer is that the mayor is speaking against EPA air limits, and that's the difficultly of this whole subject. Because if you say anything negative about current or future air limits, you give the impression that you are against trying to improve our air quality, when that's just the opposite of the case. The mayors of many cities throughout the United States, and including Mayor Kirk in Dallas, Mayor Archer in Detroit, Mayor Daley in Chicago—and I am speaking on behalf of all of the mayors through the Conference of Mayors and as chairman of the Environmental Committee—are working at the grassroots level to try to improve our air and water quality through such things as mass transit, busways, light rail, HOV lanes, pedestrian-friendly things, through sidewalks, greenways, bike paths.

In Charlotte, for example, I've taken a very controversial stand as a Republican, pushing for a half-cent sales tax to help improve our air quality through mass transit, similar to what Mayor Kirk in Dallas and others are doing in the Texas area. I know that Dallas, for example, has been very successful with the DART program, converting buses to natural gas, I think by the year 2002, which will have a tremendous positive impact on that area.

But the fact of the matter is we are at the grassroots level working very hard to clean our air and water. However, representing all of the mayors, we do have some very serious concerns about the new, more stringent standards that are being proposed by EPA, and we fully support the legislation which will delay or will further study those standards for up to 4 years, so we can get some clear facts to help us in that implementation and to improve the partnership between local, State, and the Federal Government.

One of the immediate concerns we have is that we are concerned the whole EPA strategy with these new air regulations is not consistent with other strategies, especially brown fields. For example, if we further invest in brown fields, and many of us are asking for further Federal moneys to help us in brown fields, that may in fact contradict the effort to control suburban sprawl, which will increase air pollution. But, in fact, the new air standards, may, in fact, encourage suburban sprawl and not encourage in-field development and the recycling of pollution land, which we're attempting to do in Charlotte, in Dallas, in Chicago, and other major urban centers.

Another major concern we have is that the boundaries within these regulations are basically political boundaries. And as you know, air and water does not recognize political boundaries. And I think it's been recognized much of our pollution is out of our control and is determined by weather patterns. In fact, in our measurements for 3 years, most of the controlling factor is not the work that we do to help improve our air, but is impacted by the weather patterns in not only in our State, but up and down the eastern coast.

The other concern that we would like to State is, of course, the real benefits of this and the science of this, but we don't want to get into those details. I know that you're also debating that issue.

We would also like to make two major points with respect to the cost, as the Conference of Mayors in the National League of Cities have worked very closely with Congress to deal with the issue of unfunded mandates and the cost of unfunded mandates. And there are two issues related to this. We do not know what the cost of these new regulations are going to be on either the State or cities. Yes, there is some money budgeted for the bureaucracy to push the paperwork to help us in this effort, but there is no budgeted money to assist in implementing and helping us with such things as mass transit, to help us meet these new, more stringent air standards. And that is a major concern that we feel, and we feel the legislation, which many of you are sponsoring, helps get a better control and a feel of what the true cost of the new air standards will be, and we would strongly recommend that we get a control of what the costs are and who pays for it. As you know, at the local level, we're paying a large portion of local tax dollars to try to improve our current pollution standards.

With respect to unfunded mandates, there is no doubt the Conference of Mayors strongly believes that this goes against the grain of recent legislation which forbids new Federal legislation of unfunded mandates. I think there is a loophole that we do not agree with from the EPA, which is bypassing recent legislation regarding unfunded mandates.

Again, I would like to just emphasize that, on behalf of the Conference of Mayors, that we do delay this for 4 years and support legislation which would verify the science and the benefits. This will help determine the cost, which will determine the resources to help pay for this, consistent with the overall goals of the EPA, and will help us further develop partnerships between the cities, State, and Federal Governments. And we are very interested in further developing these partnerships and negotiating because, again, it comes down to the basic implementation. It comes down to the cities and to the States, regardless of what people are doing in Washington, DC.

We look forward to working with you and also supporting legislation regarding ISTEA and other types of Federal legislation which will help the mayors throughout this country.

[The prepared statement of Hon. Pat McCrory follows:]

PREPARED STATEMENT OF HON. PAT MCCRORY, MAYOR, CHARLOTTE, NORTH CAROLINA, ON BEHALF OF THE U.S. CONFERENCE OF MAYORS

Mr. Chairman and committee Members, I am Pat McCrory, Mayor of Charlotte, North Carolina. I am here today on behalf of The U.S. Conference of Mayors where I serve as the Chair of the Energy and Environment Committee.

I appreciate the opportunity to provide testimony in response to the U.S. Environmental Protection Agency's (EPA) proposed implementation plan for the new ozone and particulate matter air quality standards, especially because implementation of the new standards is at the heart of the Conference's concerns.

I would like to begin by emphasizing that The U.S. Conference of Mayors wants cleaner air for our cities and our nation. This is perhaps best illustrated by the fact that many mayors have long been involved in efforts to improve air quality, even apart from what is required by federal law. Cleaner air is a critical goal for us because we recognize it improves not only the physical health of our citizens but also the environmental and economic well-being of our communities. This is more than just a perception; just two weeks ago *The Wall Street Journal* reported that Asia's pervasive air pollution problem is a primary reason for its suffering economy.

Despite a desire for cleaner air, The U.S. Conference of Mayors is opposed to EPA's proposed implementation package for the new air quality standards. While EPA characterizes the package as "common sense," for a number of reasons, we are at a total loss as to how the plan could be characterized as such, or, more importantly, how it could be promoted as a plan to improve air quality, public health, and the environment, when it appears it will accomplish none of these objectives.

From our perspective, the implementation package defies logic. Among its most serious shortfalls is that is fails to correct recognized shortcomings of the existing implementation plan that undermine public health and the environment. Instead, EPA's proposal retains and perpetuates these problems by adding additional layers to an already unnecessarily complex and counterproductive regulatory framework.

Overview of Conference's Actions on New Air Standards

Let me begin by discussing briefly the Conference's most recent actions on these issues, I have attached to my testimony our recent policy statements which were adopted by our full membership in June at our Annual Meeting in San Francisco.

At our San Francisco Meeting, the nation's mayors recorded their opposition to the proposed standards which U.S. EPA subsequently issued in July. Mayors also believe that one of our most pressing concerns with the new standards is that we have not yet built the partnership—among federal, state and local governments—that can produce the results these new standards envision.

We debated these issues and among the mayors there was a clear expression that we are not ready to embark on such a broad-based national effort to control ozone and fine particulates to the levels set forth in U.S. EPA's revised standards.

The Conference and its member mayors are now expressing their support for the Klink/Boucher/Upton legislation as one course of action to deal with the most immediate issue before the Congress and that is EPA's recent action adopting new standards for ozone and fine particulate matter.

And, we have a number of mayors now in nonattainment or maintenance areas who continue to confront unreasonable constraints and challenges in satisfying currently applicable standards. We believe there are issues pertaining to areas now in non-attainment or maintenance that also need attention when you move the Klink/Boucher/Upton legislation forward.

Overview of Key Concerns

Let me first discuss some of the overriding issues that mayors have raised about the new standards and the implementing policies.

It is certainly a challenge for us at the local government level to fully characterize the myriad of issues associated with how state and local governments will implement these new standards, given that most days it appears that we are dealing with a moving target.

These standards are not ready for prime time. The entire process of development of these new standards does not increase our comfort level at the local level that many of the daily issues we confront in trying to find ways to improve air quality are even in play.

We have a process that was driven by litigation, an effort by a party to force U.S. EPA to implement directives under the Clean Air Act. We know that the Agency failed to perform its duty, and there are many reasons for missing the statutorially-prescribed deadlines, but nonetheless the burden then shifts out to the state and local government sector to absorb the shortfalls of the process. This is not the way to attack this complex problem and it strains credibility in the system.

Consider the standards from our vantage point at the local level. Over the years, we have seen Congress impose requirements and deadlines on the Agency that can't be met. Deadlines often are simply an expression of a "federal desire" to get things done, with the full knowledge by lawmakers that the deadlines would not be met. Once the Federal Government overcomes its own hurdles, those of us at the state and local level must conform to constraints and timetables that have been largely pre-determined. It presumes a level of subordination to the federal government that is simply out of step with the voters and the taxpayers.

Last November, U.S. EPA under a court-ordered schedule was directed to issue a new PM standard, allowing a very brief public comment period. The Agency elected to include revisions to the ozone standard in this package. Local officials, most of whom do not even know if they are affected by the proposal, were given 45 days to comment on EPA's proposed standards. Even if they did know that they were affected, they had 45 days to comment on federal rules that many communities will spend up to ten years working to comply with and another decade or more trying to maintain compliance with.

The Conference and the other local and state government organizations strenuously objected and urged the Administrator to request an extension of the public comment period, with the court ultimately agreeing to slip the schedule by three weeks.

It is this continuing internal conflict at the federal level that sends the wrong signal to state and local governments and their decision-makers, ironically the levels of government that often must reconcile federal standards with reality.

No unfunded mandates report was prepared. Despite our numerous requests to the EPA Administrator and others for an unfunded mandates report, we have been advised that the U.S. EPA is not required to undertake such a study. We strongly disagree.

I just mentioned that our ability to participate in shaping decisions affecting our communities was fairly limited in the first instance. To compound matters, our federal partners won't even undertake the analyses to advise us of the likely cost impacts on our taxpayers.

U.S. EPA missed the point on why this study was so important to us. The state and local government organizations never suggested that a mandates cost report should drive EPA's decision on where to set the standard. Rather, we believe that the "Unfunded Mandates Reform Act" directs that such a report be prepared.

We are well past the days where we simply change federal rules and requirements, and simply impose unknown costs on state and local taxpayers. This is a threshold issue for mayors and other local and state officials.

We had believed that this practice had come to an end, when the Congress acted on this legislation in the last Congress and the President signed this legislation and issued two related Executive Orders.

If we, at the local government level and in the states, are going to partner effectively with the federal government, we must have a fuller description of the potential costs of meeting these standards. In this way, we can consider what investments are needed and how we partner with the federal government to shoulder the costs of meeting these federal standards.

To illustrate this point, I checked the *Register Register* on revisions to existing air plans. In each case, EPA announced that it had not performed a mandates study on a State's revisions because it was an existing mandate and the Agency wasn't required to conduct such a study. I hope that after we skipped the mandates report here, we don't read in the *Federal Register* sometime in the future that State plans implementing these rules are existing mandates and therefore don't require any cost estimate.

If we are going to meet these standards, it is essential that we first know what will be required in terms of investment. We can then work to secure the resources at the federal level and among state and local governments to successfully achieve what we have set forth in these new standards.

Brownfields redevelopment is threatened by new standards. Despite EPA's continuing efforts to work with communities on brownfields, the new air standards and the accompanying implementation policy ignore the impact these standards will have on city and county strategies to redevelop brownfields.

As Members of this Committee know, the nation's mayors are continuing to press for a broader federal response to the growing inventories of vacant, abandoned and underutilized properties in cities and counties all across the nation where the threat of Superfund liability has choked off redevelopment of these sites.

At the very time mayors are finally focusing national attention on the need for action on brownfields legislation, we must confront new air rules that we believe will layer new burdens on cities and counties seeking to recycle these sites.

Many mayors have also expressed concerns about the public health implications of further stalling redevelopment of these sites, many of which can be found in neighborhood of economic disinvestment with prevailing unemployment and below average access to health services. Without cleanup and redevelopment of these sites, there will be persistent and significant public health risks to populations within these areas where employment opportunities would otherwise lower these risks.

Mayors also assert that brownfields redevelopment is likely to produce many environmental gains, as sites which are concentrated in higher density and established communities are redeveloped. These sites are more likely to utilize incumbent public transportation capacities or support transit corridor development, shorten work trips and offer alternatives to lower density, exurban sprawl.

Process is flawed. Even today as we meet for this hearing, we still don't even know what areas of the country are affected by the new standards. Consequently, there are many officials who will not be fully engaged in these deliberations, and many subsequent efforts at U.S. EPA and in the states, that will direct how these standards are implemented in future years. It is simply unfair to affected parties that the scope and applicability of these standards are not better described and analyzed. As a result, there will be many local elected officials who will soon find themselves confronting burdens under these standards and operating under requirements that were developed without their full participation.

This is wrong. We should not be approaching a problem of such importance without a better understanding of the full scope and coverage of these new standards.

Specific Concerns with the Proposed Implementation Plan

Under the proposed implementation plan for ozone, areas categorized as "nonattainment" under the current 1-hour ozone standard, and only those areas, will remain subject to the existing implementation plan. That existing plan, which singles out individual communities—primarily the nation's cities—for a sanction-based scheme of emission control measures, has proven over the past few years to be a faulty approach for improving air quality, and should be eliminated under a new implementation plan, not retained.

The existing plan creates a slew of public health, environmental, and economic problems both in and around those areas by encouraging businesses to leave nonattainment areas in favor of outlying areas where stringent emission controls and offset requirements can be avoided. Driving jobs out of our cities strips individuals of livelihoods and job-related benefits, such as effective health care, that are critical to maintaining better public well-being. It results in more brownfield sites, leaving nearby residents with neighborhood property that is abandoned as well as a potential risk to health and the environment. It intensifies urban sprawl, thus increasing auto emissions, degrading natural resources in outlying areas, and destroying farmland critical to the nation's well-being.

Some EPA officials have even recognized the detrimental consequences of its existing plan. They have claimed that the new standards would eliminate these detriment consequences by throwing more areas into the nonattainment category. Their notion appears to be that this expansion of the detrimental, sanction-based scheme would "level the playing field" among different geographic areas. In itself, this would merely spread the problems rather than alleviate them. This scheme, however, has not been carried forth in the proposed implementation plan. Instead, an even worse unfair regulatory apartheid is proposed: areas not meeting the new ozone standard will be designated "transitional" rather than "nonattainment" so that they can avoid what EPA itself has called the "burdensome local planning requirements" and "stigma" that accompany the nonattainment designation under current law. This is a confusing and incoherent approach. It means that in some areas where the ozone standard is not met, the traditional nonattainment requirements would apply, while in other areas where the standard is not met, no nonattainment requirements would apply.

This precarious proposal for "transitional" areas is a clear acknowledgment by EPA that the current framework for improving air quality no longer works. The public would be far better served if EPA simply admitted this, and worked for real and effective reform, instead of engaging in legal maneuvering and linguistic games in an effort to get around its legal obstacles.

Moreover, the "transitional" category will place some communities in a state of ambiguity, unsure of what actions they may be required to take, unsure of what the penalties will be for taking them, and unsure of whether to make local economic investments for fear of degrading air quality. Not only with this regulatory uncertainty will have a chilling effect on local economic development, but it will also create uncertainty about future air quality planning activities.

If we truly want to achieve cleaner air, a better environment, and improved public health, new implementation policies should be crafted, mindful of the lessons we have learned under the existing implementation policies, rather than ignoring them. Cleaner air is critical, but it defies common sense to pursue it to the detriment of other equally critical public goals. Clean air is but one important public health and environmental goal among numerous others, such as clean water, clean land, freedom from illness and disease, etc. If the solution to air pollution comes at the expense of achieving other equally important public health and environmental benefits, it is in fact no solution at all.

Conclusion

The U.S. Conference of Mayors believes strongly that air quality, public health, and the environment can be improved simultaneously, but first, it is essential to climb out of the existing regulatory box. Past efforts should be the guide for eliminating or restructuring policies that are no longer effective. Simply placing a moratorium on EPA's proposals will not do the trick. New policies and strategies that are truly flexible, affordable, and rational should be created;. Most importantly, they should be based on reality and be effective.

A solid start in this direction would be for EPA to forego its plan to spend the next year developing additional implementation strategies and promulgating more rules. Instead, the Agency should focus on immediately implementing programs that will actually bring about the improvements we are seeking. Examples include: development of a nationwide infrastructure for alternate fuels; increased investment in research and development for alternatives to existing products and processes that pollute the air; and a solid plan for improving indoor air quality.

In addition to EPA's work in these areas, we believe that there is much more that can be done to address mobile sources. The Conference has worked extensively on the renewal of the Intermodal Surface Transportation Efficiency Act of 1991 (ISTEA), seeking to preserve and increase investment in the Congestion Mitigation and Air Quality program (CMAQ) and expand federal commitments to public transportation and other alternatives to our growing reliance on single occupancy vehicles. We want to see a broader federal commitment to helping us deal with emissions from automobiles and trucks by extending and expanding ISTEA's commitment to public transportation and other investments that over time will prove beneficial to our collective efforts on clean air.

The ultimate outcome of the current debate, whether legislative or regulatory in nature, will determine air quality policy at lease for the foreseeable future, and may have environmental and economic repercussions even beyond that. Accordingly, it is imperative to move forward carefully and effectively, and with greater certainty about the costs and the investments that are needed to achieve standards.

Mr. BARTON. We thank you, Mayor.

The Chair will now recognize State Senator Walaska. Your entire statement will be put into the record and we ask that you would summarize in 5 minutes.

STATEMENT OF HON. WILLIAM A. WALASKA

Mr. WALASKA. Thank you, Mr. Chairman. Thank you for allowing me to testify this morning—or this afternoon—before this committee.

My name is Bill Walaska. Today I represent two constituencies. First, I'm a Democratic senator from the State of Rhode Island, and second, I am a member of ALEC's Task Force on the Energy, Environment and Natural Resources. As a representative of these two constituencies, I feel compelled to discuss with you the seriousness of the upcoming EPA proposals.

As for the issue of science behind the new standards, I feel that conclusive evidence does not warrant them. A view of ALEC's membership of over 3,000 State legislatures holds this view. Last year ALEC's Task Force passed a resolution on ozone and particulate matter, submitted to this committee by State Representative Tom Alley of Michigan. It opposed the new standards. Last spring, Representative Alley won passage of a resolution based on this ALEC model, which was submitted to Congress from the State of Michigan. I might add that we passed a similar resolution in the Rhode Island Senate this past session.

I feel that the EPA has not taken into consideration the difficulties in implementing the standards already required by the States. Simply put, the science is inconclusive. But I know that I'm not here to discuss that issue, but rather the implementation of the new standards.

With the current standards, Rhode Island is already under sanctions from the Clean Air Act for failure to implement certain control strategies. We have reformulated fuels, installed vapor recovery systems at all of our gas stations, and require new source standards at all power plants, but we have yet to implement an inspection and maintenance program for automobiles. Regulations are currently being drafted in these areas. Implementation of new ozone standards, therefore, is something of a hollow issue, as we can't get into any worse trouble with the EPA.

Winding the net of non-compliance across a larger land mass is favored by the EPA. The control strategies that we are lacking will need to be in place before we focus on any new strategies. Several practical questions exist as to how the new standards will be implemented: What controls will be required if the States are shown to be out of compliance? Who will pay for the new particulate matter monitoring systems? Will the particulate controls include, for example, new equipment for trucking concerns? Will public buses be upgraded? At what cost? Who's to pay?

Rhode Island, for one, knows the difficultly in answering these questions with regard to the current standards. And these difficulties would only be compounded with new standards.

Timing is also critical. EPA is well-known for allowing slippage on timing mandates. From a State's perspective, this may be favorable to allow more time to achieve compliance with the new standards. However, such groups as the American Lung Association may oppose a more flexible implementation schedule. Such a group might sue EPA to achieve rapid implementation, leaving States and localities with no practical ability to meet the standards. In order to assist the States in meeting the standards, Congress should specify the timing of implementation rather than leave it to the EPA.

We must also answer the question, implementation of what? What will the control measures be? EPA is well-known for overreaching in the creation of control measures. The failure of mandated centralized IM programs and EPA's recent shift to acceptance of decentralized programs are extremely costly backtracking examples.

With the transitional plan for implementation at the will of the EPA, any unilateral changes in the plan would have major consequences for the States. Congress should specify the terms of the transition plans to prevent substantive changes during agency rulemaking. The lesson to be learned is reasonableness will show faster results at a lower cost.

As we look at compliance under title V of the Clean Air Act, it is important to note the enforcement measures the EPA can utilize. The EPA has discretion to remove title V permitting, if they feel the State is not adequately enforcing the laws. However, the EPA is the sole arbiter of what's adequate, leaving States to shoot at a moving target.

As a case in point, ALEC has led the way on legislation designed to encourage voluntary compliance of environmental regulations, the Environmental Audit and Qualified Disclosure Act. It allows public and private entities to conduct environmental audits to correct problems without penalty as long as those problems do no harm to third parties, are done in a timely manner, and aren't criminal in nature. Over 20 States have enacted this legislation the past 4 years.

This legislation, which I sponsored in the State of Rhode Island this year—and it passed in the State of Rhode Island—creates far greater compliance with environmental regulations than the EPA and the State environmental agencies put together, at a much lower cost.

The EPA has gone on the offensive with this law, testifying before State legislatures against the bill and attacking businesses in the districts of legislators who sponsor the legislation, as is happening to Representative Scott Orr of Montana. EPA threatened title V permitting in Idaho, Texas, and Michigan, and has threatened to nationalize Virginia's environmental programs this year.

Mr. BARTON. The gentlemen's 5 minutes has expired, if you could summarize in the next minute, the rest of your testimony, please.

Mr. WALASKA. Well, it's no exaggeration that the EPA has significant effects on the American economy and the economy of each State.

Mr. BARTON. Right.

Mr. WALASKA. It's incumbent upon the Congress to legislate these regulations. To leave them arbitrarily up to the EPA—and I think we saw that by the questioning, the questions and answers this morning—is dangerous. We know the Congress is liable to the public, to its constituency. Who's the EPA liable to; who do they answer to?

Thank you very much.

[The prepared statement of Hon. William Walaska follows:]

PREPARED STATEMENT OF HON. WILLIAM WALASKA, A STATE SENATOR FROM RHODE ISLAND

Good Morning, my name is Bill Walaska. I am here today representing two constituencies: first as a State Senator in the majority party of Rhode Island, and secondly, as a member of the American Legislative Exchange Council. Within ALEC, I serve as an executive committee member of the Energy, Environment, Natural Resources and Agriculture Task Force. As a representative of these two constituencies, the citizens of Rhode Island and the legislators of ALEC, I feel compelled to discuss with you the seriousness of the upcoming proposal by the EPA for a transitional plan for implementation of the new National Ambient Air Quality Standards and Particulate Matter standards.

As for the issue of the science behind the new standards, I feel that conclusive evidence does not exist to warrant them: a view held by ALEC's membership of over 3,000 state legislators. Last year, ALEC's EENRA Task Force passed the Resolution on Ozone and Particulate Matter NAAQS Revisions, opposing the new standards.

The ALEC resolution was submitted to this committee by State Representative Tom Alley of Michigan, along with a resolution passed in Michigan. Rhode Island also passed a resolution, which I cosponsored, based on the ALEC model. In it, Rhode Island asks EPA to quantify health benefits and economic consequences before implementing new standards. I feel that the EPA has not taken into consideration the difficulties in implementing the standards already required of the states. Simply put, the science is inconclusive. But I understand that I am not here today to address this issue, but rather the implementation of the new standards.

With the current standards, Rhode Island is already under sanctions from the Clean Air Act for failure to implement certain control strategies. We have reformulated fuels, installed vapor recovery systems at all our gas stations, and require new source standards at all power plants; but we have yet to implement an inspection and maintenance program for automobiles. Regulations are currently being drafted for these areas. Implementation of new ozone standards, therefore, is something of a hollow issue. As is, we can't get in any worse trouble with the EPA. Widening the net of non-compliance across a larger land mass, however, is favored by the EPA. The control strategies that we are lacking will need to be in place before we focus on any new strategies. Several practical questions exist as to how the new standards would be implemented. What controls will be required if the states are shown to be out of compliance with the particulate standards? Who will pay for the new particulate matter monitoring systems? Will the particulate controls include for example, new equipment for trucking concerns? Will public buses need to be upgraded? At what cost? Who is to pay? Rhode Island, for one, knows the difficulty in answering these questions with regard to the current standards. And these difficulties will only be compounded with the new standards.

Timing is also critical. EPA is well known for allowing slippage on timing mandates. From a state's perspective this may be favorable, to allow more time to achieve compliance with the new standards. However, groups such as the American Lung Association may oppose a more realistic and flexible implementation timetable. Such a group might sue EPA to achieve rapid implementation, leaving many states and localities with no practical ability to meet the standards. In order to assist the states in meeting the new standards, Congress should specify the timing of implementation, rather than leave it to the EPA.

We must also answer the question; implementation of what? What will the control measures be? EPA is well known for over reaching in the creation of control measures. The failure of mandated centralized I/M programs and EPA's recent shift to acceptance of decentralized programs are extremely costly examples of backtracking. With the transitional plan for implementation at the will of EPA, any unilateral changes in the plan would have major consequences for the states. Congress should specify the terms of the transition plan, to prevent substantive changes during agency rule making. The lesson to be learned is that "reasonableness" will show faster results, at a lower cost.

As we look at compliance under Title V of the Clean Air Act, it is important to note the enforcement measures the EPA can utilize. The EPA has the discretion to remove Title V permitting if they feel the state is not adequately enforcing the laws. However, EPA is the sole arbiter of what is "adequate," leaving states to shoot at a moving target. As a case in point, ALEC has led the way on legislation designed to encourage voluntary compliance of environmental regulations. The Environmental Audit Privilege and Qualified Disclosure Act allows private and public entities to conduct environmental audits and correct problems without penalty, as long as those problems do no harm to third parties, are done in a timely manner, and aren't criminal in nature. Over 20 states have enacted this legislation in the past four years. This legislation, which passed in Rhode Island this year, creates far greater compliance with environmental regulations than the EPA and state environmental agencies could together, at a much lower cost. The EPA has gone on the offensive to kill this law, testifying before state legislatures against the bill, and attacking businesses in the districts of legislators who sponsored the legislation, as is happening to Rep. Scott Orr of Montana. EPA threatened Title V permitting in Idaho, Texas, and Michigan, and is threatening to nationalize Virginia's environmental programs this year.

It is this sort of attack on innovative legislation which help states meet the burden of complying with environmental regulation that is so disheartening. The Environmental Audit Privilege and Qualified Disclosure Act, supported by members of the public sector such as Gov. Roy Romer of Colorado, is one approach that could help states meet new NAAQS and PM standards. Yet EPA stands foursquare against this approach. States are asked to enforce these new standards, regardless of cost, yet are not allowed the tools with which to do so.

It is no exaggeration to state that the EPA touches every part of our economy through the Clean Air Act. We in Rhode Island recently deregulated electricity, yet EPA mandates may well eat up much of the benefits to consumers. The EPA puts regulations on our businesses and farms and drives up prices. And just as the EPA argues that public health will benefit from the new PM standards, we read in the Wall Street Journal that the EPA implemented a ruling taking asthma inhalers out of the hands of children, for violating the CFC ban! Congress needs to either repeal the new standards, or pass legislation to safeguard the rights of states during the implementation process. EPA has truly made a Browner environment for all.

Mr. BARTON. Well, they answer to the Congress, ultimately.

The situation on the floor is that there are 14 votes, each of which could be called for a 5-minute vote. I've stayed to hear the testimony. We're into the second vote. There is a chance an agreement may be made, so that the rest of the votes may be taken on a voice vote. Therefore, we are going to continue the panel. Hopefully, we'll begin to see other members trickle back in.

If they are going to have record votes all 14 votes, however, especially those of you who are State senators and mayors will understand that I will have recess to go vote; I can't miss 14 votes. I can miss 1 or 2 but I can't miss 14.

So we are going to continue with State Senator Morse from the great State of Minnesota, He's representing the National Conference of State Legislatures. Your written statement is in the record; try to summarize in 5 minutes, please.

STATEMENT OF HON. STEVEN MORSE

Mr. MORSE. Thank you Mr. Chairman.

Mr. BARTON. Use the microphone, please.

Mr. MORSE. Thank you, Mr. Chairman, and I appreciate your staying with us here and hearing us out.

I am, as you said, Minnesota State Senator Steve Morse. I serve as vice chair of the Assembly on Federal Issues of the National Conference of State Legislatures or the NCSL. I'm also the immediate past-chair of the NCSL's environment committee. I offer testimony today on behalf of NCSL, and I appreciate this opportunity.

NCSL is a bipartisan organization that represents America's 7,541 State lawmakers. One of the most important functions of our organization is to analyze Federal legislation and regulations to ensure that the State and Federal roles are clearly and equitably defined. NCSL facilitates successful development and implementation of Federal laws and regulations by expressing to Federal officials the needs and concerns of the States.

Now, I would also like to point out that, although the EPA Administrator may trust the Governors, it is the legislatures that appropriate the money and set up the policies that the Governors follow.

Now to address the issue of the Clean Air Act: NCSL continues to be a strong supporter of the principles underlying the Clean Air Act. NCSL fully supports the goals of the 1990 Clean Air Act Amendments and believes EPA should proceed diligently with full implementation of the law to achieve the goal of clean air.

At the recently completed 23rd Annual Meeting of the NCSL in Philadelphia in August, State legislators from around the country reaffirmed their solid commitment to the Clean Air Act and voiced their support for continued efforts to improve our Nation's air quality in a newly adopted air quality policy. This policy is attached and has been submitted.

NCSL lauds the public outreach efforts of the EPA during the development of the final rules of the National Ambient Air Quality Standards for particulate matter and ozone. EPA's attention to the concerns of the States during this rulemaking process is very important and very much appreciated. My colleagues and I look forward to working with EPA, and to continue working with EPA and Congress to ensure that the needs and concerns of the States are addressed during implementation.

NCSL also greatly appreciates the flexibility offered in the implementation plan that has been outlined in the Presidential memo of July 16. We strongly support this flexibility in the funding offered by the EPA to the States and evidently agreed to last night by Congress. NCSL believes this flexibility will ease the administrative and financial burdens that may be placed on the States by the new standards. We also support EPA efforts to provide informational resources to the public, so that the public, too, understands it's role in achieving Clean Air Act goals.

On behalf of the NCSL, I'd like to focus the rest of my comments on some of the concerns we have vis-a-vis the flexibility and funding offered by the EPA. Namely, EPA—or excuse me—NCSL has some concerns about the authority, as have been expressed at this meeting earlier today, the authority for the EPA to do the flexibility that has been that has been proffered. NCSL urges Congress to ensure that the EPA has the authority and the resources to provide States with the flexibility and funding promised in the administration's plan.

Specifically, NCSL is concerned about the legal authority that the EPA Administrator may have, and that these may be challenged in court. Furthermore, NCSL is concerned that the implementation plan is a Presidential memorandum rather than a Federal regulation. Because the plan is essentially a guidance document, States may face direct legal challenges while carrying out the flexible provisions of the implementation plan.

There are three specific areas of concern, some of which we've heard about earlier today. The first would be to phaseout of the 1-hour ozone standard. The Clean Air Act does not specifically authorize EPA to allow certain areas to meet the current standard before being required to develop plans and impose the new programs to meet the new 8-hour standards. Regarding the implementation of the 8-hour standards, the administration's implementation plan, as we heard today, calls for designations in the year 2000 and nonattainment revisions of the SIP plans by the year 2003. It appears under the administration's plan that the State's action to adopt the new ozone standard is not required for 6 years. However, under section 110(a) of the Clean Air Act, States may need to update their authorities by the legislature, if it is not already in place, within 3 years. We are concerned that this will lead to confusion on the parts of States in how they should proceed with implementation.

Second, transitional classification. Section 172(a)(1) of the Clean Air Act does not specifically authorize EPA to create transitional zones, as described in the Presidential memo, or to exempt such newly created areas from traditional non-attainment requirements.

The third major area is monitoring costs, and this, too, has been addressed, but NCSL is concerned that the States remain ultimately responsible. We appreciate the efforts that have been made to secure monitoring so far; I want to ask that those continue.

In conclusion, I would reiterate that the NCSL supports the goals and principles underlying the Clean Air Act. NCSL appreciates the flexibility in funding, and we pledge to work with EPA and Congress to make sure that these are fully implemented. However, our concerns remain that these may not be realized and that we would be faced with additional financial/administrative burdens, if that were to occur.

And with that, I'll hold the rest of my comments. Thank you.

[The prepared statement of Hon. Steven Morse follows:]

PREPARED STATEMENT OF HON. STEVEN MORSE, MINNESOTA STATE SENATE, ON BEHALF OF THE NATIONAL CONFERENCE OF STATE LEGISLATURES

Mr. Chairman and members of the committee, I am Minnesota State Senator Steven Morse. I serve as a vice chair of the Assembly on Federal Issues of the National Conference of State Legislatures (NCSL). I am also the immediate past-chair of the NCSL Environment Committee. I offer testimony today on behalf of NCSL. Thank you for the opportunity to discuss the issues surrounding implementation of the new National Ambient Air Quality Standards for particulate matter and ozone.

NCSL is a bipartisan organization that represents America's 7,541 state lawmakers. One of the most important functions of NCSL is to analyze federal legislation and regulations to ensure that the state and federal roles are clearly and equitably defined. NCSL facilitates successful development and implementation of federal laws and regulations by expressing to federal officials the needs and concerns of the states. Another critical function of NCSL is to provide state lawmakers with accurate and timely information regarding federal policy and initiatives.

NCSL continues to be a strong supporter of the principles underlying the Clean Air Act. At the 23rd Annual Meeting of NCSL held this past August in Philadelphia, state legislators from across the country reaffirmed their solid commitment to the Clean Air Act and voiced their support for continued efforts to improve our nation's air quality in a newly adopted official Air Quality policy. This policy is attached to my testimony and I ask that it be entered into the record of today's hearing.

NCSL lauds the public outreach efforts of the U.S. Environmental Protection Agency (EPA) during the development of the final rules on the National Ambient Air Quality Standards for particulate matter and ozone.[1] EPA's attention to the concerns of the states during this complex rulemaking process was very important and much appreciated.

My colleagues and I look forward to working with EPA and Congress to ensure that the needs and the concerns of the states are addressed during implementation of the new standards. NCSL greatly appreciates the state flexibility offered in the implementation plan that was outlined in a presidential memorandum of July 16, 1997.[2] NCSL believes this state flexibility will ease the administrative and financial burdens that may be placed on the states by the new standards.

My testimony on behalf of NCSL will focus on the concerns of state legislators regarding the flexibility and funding offered to states by the EPA in the implementation plan for the recently promulgated standards for ozone and particulate matter. Because the official NCSL Air Quality policy does not specifically address the National Ambient Air Quality Standards, NCSL neither supports nor opposes the new air quality standards. However, NCSL does support the federal government providing states with adequate funding, complete technical assistance, and as much administrative flexibility as the law allows for carrying out programs required by federal law. Therefore, NCSL supports the flexibility and funding offered by EPA to states in the Administration's implementation plan.

---

[1] *Federal Register,* pg. 38856-38896, 38652-38760, 38764-38854, 38762, Vol. 62, No. 138, July 18, 1997.

[2] *Federal Register,* Vol. 62, No. 138, July 18, 1997.

NCSL believes the following provisions are necessary for successful implementation of the new standards:

- A provision phasing-in the one-hour ozone standard to allow current ozone nonattainment areas to achieve the current one-hour ozone standard before being required to implement new measures to achieve the new, stricter eight-hour ozone standard.
- A provision implementing the new eight-hour ozone standard to allow three years for nonattainment area designations and an additional three years after designation for states to submit state air quality plans, also referred to as "SIPs," that demonstrate how the new standard will be met.
- A provision creating "transitional areas" to relieve qualified areas[3] from certain "unnecessary local planning requirements"[4] for ozone nonattaimnent areas.
- A pledge by the President and EPA to provide full federal funding for the purchase, installation, and operation of the new particulate matter monitoring networks.

While appreciative of the flexibility and funding, NCSL has some concerns about the provisions listed above. NCSL looks, forward to working with EPA and Congress to address these concerns. NCSL urges Congress to ensure that EPA has the authority and the resources to provide states with the flexibility and funding promised in the Administration's plan for implementation of the new standards. NCSL's concerns include:

I. LEGAL AUTHORITY

NCSL fears that the provisions intended to provide states with flexibility may be removed from the plan if the EPA's legal authority to offer such flexibility is challenged in court. There is concern that these attempts by EPA to offer flexibility to states may be blocked if the courts rule that the Clean Air Act does not specifically authorize such EPA actions.

Furthermore, NCSL is concerned that the implementation plan is a presidential memorandum rather than a federal regulation. Because the plan is essentially a guidance document, states may face direct legal challenges while carrying out the flexible provisions of the implementation plan.

a) Phase Out of One-Hour Standard

The Clean Air Act does not specifically authorize EPA to allow certain areas to meet the current one-hour standard before being required to develop plans and impose new programs to meet the new eight-hour standard. The Clean Air Act is silent on the issue of if and how states may progress from an existing standard to a revised standard. One interpretation of that silence is that revised air quality standards simply replace existing standards. This would require states to immediately begin the process to adopting and planning to meet the revised standards. Under this interpretation, states that continue programs toward meeting the old standards may face legal challenges and, possibly, sanctions under the Clean Air Act.

b) Implementation of the New Eight-Hour Ozone Standard

Section 107 allows up to three years, after promulgation of new or revised standards, for area designation. Specifically, governors have up to one year to submit a list of potential nonattainment areas. EPA then has two years to make final nonattainment area designations. Section 172(d) of the Clean Air Act specifies that states must submit to EPA an air quality plan, or "SIP," revision, that demonstrates how the air quality standards will be met. This SIP revision must be submitted within three years after area designation. As the Administration's implementation plan for the new standards specifies, the deadline for area designations is July 18, 2000 and the deadline for nonattainment SIP revisions is July 18, 2003.

It appears under the Administration's implementation plan that state action to adopt the new ozone standard is not required for six years. However, Section 110(a) of the Clean Air Act requires state action within three years. Section 110(a) requires states to submit SIP revisions to EPA within three years after *promulgation* of a new or revised national ambient air quality standard. SIP revisions must provide for the implementation, maintenance, and enforcement of the new standard. Specifically, these revision must include:

---

[3] Areas that comply with current ozone standard but will not comply with the new ozone standard may be designated as transitional area, provided the new NO_x regulations that are expected this Fall will bring the areas into or close to compliance with the new standard by the year 2000.

[4] *Presidential Memorandum* dated July 16, 1997 published in *Federal Register,* Vol. 62, No. 138, July 18, 1997.

- Emission limitations and other control measures and compliance schedules.
- Programs to monitor, compile, and analyze air quality data.
- Enforcement provisions.
- Provisions to prohibit sources from emitting air pollutants that will greatly contribute to violation of the standards.
- Demonstration of adequate funding, resources and legal state authority to carry out the SIP.
- Provisions to ensure monitoring of emissions from stationary sources by owners or operators of such sources.
- Emergency powers and contingency plans.
- Provisions for periodic review and update of the plan.
- Air quality modeling.
- Stationary source permitting program.
- Provision to ensure consultation with local officials.

Section 110(a) gives specific authority for the Administrator to require submission earlier but not later than the three-year period after promulgation. Under this section, the deadline for states to submit SIP revisions for the new particulate matter and ozone standards is July 18, 2000.

Section 110(a) requires state action within three years and Section 172(d) requires state action within six years. NCSL is concerned that states may be confused by the President's implementation plan, as it only discusses the states' responsibilities under Section 172(d). States may not understand that they must submit SIP revisions to comply with Section 110(a) by July 18, 2000. States that fail to comply with Section 110(a) by the deadline are likely to face legal challenges and sanctions under the Clean Air Act.

c) Transitional Classification

Section 172(a)(1) of the Clean Air Act does not specifically authorize EPA to create "transitional areas,"[5] as described in the presidential memo, or to exempt such newly-created areas from the traditional nonattainment area planning requirements.[6] Under the Administration's plan, transitional areas are a classification of ozone nonattainment. These areas will be required to fulfill, at a minimum, the basic nonattainment requirements listed in Sections 172(a)(1), 173 and 174. The relief offered under the guidance appears to be limited to promised relief in soon-to-be revised rules for new source review and conformity. There do not appear to be very many planning and regulatory burdens lifted from states with areas classified as transitional ozone nonattainment areas.

II. MONITORING COSTS

NCSL is concerned that states remain ultimately responsible for the purchase, installation, and operation of fine particulate matter monitoring stations. The following is a table based on EPA estimates of the FY 1998 through FY 2001 monitoring costs.

Estimated Costs of New Particulate Matter Monitoring Networks

	1997	1998	1999	2000
PMFine	$4,500,000	$18,225,000	$38,143,110	$37,502,023
PM10	$15,860,689	$13,494,542	$12,035,187	$9,292,983
Total	$20,360,689	$31,719,542	$50,178,297	$46,795,006

This chart was produced by NCSL. The U.S. EPA is the source of the estimated monitoring costs.[7]

Federal funding for the monitoring networks, as promised by the Administration, is crucial to states' efforts to meet the monitoring requirements of the new particulate matter standards. Though the FY 1998 appropriations for EPA are not yet fi-

---

[5] Section 185(a) of the Clean Air Act created a very specific definition of "transitional area" by authorizing EPA to suspend until December 31, 1991 the applications of certain requirements in ozone nonattainment areas that did not violate the ozone standard for the period of tune between January 1, 1987 to December 31, 1989. The Administrator was required to review the attainment status of these transitional areas by June 30, 1992.

[6] Section 172(c) requires all nonattainment plans to provide for: implementation of control measures; reasonable further progress; inventory of pollution sources; identification and quantification of emissions; permits; emissions limitations and other control measures; compliance with SIP; planning techniques; and contingency measures.

[7] *Ambient Air Quality Surveillance for Particulate Matter.* U.S. EPA, Office of Air Quality Planning and Standards, Emissions, Monitoring, and Analysis Division. May 22, 1997.

nalized, Congress has agreed to provide states with funding to cover the FY 1998 monitoring costs. NCSL looks forward to working with EPA and Congress to ensure the monitoring costs are completely covered in FY 1998 and in future years.

Federal funding is particularly important for FY 1998 given that many states' legislative schedules and budgetary practices cannot accommodate appropriation of funds on short notice. Many states completed their FY 1998 budgets and concluded their 1997 legislative sessions prior to promulgation of the final rule on the particulate matter standards. State budgetary practices cannot accommodate the costly FY 1998 requirements within such a short time.

In conclusion, I would like to reiterate that NCSL supports the goals and principles underlying the Clean Air Act. NCSL appreciates the flexibility and funding offered to states in the implementation plan for the new standards for particulate matter and ozone. NCSL pledges to work with EPA and Congress to ensure that the flexibility and funding is realized during implementation of the new standards.

NCSL is concerned that this flexibility and funding may not be realized and that an enormous financial and administrative burden may fall on state and local governments. In addition, states that implement the flexible provisions of the Administration's plan may face sanctions for failure to comply with the Clean Air Act. NCSL urges Congress to take the necessary steps to ensure that EPA has the authority and resources to give states the flexibility and funding promised in the Administration's Implementation plan. Thank you for this opportunity to testify.

Mr. BARTON. The bells going off is a motion to adjourn on the House Floor. Now after the motion to adjourn, there could be all heck breaking loose on the House floor, just to be blunt about it.

So, unfortunately, I'm going to have to recess. I hate to recess in the middle of this panel. I can't tell you exactly when we will be back. I'm told that it's a 15-minute vote. So I'm going to hear Mr. Frosh and Mr. Heilman, and then I'll probably recess between Mr. Hawkins and Mr. Ottenberg. So I've got 15 minutes. So, Mr. Frosh, your entire statement will be entered in the record, and you're recognized for 5 minutes.

STATEMENT OF HON. BRIAN FROSH

Mr. FROSH. Thank you, Mr. Chairman. First of all, perhaps I'm sort of a thorn among the roses here, but I support the NAAQ standards. I believe they will yield important health benefits to most Americans. I say that as a representative of Montgomery County, Maryland, which is a non-attainment area, and also as the sponsor of Maryland brown field's law.

Mr. BARTON. I believe Mrs. Browner is a constituent of yours then?

Mr. FROSH. So are some of the other panelists——

Mr. BARTON. You're only doing what your constituents want you to do?

Mr. FROSH. Yes.

Mr. BARTON. All right.

Mr. FROSH. We all do that.

I'm told that Maryland ranks second in the Nation in hospital visits during bad air days, and we do suffer from some of the worst air in the country. I think that we need to set a goal and we need to set it now; otherwise, we don't know where we're going. And I think that, for Maryland, it's critical to be able to achieve that, to address sources of pollution that are outside our State, and the new NAAQ standards do that. It's an opportunity to clean up sources of pollution that we can't control.

Pollution sources from the Midwest are a problem for us. The regional approaches laid out in the implementation strategy will give some relief to Maryland communities that are doing their part to reduce pollution, only to find that out-of-State sources continue to contribute to ozone non-attainment.

For our cities and our environment, there will now be some recourse that identifies and stops transient sources of pollution. I think that the health impacts that the new standards will have are, by far and away, the most important reason to stick with them, but I want to offer the committee and the Chair a slightly different perspective that I think also lends support to these standards. And that has to do with the Chesapeake Bay.

The Chesapeake Bay is a huge capital asset for the State of Maryland, and it brings in hundreds of millions of dollars every year in commercial fishing and billions of dollars a year in tourism. We have more than a million people a year who go out to fish on the Chesapeake Bay.

It's absolutely critical for us to maintain to Chesapeake Bay as a viable resource. The EPA has estimated that the NAAQ standards will result in the reduction of more than 17 million pounds of nitrogen going into the Chesapeake Bay every year. As the committee is probably well aware, we have suffered in recent weeks and months with a micro-organism called pfiesteria that has killed of tens of thousands—hundreds of thousands—of fish in the Bay and poses a threat to human health as well, to the point that our Governor has been forced to close several of the major tributaries to the Bay.

This pfiesteria organism can cause severe neurological problems for humans, including such severe loss of memory that people forget their own names. It is critical, I think, to the help of the Chesapeake Bay and to the health of all Marylanders that we clean up the air.

I would like to close by offering the Chair an excerpt from the diary of Captain John Smith, who explored the Chesapeake Bay in the early 17th century. He wrote, "We found, and in diverse places, that abundance of fish lying so thick, with their heads above water, that for want of nets, our barge driving amongst them, we attempted to catch then with a frying pan, but we found it a bad instrument to catch fish with. Either better fish, nor more plenty, nor more variety for small fish had any of us ever seen so swimming in the water, but they are not to be caught with frying pans."

Well, I think that it's clear that we are never going to get back to the pristine state of the Chesapeake Bay, never going to get back to the point where it looks like you can just reach down into the water and grab a fish, and we're certainly never going to get to the point where you can catch them with frying pans. But I think that the NAAQ standards offer some hope that we can begin to address the most serious problems that afflict the Bay, and perhaps bring it back to a state of help.

Thank you, Mr. Chairman.

[The prepared statement of Hon. Brian E. Frosh follows:]

PREPARED STATEMENT OF HON. BRIAN E. FROSH, A STATE SENATOR FROM MARYLAND

Thank you Chairman Bilirakis, Chairman Barton, and members of the subcommittees. I appreciate this opportunity to testify regarding the implementation of new national ambient air quality standards for ozone and fine particulate matter. The implementation of these standards will be the responsibility of the states, and as the Chairman of the Maryland Senate Subcommittee on the Environment, I am very interested in the strategy put forth by the Environmental Protection Agency.

First, let me state that I support these new standards; I believe that they will improve the quality of the air throughout Maryland and can be implemented in a cost-effective manner. As members of this committee know, state administrations will be responsible for modifying state implementation plans to ensure that counties state-wide can attain the new standards. State legislatures will be called upon to establish new emissions policies in order to comply with the implementation plans, and speaking on behalf of state representatives who are looking for solutions to air quality problems, I am glad that we have the incentive to work locally, with the added assurances that regional solutions must be devised to reduce the incidences of pollutant transfer.

I believe we all are well aware of the health implications of poor air quality. The health concerns alone are sufficient reason to move forward in order to ensure that hundreds of thousands of people in the state of Maryland do not suffer from poor air quality. Ten counties of the twenty-three in Maryland likely will be in violation of the new ozone standards. But I do not see this as a doom and gloom scenario that bodes poorly for Maryland's economy. Quite the contrary, I see the implementation of the new rules as an opportunity to plan for Maryland's continued prosperous future.

Mr. Chairmen, I say this for a couple of reasons. For one, the sooner we begin planning for the new standards, the better off we will be in planning for our economic future. Local improvements in air quality will allow regions of my state to continue to recruit new business, with the knowledge that early pollution prevention efforts will provide a more sustainable economy, not the boom and bust economies that some regions exalt and then suffer through.

Delays in implementing these standards could mean that current improvements in air quality will be set back, and attainment will be much harder to reach for many areas. Such delays would represent short term environmental and economic planning.

Secondly, this is an opportunity to clean up sources of pollution that we in the Maryland legislature do not control. Pollution sources from the midwest, specifically the Ohio Valley, are a problem that we cannot solve easily. The regional approaches laid out in the implementation strategy will give some relief to Maryland communities that are doing their part to reduce pollution, only to find that out of state sources continue to contribute to ozone nonattainment. For our cities and for our environment, there now will be some recourse that identifies and stops transient pollution.

So much of Maryland's economy is built on a healthy environment. I cannot overstate the importance of the Chesapeake Bay to Maryland and its economy. As a member of the Chesapeake Bay Commission, I am very concerned about the Pfiesteria outbreak along the Pocomoke River and the implications about the health of the Bay. While Pfiesteria may not be linked directly to air pollution, excess nutrients in the Bay are exacerbating the problem. The continuing deposition of pollutants from the atmosphere into the Bay is just as alarming as the bacterial threat.

A recent report about the impacts of nutrient loading on the Chesapeake Bay shows that atmospheric deposition is responsible for nearly twenty-five percent of nitrogen entering the Bay. Much of this nitrogen comes from sources that we cannot control in Maryland. No fewer than fourteen states are included in the airshed that contributes pollution to the Bay.

With a new regional approach to controlling nitrogen oxides, we have an opportunity to eliminate one of the primary sources of excess nutrients entering the Bay. The Environmental Protection Agency estimates that the new standards will prevent 17 million pounds of nitrogen from entering the Bay on an annual basis. That is the equivalent of the annual nitrogen loads from the York and Rappahannock rivers combined as reported by the *Bay Journal*. The money spent to prevent that pollution will save us millions of dollars in clean up costs.

In short, improved water quality means enhanced tourism and a stronger fishing based economy. That is one of the primary reasons I say these new rules represent an opportunity, not a threat.

Tourism is a significant part of the economy in eastern Maryland. Clean air and improved water quality are important ingredients to making that economic recipe work. In addition to protecting the Bay, improved summer air quality enhances the experience for visitors, whether they are visiting one of Maryland's beautiful beaches or attending a baseball game at Camden Yards. Can you imagine either of these experiences being taken away because the air quality is so poor that it becomes unhealthful to be outside?

A clear link can be made between the health of our environment and our quality of life. Governor Glendening has recognized this and is encouraging Marylanders to consider water quality and growth issues in their decision making. We should take a similar approach to ensure that future progress does not come at the cost of clean air.

In addition to ozone problems, there are questions about the impacts of fine particulate matter. The funding envisioned to set up monitoring and data collection will be extremely valuable to Maryland. I know that our farming communities have some concern about these particulate standards. But those concerns are based on a number of unanswered questions. The time frame set up for implementing the new particulate matter standards appears to give those of us at the state level the financial assistance, resources and information necessary to make good policy decisions about controlling fine particulate matter. I believe these standards ultimately will be another tool in protecting watersheds as well as human health.

The implementation schedule is a good example of federal regulators working hand in hand with local and state parties to address a problem. I am pleased that as a legislator, I will have an opportunity to determine how Maryland protects its air quality. But, I also know that air pollution does not recognize state boundaries. That is why we need regional and national approaches to improving air quality, and that is why I support the new standards and the implementation schedule put forth by the EPA.

I also want to note for the record a letter of support for the new standards that is signed by twelve environmental and public health organizations from Maryland. This letter indicates a broad base of support for the standards and the implementation schedule. I believe there will be a cooperative effort on the part of all interested parties to see that the new standards are fully implemented with the greatest flexibility.

Mr. Chairmen, that concludes my testimony. I hope this committee will not act to overturn the new standards, because for many of us in the eastern United States, they offer a much needed remedy to a regional and national problem. I will be happy to answer any questions from the committee.

Mr. BARTON. Thank you, Senator. We use frying pans sometimes to hit people up the side of the head in Texas with, but we've never tried to catch a fish with one.

We'll now hear from—I believe that's all of our elected officials— we're now going to go to some of our private sector representatives: Mr. Glenn Heilman, representing the NFIB. He is the vice president of Heilman Pavement Specialties. Again, your written statement is in the record; please allot 5 minutes or less to your oral statement, and your summary, please.

STATEMENT OF GLENN HEILMAN

Mr. HEILMAN. Thank you, Mr. Chairman, and good afternoon. My name is Glenn Heilman. I'm the vice president of Heilman Pavement Specialties. We're a small family business that's been in operation for over 41 years. We're located in Freeport, Pennsylvania, which is just above Pittsburgh, and I want to thank you for giving me the opportunity to testify on behalf of the National Federation of Independent Business regarding the revisions to National Air Quality Standards for Ozone and Particulate Matter.

In addition to being a small business owner, I volunteer and serve as chairman of Pennsylvania's Small Business Compliance Advisory Panel. Now this panel is mandated by section 507 of the Clean Air Act Amendments of 1990 to help small businesses as part of the Small Business Stationary Technical and Environmental Compliance Assistance Program. This program has been successful, despite lack of funds, and has become a model for small business programs and other environmental legislation.

In my position as chairman, the panel has identified over 86,000 small businesses in Pennsylvania that are subject in some way to the existing clean air regulations. Through the continued work of the Compliance Advisory Panel and our program called AIRHELP, over the past 4 years we've provided compliance assistance to small businesses specializing in wood furniture manufacturing, autobody repair, printing, baking, dry cleaning, and metal fabricating. Yet, despite 4 years of outreach, we estimate that only about 30 percent of the small businesses affected by the Clean Air Act Amendments of 1990 have been reached in Pennsylvania. Clearly, 70 percent of the small businesses that could be affected don't even know about our AIRHELP Program.

These statistics reinforce the fact that the 1990 Clean Air Act Amendments are far from being fully implemented. At the same time, EPA has pointed out that the air has increasingly gotten cleaner. Given this progress, only requirements that are essential should be mandated.

As a small business owner, the economic impact and burdensome requirements of the new standards would significantly affect and threaten the livelihood of my business. As a manufacturer of road pavement, my business operates asphalt plants and hauls stone as raw material. The moving of equipment and materials creates a minor particulate matter. I also have air emissions from my off-highway road equipment and heavy trucks. Some of this equipment is old but works very well, and I simply cannot afford to buy new equipment to comply with new regulations.

As a small business owner, I'm actively involved because I have to be. Misguided regulations will put me out of business. Not only will small business owners lose lifesavings and investment, but our employees lose their jobs and communities suffer economically.

For that reason, I'm shocked and disappointed that EPA has gone forward and declined to consider the effects of this rule on small business. Last week, I was invited to attend and speak at the EPA's Small Business Compliance Advisory Panel training program in Phoenix, Arizona. Representatives from 21 States and Puerto Rico were in attendance. Many State representatives posed questions during the conference to EPA senior staff about the new standards for ozone and particulate matter. We expressed our concerns about how small businesses would know whether they would have to comply, what the cost would, and who could assist them with implementation. EPA said that there is essentially no guidance for small business in this area. Needless to say, the business owners left with a certain degree of apprehension and desperation.

My industry is not glamorous, but it is my family's livelihood. It distresses me, and small business owners like me, that EPA has not considered the potential consequences on our lives or our communities or the entire economy before imposing this new standard.

Rather than implementing new standards for clean air, I recommend utilizing and encouraging the use of present means to achieve air quality improvements. There are technologies presently available to help clean our air. In our company, we voluntarily look for them. In 1980, my father developed a new ozone-friendly way to improve asphalt roads. This technology is exemplified in a material called Highway General Purpose Material or HGP. A 2-year university study documents that HGP emits seven times less Volatile Organic Compounds, VOC, in the form of low molecular weight normal and branched alkane hydrocarbons, than the present technology used to pave roads.

Under standard technology, present road paving methods allow more than 1,000 gallons or 3 tons of gasoline-type VOC to evaporate into our troposphere for every mile paved. HGP reduces this VOC pollution by 85 percent. Next week I will be receiving a Governor's Environmental Excellence Award for this HGP technology.

In closing, it's important to keep in mind the unique nature of a small business owner when examining our reaction to environmental legislation and regulations. Small business owners wear many hats—we drink the water; we breathe the air; we fish the lakes. We want to help the environment for ourselves and our children and expect the government to be fair and responsible.

The new regulations for ozone and particulate matter are unnecessary and will result in an enormous regulatory burden and threaten the business that my family has spent 41 years to build. A viable framework is in place and consists of new environmentally friendly technologies, such as HGP, and couples these technologies with existing programs, such as the AIRHELP program. The system is working; please, let's use what we have. Thank you.

[The prepared statement of Glenn Heilman follows:]

PREPARED STATEMENT OF GLENN HEILMAN, VICE PRESIDENT, HEILMAN PAVEMENT SPECIALTIES INC.

Good Morning. My name is Glenn Heilman. I am the Vice President of Heilman Pavement Specialties, Inc., a small family owned business that has been in operation for 41 years. We are located in Freeport, Pennsylvania which is just above Pittsburgh. Thank you for giving me the opportunity to testify on behalf of the National Federation of Independent Business regarding the revisions to the national air quality standards for ozone and particulate matter.

NFIB is the nation's largest small business advocacy group representing 600,000 small businesses in all fifty states. A typical NFIB member has five employees and grosses $350,000 dollars in annual revenue. NFIB's membership reflects the general business profile by having the same representation of retail, service, manufacturing and construction businesses that make up the nation's business community.

In addition to being a small business owner, I serve as Chairman of Pennsylvania's Small Business Compliance Advisory Panel. This panel is mandated by Section 507 of the Clean Air Act Amendments to help small business as part of the Small Business Stationary Technical and Environmental Compliance Assistance Program. This program has been successful despite lack of funds and has become a model for small business programs in other environmental legislation.

Our small business program conducts seminars, offers a toll-free confidential hotline, low interest loans and many other outreach efforts for small businesses. Every state has such a program in varying degrees of effectiveness. These programs are valuable tools to improve our air quality and are overseen by the Environmental Protection Agency.

In my position as Chairman, the panel has identified over 86,000 small businesses in Pennsylvania that are subject to existing clean air regulations. Through the continued work of the Compliance Advisory Panel (CAP) over the past four years, we have provided compliance assistance and our program, called AIRHELP to small businesses specializing in wood furniture manufacturing, auto body repair, printing, baking, dry cleaning, and metal fabricating. Yet, despite four years of outreach, we estimate that only about 30 percent of the small businesses affected by the Clean Air Act Amendments of 1990 have been reached in Pennsylvania. We estimate that 70 percent of the small businesses that could be affected do not even know about Pennsylvania's AIRHELP Program.

These statistics reinforce the fact that the 1990 Clean Air Amendments are far from being fully implemented. At the same time, EPA has pointed out that the air has increasingly gotten cleaner. Given this progress, only requirements that are essential should be mandated.

As a small business owner, the economic impact and burdensome requirements of the new standards would significantly affect and threaten the livelihood of my business. As a manufacturer of road pavement, my business operates asphalt plants and hauls stone as a raw material. The moving of equipment and materials creates minor particulate matter. I also have air emissions from my heavy truck and off-road equipment. Some of this equipment is old, but works well. I simply cannot afford to buy new equipment to comply with the new regulations.

As a small business owner, I am active and involved because I have to be. Misguided regulations will put me out of business. Not only will small business owners lose life savings and investment, but our employees lose their jobs and our communities suffer economically. For that reason, I am shocked and disappointed that the EPA has gone forward and declined to consider the effects of this rule on small business.

Last week, I was invited to attend and speak at the EPA Small Business Compliance Advisory Panel Training Program in Phoenix, Arizona. Representatives from 21 states and Puerto Rico learned about facilitating the Section 507 Programs for their areas.

Many state representatives posed questions during the conference to EPA senior level staff about the new National Air Quality Standards for Ozone and Particulate Matter. We expressed our concerns that EPA had never convened a review panel to study the costs of the new regulations on small businesses as required by law under SBREFA. The EPA officials responded with the statement that they are only responsible for guidance to the states—the states are responsible for implementation. When asked how small business would know whether they would have to comply, what the cost would be and who could assist them with implementation strategies, EPA stated that there is essentially no guidance to help small business in this area. Needless to say, the small business owners who attended the conference took back to their respective states, feelings of apprehension and desperation.

My industry is not glamorous, but it adds billions of dollars and thousands of jobs a year to the U.S. economy. My small business is my family's livelihood. It distresses me and small business owners like me, that EPA has not considered the potential consequences to our lives, our communities and the entire economy before imposing this new standard.

Rather than implementing new regulations for clean air, I recommend utilizing and encouraging the use of present means to achieve air quality improvements. There are technologies presently available to help clean our air. In our company we voluntarily look for ways to improve the environment. In 1980 my father developed a new, ozone-friendly technology for asphalt roads. This technology is exemplified in a material called HEI-WAY General Purpose Material or HGP. A two year university study documents that HGP emits seven times less Volatile Organic Compounds (VOC)—in the form of low molecular weight normal and branched alkane hydrocarbons—than the present technology used to pave roads. Additionally, this technology also eliminates a significant water pollution threat to rural streams and wetlands.

Under standard technology, present road paving methods allow more than 1000 gallons (or three tons) of gasoline-type VOC to evaporate into our troposphere for every mile paved. HGP reduces this VOC air pollution by 85 percent. On a nationwide basis, of the nearly 4 million miles of roads in the country, this technology is applicable to over 60 percent of them. In Pennsylvania alone, if just 1 percent of the roads were paved each year with HGP instead of the standard technology, over 3000 tons of VOC air pollution would be eliminated. Next week, I will be receiving the Governor's Environmental Excellence Award for HGP technology. On numerous occasions, I have spoken with EPA officials about the possibilities of HGP. My efforts have been met with reluctance and disregard. My research has proven to me that there are more innovative and effective ways of doing things than undercutting small business growth and jobs.

In closing, it is important to keep in mind the unique nature of a small business owner when examining our reaction to environmental legislation and regulations. Small business owners wear many hats. Two of the most important are being both a business owner and a citizen of a community. We drink the water, breathe the air, and fish in the lakes. We want a healthy environment for ourselves and our children. However, we also expect the government to be fair and responsible.

The new regulations, signed in July, by EPA for ozone and particulate matter are unnecessary, will result in an enormous regulatory burden and threaten a business that my family has spent 41 years to build. A viable framework is in place. It consists of new, environmentally friendly technologies, such as HGP, and couples these initiatives with existing programs. The system is working. Let's use what we have.

Mr. KLINK [presiding]. Thank you very much. Mr. Hawkins.

STATEMENT OF DAVID G. HAWKINS

Mr. HAWKINS. Thank you, Congressman Klink. My name is David Hawkins. I'm here on behalf of the Natural Resources Defense Council, and I'd like my testimony to be submitted for the record and would just make a few quick points.

Mr. KLINK. Without objection.

Mr. HAWKINS. Congressman, you and others have indicated a lot of concerns about the course of implementation of the new standards, and I think the way I would like to suggest you approach those concerns is to look at them in the context of the implementation of the Clean Air Act over the last 27 years. Since 1970, the act has had a common structure: strong health-based standards coupled with prompt schedules for implementing and carrying out those standards, with responsibility shared by State and local governments as well as the Federal Government. It's a program that has worked. We've had enormous improvements in air quality and enormous reductions in emissions, and we've done it without hindering the growth in the American economy. We have a thriving economy; we have cleaner air.

Is the job done? No, it's not done. The research that the EPA looked at to revise the standards clearly shows, in our view and in the views of many, many independent scientists, that tougher standards are needed to protect public health. We need to have those standards on the books to advise the public of how clean their air is, how much cleaner it needs to be to protect their health, their kids' health, their parents' health. We need those standards on the books to structure and motivate the kind of innovation and creativity that we've seen over the last 27 years.

Sending a signal to slow down or cause a moratorium on implementing those standards, in our view, would be the wrong medicine. It would create signals that would cause industry to have second thoughts about pursuing cleaner technologies, cleaner methods. It would create signals; it would create confusion and uncertainly in the very urban development types of issues that have been have been raised.

As long as we defer getting down to the hard work of figuring out what are the emission reduction opportunities that are attractive, we will have remaining uncertainty. As I say in my testimony, Congress could enact a law declaring that communities did not have to proceed to clean up their air, but Congress cannot enact a law saying that these areas are not polluted. They are polluted. And as long as they are polluted, there will be both a public health toll and a need to take action. That need to take action will create uncertainty until we get down to the business of taking that action.

There are a couple of points that were raised earlier that I would just like to touch on.

The first is billions of dollars in highway sanctions. Just to clarify, the law does not impose highway sanctions on States for not meeting air quality standards. The highway sanctions are an available tool, if an area fails to do the planning needed to demonstrate attainment of the standards, but failure to meet those standards is not a cause for imposition of highway standards—of highway sanctions.

Brown fields have been mentioned. Brown fields—I really don't see how that is an issue with respect to the standards. New stationary sources of air pollution are required to undergo preconstruction review. Whether they are in an attainment areas or non-attainment areas, they are required to pass certain air quality tests; they are required to be equipped with best technology.

Urban redevelopment projects typically don't involve many new major sources of air pollution, but if they did, whether the area is an attainment area or a non-attainment area, it's still would be required to pass air quality tests to show it is not harming the health of residents, and it will be required to impose best technology. So if there is a policy issue here, it's important to figure out and be clear about what it is.

Litigation is another aspect that's been mentioned. I've read the EPA implementation schedule, and I can say, based on my experience, what I've seen in that implementation schedule, I don't see anything that violates the law. So to me, the real issue is a policy issue. And the policy issue is, how do we cooperate to get clean-up programs to continue to move forward? In our view, the program and structure under the existing Clean Air Act has proven itself to be workable: Set strong standards, come up with programs to make progress to meet those standards, and when there are bumps in the road, have EPA and congressional oversight operate to adjust and make changes where necessary.

Thank you.

[The prepared statement of David G. Hawkins follows:]

PREPARED STATEMENT OF DAVID G. HAWKINS, SENIOR ATTORNEY, NATURAL RESOURCES DEFENSE COUNCIL

Thank you for inviting the Natural Resources Defense Council (NRDC) to testify today on U.S. EPA's plans to implement the revised National Ambient Air Quality Standards (NAAQS) for ozone and for fine particulate matter ($PM_{2.5}$) pollution.

NRDC is a membership organization, founded in 1970, dedicated to protection of human health and the environment. NRDC has worked on behalf of its members, which number more than 350,000 people throughout the United States, to carry out goals that the American public strongly supports: environmental quality that is free from manmade threats to human health and that preserves and enhances the wonderful diversity of natural resources with which America has been blessed.

NRDC supports strong continuing efforts to reduce air pollution as called for by the Clean Air Act. I personally have worked to encourage these efforts for 27 years, beginning with the initial implementation steps under the Clean Air Act of 1970 and including a stint as the EPA Assistant Administrator for Air, Noise and Radiation during the administration of President Jimmy Carter. I have followed closely EPA's revision of the ozone and $PM_{2.5}$ NAAQS and appreciate the opportunity to present our views on implementation of these standards.

During the last year and one-half Congress and the American public have been treated to a drumbeat of claims that EPA's adoption of the revised air quality standards will impose exorbitant costs on consumers, will disrupt community lifestyles, and will threaten the existence of small businesses. Those of you who have studied the history of the Clean Air Act will find these claims familiar. That history also shows clearly that, contrary to these claims, the Act has produced major air quality gains while avoiding the adverse impacts so loudly predicted by the law's opponents.

To evaluate the likely impacts of implementing the new standards we should take stock of where we are today and how we have gotten here. The facts are clear: air quality is greatly improved in the quarter-century since the 1970 Clean Air Act was enacted. We have made dramatic gains in controlling pollution from stationary sources, from motor vehicles and from many other sectors. We have done this in the face of repeated claims that each air quality goal we set was not needed and would lead to economic and social disruption. In fact, setting sound air quality targets and pursuing them under a program that makes decision-makers accountable has focused energies and stimulated creativity in a way that has found solutions that were not easy to predict at the beginning of these programs. Emissions have been reduced, air quality has improved and the economy has grown despite confident assertions that we could not have clean air without sacrificing economic well-being. The Clean Air Act is working.

The central structure of the Act has remained intact since 1970. Congress has required EPA to establish desirable standards of air quality for the nation based on a careful assessment of the damage that air pollution inflicts on human health, the economy, and the environment. Under the law the national standards are to be set so that the air will be clean enough to prevent these damages. Since 1970 the job of developing and carrying out programs to reduce air pollution has been a joint responsibility of EPA and state and local agencies. The NAAQS air quality standards are the target that state and local programs are required to aim for, taking into account the reductions achieved by federally-adopted control programs. Since 1970 the Act has included timetables for adoption of state and local air quality plans (SIPs) and timetables for reaching the NAAQS targets.

At each stage of the Act's evolution voices have been raised to challenge the legitimacy of the standards and to assert that the costs of reaching for those standards will be unacceptably large. History has proved these claims wrong. EPA adopted the first set of NAAQS in early 1971 and the Act at that time required the standards to be met by 1975, with a possible two-year extension. The combination of a demanding air quality target and the Act's tight timetables caused nearly all states to adopt broad programs to reduce air pollution. While major air quality improvements resulted, not all areas could forecast attainment by the Act's deadlines. Accordingly, Congress conducted oversight of actual progress and, based on an understanding of several years of real-world experience with efforts to meet the standards, Congress made adjustments to the Act that reflected an appreciation of both the public's desire for cleaner air and the difficulties that more polluted areas were facing in meeting the Act's ambitious timetables. In 1977, Congress extended the attainment deadlines to 1982, with EPA allowed to grant an additional 5-year extension, and added requirements for reasonable further progress in emission reduction as well as certain other required measures that Congress concluded would help areas make clean air progress.

In the early 1980's pollution reduction progress slowed somewhat due in part to mixed implementation signals emanating from the Administration and uncertainty caused by active consideration of Clean Air Act amendments in both the House and Senate. At the end of the 80's the Bush Administration proposed comprehensive amendments to the Act, including changes to the NAAQS implementation requirements in Title I of the Act. This Committee substantially revised the Administration's proposal and adopted provisions that were passed largely unchanged by Congress and signed by the President as the 1990 Amendments to the Act.

The 1990 Amendments contain several noteworthy features. First, they retain the Act's core provisions of EPA-established national air quality standards, coupled with timetables for state and local programs designed to meet those standards. Second, the amendments classify areas that were not meeting the standards in 1990 based on the severity of their air pollution problems. Third, these classifications are the basis for timetables and pollution control requirements that are tailored to the severity of the area's pollution. Under the 1990 amendments timetables for attaining the 1-hour ozone standard were extended for an additional 3 years for the least polluted areas and for up to 20 years for Los Angeles, the most polluted area. Fourth, Congress added and enhanced provisions of the Act designed to address the interstate transport of air pollution.

With 27 years of history as a guide, we can say the Clean Air Act has worked well. The standards set by EPA have been based on health and ecosystem needs from the very beginning. And, from the beginning the standards have been challenging for badly polluted areas. Yet adoption of strong, protective standards has mobilized efforts by the public and private sectors to develop less polluting ways of doing business. Despite delays in fully achieving the standards in many areas, the process of standard-setting, implementation efforts, and congressional adjustments has worked to make substantial clean-up progress possible without hindering a remarkable record of economic growth. The health-based standards also allow the public to understand how much the levels of pollution in their community must be reduced to meet what available science indicates is good air quality.

Congress has, of course, revisited the Act in the past. Does this argue for legislation now to respond to current allegations regarding the possible impacts of the revised standards? You will not be surprised to hear me say that legislation is not warranted. While claims abound, facts are more helpful in forming good policy. Despite our large air quality gains, we still release enormous quantities of pollution every day in this country and that pollution continues to cause major air quality problems in large metropolitan areas as well as in many smaller communities. While it is difficult today to precisely define the last steps needed to attain the standards in some areas, there are numerous early reduction steps that can and should be taken promptly to make needed progress. History has shown that moving forward with the process of designing programs to reach the standards will uncover many added opportunities that are easy to miss in broad overviews made at the start of the effort.

Congress has played a critical role in the past by supporting the objective of strong standards and timely adoption of sensible measures to meet those standards. This congressional support has helped the public get closer to its desired objective of cleaner air. By requiring diligent pursuit of programs to meet the standards, Congress has made it clear to investors and business planners as well as state and local officials that serious continued efforts to reduce pollution are essential and must be a key part of company and community strategies.

Were Congress to adopt legislation requiring a delay in implementing the revised standards, the impacts would be numerous and all negative. Firms considering investing in cleaner air processes would have second thoughts, leading to delays in developing the solutions we sorely need. Those concerned about the possible impacts of implementing the standards on business planning or community development would be in a position of greater uncertainty, not less. The best way to clear away the misinformation about how the standards will be implemented is to get down to the business of determining the actual amounts of pollution reductions that are needed in specific communities and regions and to identify the most attractive opportunities for achieving those reductions. If Congress were to put implementation of the revised standards on hold, this type of work would be slowed dramatically. As a consequence, not only would our kids and parents suffer even longer from excessive levels of pollution, but communities with bad air quality and firms in those communities would go through an extended period of uncertainty about the future of emission reduction requirements.

While Congress is capable of voting to delay mandatory federal emission reduction requirements, members cannot vote away the fact that millions of people live in polluted communities today. As long as that pollution remains unabated and clean-up programs are stalled, there will be large unanswered questions about the measures that eventually must be adopted to address this important problem. Accompanying that uncertainty, however, is the high likelihood that solving the air quality problem will only get that much more difficult the longer we put off action on needed next steps.

If you favor clean air progress and a stable environment for economic development you should avoid sending a message that pursuit of cleaner air conflicts with economic goals. Congress should be active in its oversight of the implementation of the revised standards. You can help identify obstacles to progress and keep the pressure on all actors to seek out innovative and efficient ways to reduce pollution while providing the other things that Americans value. If you conduct this oversight in a way that shows the public you understand and support their desire for cleaner air you will make it clear that Congress intends to keep the Clean Air Act working. Sending this policy message will be invaluable in stimulating creative responses from industry, small businesses and agencies at all levels of government. Thank you for inviting me to testify. I will be happy to answer your questions.

Mr. KLINK. Thank you very much for your testimony, Mr. Hawkins. And now Mr. Lee Ottenberg, president and CEO of Ottenberg Bakery.

STATEMENT OF LEE OTTENBERG

Mr. OTTENBERG. Thank you, Congressman, members of the committee. I'm Lee Ottenberg, president of Ottenberg's Bakery, a business started by my great grandfather in 1869. We have two plants, one near Catholic University here in Washington; the other in Eldersberg, Maryland. I'm also here representing—as chairman of the Independent Bakers' Association, a national trade association representing over 360 small and medium-sized, mostly family owned, regional wholesale bakeries and the allied trades.

We participated, IBA has participated, in the regulatory process since the early 1980's, when the San Francisco Bay area first started to deal with regulating ethanol emissions. Ethanol is produced in baking when dough is fermented by yeast as part of the leavening process. This is what makes bread and rolls rise.

Under the 1990 Clean Air Act, ethanol emissions are recognized as VOCs. It's a low-reactive, non-toxic VOC. VOC is a component of ground-level ozone in the form of smog. It is estimated that bakery oven emissions of ethanol contribute less than one-tenth of 1 percent of the ozone in the Nation's stationary source emission inventory.

Despite such a low overall percentage contribution, bakeries are a recognized source with a Federal guideline document for abatement control. As a recognized source, States are compelled to include bakery oven controls in their State implementation plans or face the threat of losing Federal highway construction funds.

We operate in the Northeast Ozone Transport Region. Our ethanol emissions do not cross the threshold required for controls. However, three bakeries in Maryland that do meet the statutory threshold have installed oven emission controls.

I am concerned that the proposed EPA modifications will cause Maryland and the District to reconsider and reduce the current threshold requirements for baking industry abatement controls. My fears carry over to other regional independent bakeries throughout the United States.

EPA issued a comprehensive document called the Ozone Regulatory Impact Analysis this year. This RIA grossly underestimates by 25-fold the cost of controlling bakery emissions. The RIA states that a facility cost of $9,000 will do the controls for VOCs and NO_X. In fact, the cost of controlling one bakery oven is close to a quarter million dollars, with annual operating costs that go on every year approaching $100,000. Large bakeries may require controls for several ovens.

Furthermore, the RIA States that the impact will require controls in 32 bakery sources. The fact is that there are already over 80 sources that are being required to have controls now. Over 20 States consider, or are actively considering, bakery oven emissions as necessary for abatement controls. This number will increase with hundreds of new and mostly rural counties that will be subjected to the more stringent EPA regulations.

EPA has no excuse for such disparity between the RIA modeling calculations and actual industry figures, since the agency in 1992 issued an official document dictating how States should regulate the control of bakery oven emissions. The guidance document, or ACT, is already out of date with the development of new, more cost-effective process technologies for ethanol abatement.

Regional bakeries with one or two plants in non-attainment areas have few options for continuing operations in States that mandate oven control requirements. A bakery can close, and we've had bakeries that have closed over the years; bakers can install controls at great expense, or they can synthetically cap emissions. An IBA member in the New York metropolitan area was capped at 25 tons per year because it is a severe non-attainment area.

The problem with capping emissions is it hampers the company's ability to grow, without building new plants, or to respond to sudden, emergency production requirements, such as in the case of economic strikes or when hurricanes impacted the facilities in the south Florida area. And let's not forget that earthquakes in California——

Mr. KLINK. Could I ask the witness to suspend?

Mr. OTTENBERG. Sure.

Mr. KLINK. We've got a 5-minute vote over on the floor. I'm sorry to have to interrupt you, but, as you can tell, we're going through some—oh, the gentlemen's here? You just voted. There's another vote?

Mr. Ottenberg, go ahead.

Mr. OTTENBERG. I think that I can summarize our primary concern is that we see a gross underestimation of the cost of putting controls onto bakery ovens, and that's what we're concerned with. And when it's underestimated by 25-fold, it gives us a lot of pause about the whole system that we are being faced with.

Thank you very much.

[The prepared statement of Lee Ottenberg follows:]

PREPARED STATEMENT OF LEE OTTENBERG, CHAIRMAN, INDEPENDENT BAKERS ASSOCIATION

Thank you Mr. Chairman, members of the Committee, ladies and gentlemen. I am Lee Ottenberg, President and Chief Operating officer of Ottenberg's Bakery, a business started by my Great Grandfather, Isaac Ottenberg in 1869. I am a forth generation baker. Our bakery includes two plants, one here in the District close to Catholic University and another in Eldersberg, Maryland.

I am also here as the Chairman of the Independent Bakers Association, a national trade association representing over 360 small and medium sized mostly family owned regional wholesale bakeries and allied industry trades. IBA has actively participated in the clean air regulatory process since the San Francisco Bay Area first considered regulating ethanol emissions in the 1980's.

Today, we will briefly consider ethanol, a by-product of the baking process, Ottenberg's regulatory posture, the glaring inaccuracies of the Environmental Protection Agency's (EPA) Regulatory Impact Analysis for the industry and finally the impact of the proposed new ozone and particulate matter standards on the baking industry.

Ethanol is produced in baking when dough is fermented by yeast as part of the leavening process. This is what makes bread and roll rise. Under the 1990 Clean Air Act ethanol emissions are recognized as a Volatile Organic Compounds (VOCs). Ethanol itself is a non-toxic low-reactive VOC. VOCs have been found to be a component of ground level ozone in the form of smog. On a national basis it is estimated that bakery oven emissions of ethanol contribute less than one tenth of one percent of the ozone stationary source emission inventory. Despite such a low overall percentage contribution, bakeries are a recognized source with a Federal guideline document for abatement control. As a recognized source, states are compelled to include bakery oven controls in their State Implementation Plans or face the threat of losing Federal highway construction funds.

Ottenberg's bakeries operate in the Northeast Ozone Transport Region established under the 1990 Clean Air Act. Ottenberg's ethanol emissions do not cross the threshold requirement for controls in Maryland or the District. However, there are at least three bakeries in Maryland that do meet the statutory threshold and all three have installed oven emission controls. I am deeply concerned that the proposed EPA modifications in ozone measurement will cause Maryland and the District to reconsider and reduce the current threshold requirements for baking industry abatement controls. My fears carryover to many other regional independents bakeries located in attainment regions throughout the United States.

EPA issued a comprehensive document called the Ozone Regulatory Impact Analysis (RIA) this year as part of rulemaking. The RIA grossly underestimates by 25 *fold* the cost of controlling bakery process emissions. The Ozone RIA states a facility cost of $9,000.00 for baking industry process controls for ozone, VOC and NO_x emissions. In fact, the cost of controlling a bakery oven is closer to a *quarter million dollars* with annual operating costs approaching $100,000.00. Large bakeries may require abatement controls for several ovens. Furthermore, the ozone RIA states the impact of the rule change will only require 32 bakery sources to control ozone and

PM emissions. Industry figures report the actual bakery emission control installations to date TOTAL *80.* Over twenty states consider or are actively considering bakery oven emissions as necessary for abatement controls. This number will increase dramatically with hundreds of new and mostly rural counties that will be subjected to the more stringent EPA regulations. EPA has no excuse for such disparity between the RIA modeling calculations and actual industry figures, since in 1992 the agency issued an official document dictating how states should regulate the control of bakery oven emissions. It should be noted that the guidance document or ACT is already out-of-date with the development of new more cost effective process technologies for ethanol abatement.

Regional bakeries with one or two plants in clean air non-attainment areas have just a few options for continuing operations in states that mandate oven control requirements. A bakery can close, and we have lost at least two dozen independent bakers in the Ozone Transport Region in the past twenty years. Bakers can install controls at great expense with little return on investment or the bakery can "synthetically cap" its emissions by agreeing not to exceed a level of production that would create 50 tons per year of ethanol in a moderate non-attainment region. An IBA member has his bakery capped out at 25 tons per year as a baker in the New York City metropolitan area, a severe non-attainment region. The problem with capping emissions is it hampers our ability to grow as a company without building new plants and to respond sudden emergency increases in demand for our products when other regional bakery suppliers are impacted. Some examples of regional impacts include economic strikes such as the recent three week shut down of Giant Food's bakeries and natural disasters like hurricanes Andrew and Hugo which impacted regional bakeries in Florida and South Carolina respectively. Let's not forget earthquakes out west that lead to bread being shipped into San Francisco from bakeries in Washington and Oregon. Please remember independent bakers often do not have the access to the capital necessary to purchase control equipment and the additional plants to allow production shifts in order to accommodate temporary regional shortfalls.

IBA is also very concerned about the RIA for Particulate Matter (PM). grain handling and bulk sugar/confection transfer operations face significant costs in controlling particulate matter emissions under the PM RIA. These Standard Industry Codes occupy two of the top five identified industry sectors according to EPA that will need "extraordinarily expensive PM controls." EPA establishes in the RIA a three percent cost of sales threshold as an acceptable amount for companies or an industry to spend on pollution abatement. We are concerned that despite the higher costs EPA estimates for grain and sweet goods bakers, well above 3 percent, the agency fails to show an appreciable or measurable drop in overall PM emissions from regulating our industry segment and our allied trades which supply the necessary ingredients for bread and sweet goods.

Expensive PM command and control regulations could be required of all bakery operations with compressed air driven bulk powdered commodity storage of ingredients such as flour and sugar.

- Several states already require bakeries in non-attainment areas to file extensive paperwork necessary for Permits to Operate exterior flour storage silos.
- Grain handling operations and bulk powdered commodities could face control requirements including fabric filters, cyclones, vented hoods, covered conveyors, covered bucket elevators and application of oil sprays to grains for dust control.

The baking industry operates one of the largest fleets of delivery vehicles. Bakeries deliver baked goods through "direct store delivery" or in Ottenberg's case our trucks deliver to the restaurants, institutions or stores for retail sale to consumers. Our bakery operates 65 routes in two states and the District of Columbia. Besides ozone requirements for oven emissions, the PM proposals could force the control of diesel emissions for bakery fleet operations and stationary co-generation engines with reformulated diesel fuels, adding an estimated $.30-$.50 per gallon to our delivery costs.

- PM controls for delivery vehicles could force retrofit requirements for hydrocarbon absorbers, ceramic engine coatings, onboard diagnostics, upgrades of evaporate emission systems, and forced reductions in delivery vehicle miles traveled.
- Employers could be mandated by states to set up employee trip reductions; subsidized van-pools; no-drive days; parking user fees and other restrictive commuter regimes.

In conclusion Mr. Chairman the baking industry will continue to do its part to improve air quality. As responsible corporate citizens we will work with states to negotiate cost effective approaches for bakers and consumers in controlling stationary and mobile source pollutants. Today our concerns are with the ever changing

Federal requirements that states must meet to prevent the loss of important highway construction funds. The changing Federal guidelines could cause states to reconsider negotiations that lead to significant investment in abatement controls and develop more stringent regulations for the baking industry. My testimony today outlines the independent segment of the baking industry's problems with EPAs proposed rulemaking and the possible impact on my forth generation family business. Thank you again for the opportunity to appear today on behalf of our company and the Independent Bakers Association. I welcome your questions.

Mr. BARTON. Just out of curiosity, that's the first time in 3 years a Democrat has chaired a committee of this Congress. Were there any suspensions passed in my absence?

Well, we have put your testimony into the record. We're going to recess. I hope that I come back very quickly. There are now 12 votes that we could have in succession. Each would be a 5-minute vote. If that would to occur, we won't come back for questions until approximately 3 p.m. So if you need to take a personal convenience break, do so, but be in the general area, because I know that there are going to be some members that want to ask questions, and it's very important to get the State and local perspective, and, obviously, the private perspective.

I want to apologize for the committee that you've had to come on a day that we're trying to get out of town and we're having parliamentary problems on the House floor.

So the committee stands in recess, subject to call of the Chair. It could be as long as 3 p.m. Hopefully, it will be sooner than that.

[Brief recess.]

Mr. BARTON. It's apparent that we're not going to get any kind of an agreement on the floor on the suspension votes, and so, very reluctantly, I'm going to adjourn the hearing. We will have written statements. We do apologize. This is somewhat unique in that we are having this kind of an issue on the floor on the eve of a religious holiday.

We will keep the record open for any member who wishes to submit written questions to the panel. We appreciate your time. It was extremely useful.

And with that, this hearing is adjourned.

[Whereupon, at 2:20 p.m., the subcommittee adjourned subject to the call of the Chair.]

[Additional material submitted for the record follows.]

PREPARED STATEMENT OF JIM MARTIN, EXECUTIVE VICE PRESIDENT, MARTIN'S FAMOUS PASTRY SHOPPE, INC.

I am Jim Martin, an owner of Martin's Famous Pastry Shoppe, Inc. We have bakeries in Chambersburg and Shippensburg, Pennsylvania. Ours is a family business started in 1955 by my parents—perhaps you have seen our "Martin's Famous Potato Rolls" in the grocery store bread aisle.

I am representing myself and the members of the American Bakers Association. The American Bakers Association represents 80% of the wholesale baking business which includes small businesses like ours and the companies with national brand names that you may know. Bakeries are captured by the Clean Air Act because the natural process of yeast fermentation used to make bread and rolls produces ethanol, an alcohol, which is considered a volatile organic compound (VOC). Ethanol from bakeries is nontoxic and low reactive, probably eaten by bacteria in the atmosphere.

The American Bakers Association strongly supports H.R. 1984 to affect the implementation of the new air quality standards. This bill will ensure that sound science and good planning have the time to develop before costly and potentially inappropriate controls are required. The baking industry has already spent $28 million to comply with existing requirements. We estimate that the baking industry will spend

$236 million to implement the new standard. The equipment to control bread aroma from a bakery costs about $500,000 to install and costs between $35,000 and $100,000 each year to operate. That is about $12,000 per ton of VOC controlled, not the $1,400 per ton estimated by the US EPA.

The American Bakers Association has worked with the bakers in each state and with the state officials to develop reasonable rules to bring the states and industry into compliance. The states have worked hard to implement the Clean Air Act under demanding and often changing EPA policies. We offer our compliments to Pennsylvania and many of the other states where hard work has put the most cost-effective control requirements in place. However, state officials are struggling to further reduce emissions and are being forced to look at smaller and smaller sources, that is, small businesses.

Bakeries are low volume, low profit margin businesses. Let me give you an example of how a small baker might be affected by control requirements. Imagine Joe's Bakery in your state. Joe is operating the white bread bakery he inherited from his father. The bakery has been in business for 80 years with a profit margin of 1.7%. His bakery, like others, is labor intensive, providing jobs at a relatively high pay rate and supporting many families in the neighborhood. No one is getting rich but everyone is being paid and the local area has good tasting fresh bread. His potential emissions of ethanol trigger control requirements. Joe goes to his bank and asks for a loan for the $500,000 to install controls and cut his profit margin by 24%. The bank denies his request, which is good business practice for the bank. If he can find a buyer, he can sell his bakery or Joe can close. Sadly, some bakeries have closed during the implementation of the current standard.

Now let me share with you the situation at our bakeries. We emit enough ethanol to trigger control requirements. Stack testing, permitting, engineering reports, communication, and costs associated with investigating new technology have exceeded $250,000 to date at our bakeries. Our engineers calculate that adding a control device will increase our production energy consumption by 50% per package of rolls. However, rather than purchase the control equipment we have chosen to limit our production and the growth of our business. This means our ability to serve the region with the maximum variety of product or to respond quickly to changes in consumer taste and demand is limited. This is not my understanding of free enterprise. And I've wondered, is less bread on the table a good trade for no measurable improvement in air quality?

I and the bakers I represent urge you to pass H.R. 1984 to make clear legal statements about the funding of the research and science and the timing of the implementation of the air quality standard. Please help us do the right thing to protect our air quality and the earth!

U.S. House of Representatives
Committee on Commerce
Room 2125, Rayburn House Office Building
Washington, DC 20515–6115

September 22, 1997

The Honorable Carol M. Browner
Administrator
Environmental Protection Agency
401 M Street, S.W.
Washington, D.C. 20460

Dear Administrator Browner:

As you know, the Committee on Commerce has been following closely events related to the revisions to the National Ambient Air Quality Standards ("NAAQS") for particulate matter ("PM") and ozone. EPA's September 9 and September 12, 1997 letters to me raised several important issues regarding EPA's implementation plan. Accordingly, pursuant to Rules X and XI of the United States House of Representatives, please provide written responses to the four questions set forth below. As you know, the Committee has scheduled an October 1, 1997, hearing regarding implementation issues. Responses to the four implementation questions below are necessary for hearing preparation and should be provided no later than close of business September 26, 1997.

1.   Title I of the Clean Air Act contains mandatory air pollution control requirements for all nonattainment areas. For example, Section 172 requires that existing sources in nonattainment areas apply "reasonably available control technology" and that nonattainment areas achieve "reasonable further progress." Section 173 requires new or modified sources, which includes new or expanding businesses, to obtain "offsetting" emission credits. Section 173 also requires that these same new or modified sources install the more stringent "lowest achievable emission rate" technology and requires that EPA review other sources of air pollution located at the site. Given these mandatory requirements, it is understandable that an EPA implementation fact sheet referred to the "stigma" associated with being a nonattainment area.

That same fact sheet stated that certain new, "transitional" nonattainment areas "will avoid burdensome new local planning requirements and restrictions on economic growth." While I applaud this general goal, I am concerned that the Clean Air Act's

The Honorable Carol M. Browner
Page 2

mandatory control requirements for nonattainment areas prevent EPA from granting any meaningful relief to the new transitional nonattainment areas. Accordingly, does EPA believe that it has the legal authority to exempt transitional nonattainment areas from mandatory Clean Air Act pollution control requirements? EPA's implementation plan states that Clean Air Act Section 172(a)(1) is the legal authority for transitional nonattainment areas. The relevant portion of Section 172(a)(1) reads:

> [T]he Administrator may classify the [nonattainment] area for the purpose of applying an attainment date . . . and <u>for other purposes</u>.

(emphasis added). It is well-established that exceptions to statutory law are not to be implied and cannot be created by artful construction. Nor can a general phrase such as "for other purposes" be read to invalidate mandatory Clean Air Act requirements. <u>See Sutherland Stat. Const.</u> Sections 47.11, 47.18-.19 (5th Ed.). Would it be EPA's interpretation that the general phrase "for other purposes" allows EPA to exempt transitional nonattainment areas from the statutory air pollution control requirements of the Clean Air Act? Please list the federal court opinions, if any, that interpret Section 172(a)(1) in this manner. Please also provide any writings or legal opinions developed by EPA or the Department of Justice that interpret Section 172(a)(1) or any other purported legal authority for EPA's implementation of the revised NAAQS for ozone and particulate matter.

2.   State and local governments have expectations of certainty regarding Clean Air Act implementation schemes. as do the businesses that often must make operational decisions based on Clean Air Act requirements. I am concerned that EPA did not issue its ozone NAAQS implementation plan as a rule. although it proposed such a rule in November 1996. Moreover, the President's July 16, 1997, memorandum to you states that EPA's implementation plan is not judicially enforceable against the agency. Accordingly, it seems that there is no legal mechanism by which states could hold EPA to the specific terms of its July 19, 1997 implementation plan. Rather, EPA is free to change the plan at any time. If EPA decided to change the terms of its implementation plan, what legal recourse is available to states and businesses that are regulated under the Act?

3.   I am also concerned that EPA's implementation plan will give courts an unnecessary opportunity to establish implementation policy. Section 304 of the Clean Air Act authorizes citizen suits to enforce certain mandatory provisions of the Act. For example. despite EPA's interpretation otherwise. a citizen group could conclude that Section 172(a)(1) does not authorize EPA to exempt transitional nonattainment areas from the more stringent "lowest achievable emission rate" technology required by Section 173. If a court sided with the citizen group in such a suit, couldn't the court require transitional nonattainment areas to use lowest achievable emission rate technology (or other mandatory controls for which EPA exempted transitional nonattainment areas)?

The Honorable Carol M. Browner
Page 3

4.   The PM2.5 NAAQS became effective on September 16, 1997. It is a new standard: there
        was not an existing NAAQS for PM2.5. Recently published reports indicate that PM2.5's
        new regulated status has immediate compliance effects under Titles I and V of the Clean
        Air Act, despite EPA's claim that nothing will happen with regard to PM2.5 for at least
        five years. For example, these reports indicate that under Title V, regulated sources must
        obtain permits for emissions of all pollutants regulated under the Act, which now
        includes PM2.5. Additionally, Part C of Title I requires that new sources, including new
        or expanding businesses, located in attainment or "unclassifiable" areas prove that their
        operations will not impermissibly increase total emissions of regulated pollutants, which
        now includes PM2.5. Both of these provisions arguably could require a source to control
        PM2.5 emissions. How does EPA plan to deal with these important and immediate Clean
        Air Act permitting issues? Does EPA plan to issue a rule addressing these matters?
        Because statutory requirements of the Clean Air Act are at issue here, couldn't a
        businesses' permit be challenged in a citizens suit under Section 304 of the Act, even if
        EPA chooses not to enforce the Title I and V requirements discussed above?

        Although the Committee is at this time limiting its request for record to the documents
described in (1) above, please preserve all EPA records relating to implementation of the recently
revised ozone and particulate matter NAAQS (see Attachment for definition of the term
"records").

        If you have any questions, please do not hesitate to contact Mr. Joseph Stanko,
Committee counsel, at (202) 225-2927. Thank you for your prompt attention to this matter.

                                        Sincerely,

                                        Tom Bliley
                                        Chairman

cc: The Honorable John D. Dingell

ATTACHMENT

For the purposes of this request, the word "records" shall include but shall not be limited to any and all originals and identical copies of any item whether written, typed, printed, recorded, transcribed, punched, taped, filmed, graphically portrayed, video or audio taped, however produced or reproduced, and includes but is not limited to any writing, reproduction, transcription, photograph, or video or audio recording, produced or stored in any fashion, including any and all computer entries, memoranda, diaries, telephone logs, telephone message slips, tapes, notes, talking points, letters, journal entries, reports, studies, drawings, calendars, manuals, press releases, opinions, documents, analyses, messages, summaries, bulletins, e-mail, disks, briefing materials and notes, cover sheets or routing cover sheets or any other machine readable material of any sort whether prepared by current or former employees, agents, consultants or by any non-employee without limitation. "Records" shall also include redacted and unredacted versions of the same record. For purposes of this request, the terms "relating" or "relate" as to any given subject means anything that constitutes, contains, embodies, identifies, deals with, or is in any manner whatsoever pertinent to that subject, including but not limited to records concerning the preparation of other records.

U.S. House of Representatives
Committee on Commerce
Room 2125, Rayburn House Office Building
Washington, DC 20515–6115

September 22, 1997

The Honorable Carol M. Browner
Administrator
Environmental Protection Agency
401 M Street, S.W.
Washington, D.C. 20460

On November 18, 1996, Health and Environment Subcommittee Chairman Michael Bilirakis wrote to Assistant Administrator for Air and Radiation, Mary Nichols, concerning pending plans to issue State Implementation Plan (SIP) calls based on the technical information generated by the Ozone Transport Assessment Group (OTAG).

In his letter, Chairman Bilirakis asked Assistant Administrator Nichols to address several questions concerning the scientific information which would be used to issue the OTAG SIP calls and questioned whether these SIP calls would take into account information generated through ongoing efforts to revise the National Ambient Air Quality Standard (NAAQS) for ozone. In specific, in question four of his November 18, 1996 letter, Chairman Bilirakis asked what ozone NAAQS would be used to establish individual state NOX and VOC reductions required in the SIP call and to what extent specific reductions required in the contemplated SIP call would be adjusted upward or downward based on the final ozone NAAQS.

The response received from Assistant Administrator Nichols on December 17, 1996 indicated that EPA planned to issue SIP calls and publish a Notice of Proposed Rulemaking by March, 1997, and publish a final rulemaking by the summer of 1997. Moreover, in response to question four noted above, it was stated that:

> EPA does not believe that the Notice of Proposed Rulemaking on regional emission reductions should take into account public comments on the proposed NAAQS. EPA will use the *current NAAQS* to establish the individual state NOx and VOC reductions because we cannot prejudge the outcome of the ozone standard revision process. *Furthermore, OTAG is using the current standard in its analyses. And, for the reasons cited above, the reductions indicated in the*

The Honorable Carol M. Browner
September 22, 1997
Page 2

> *Notice of Proposed Rulemaking will not be adjusted based on a revised ozone*
> *NAAQS. . .* (Emphasis added).

In an August 1, 1997, letter to Representative John Dingell, however, Assistant Administrator Nichols indicated that "After an independent analysis of the recommendations and *considering the reductions necessary for both the 1-hour and 8-hour ozone standards,* EPA will issue this September a proposed rule calling on States to reduce significantly NOx emissions. After an opportunity for public comment, including additional input from affected states, EPA will issue a final rule by September 1998." (Emphasis added). In addition, the July 16, 1997, Memorandum for the Administrator of the Environmental Protection Agency on Implementation of the Revised Air Quality Standards for Ozone and Particulate Matter (hereinafter referred to as the "July 16, 1997 Memorandum") indicates that a regional control strategy for ozone will be an integral part of efforts to implement the newly revised ozone standard.

The disparity between representations made in December, 1996 correspondence to Chairman Bilirakis and the present implementation plans being advanced by the Environmental Protection Agency leads me to ask for several clarifications:

(1) Please explain what is meant by the statement to Representative Dingell which indicates that the proposed rule regarding the OTAG SIP call will "consider" the reductions necessary for both the previous 0.12 ppm and newly established 0.08 ppm ozone standard. Is it the intent of the EPA to require that SIPs in states in the OTAG region provide NOx reductions that may be in excess of the reductions necessary to achieve a 0.12 ppm ozone standard?

(2) If the EPA intends to require or approve SIPs to achieve reductions beyond those required to meet a 0.12 ppm standard in downwind states, on what specific modeling will a requirement for such NOx reductions be based? If EPA has made a decision to "adjust" or "consider" NOx reductions beyond those necessary to achieve a 0.12 ppm standard in its September 1997 proposed rule, what is the extent of the incremental tonnage "adjustment" or "consideration" that will be required in each of the states affected by the notice? What is the legal basis for including emission reductions for the purpose of attaining the new 0.08ppm standard in the upcoming SIP calls?

(3) To what extent, if any, did OTAG modeling analyses address strategies for attaining an 8 hour ozone standard of .08 ppm as the average 4th highest value over three year averaging periods? To what extent did OTAG's assessment of the causes and nature of ozone transport distinguish between transport contributing to exceedences of the previous 0.12ppm ozone standard and the newly promulgated 0.08ppm standard?

(4) It is my understanding that a central recommendation of the OTAG process was that

The Honorable Carol M. Browner
September 22, 1997
Page 3

individual states should have the opportunity to conduct more refined subregional
modeling before reaching judgements on what remission reductions might be sought from
specific sources. Please explain what efforts EPA and OTAG states have made, to date,
in order to implement this recommendation and actually conduct more refined
subregional modeling. In addition, please explain the relationship between state-specific
emission reductions to be proposed by EPA in its September 1997 rulemaking and the
results of such subregional modeling efforts.

(5) As you know, the July 16, 1997 Memorandum indicates that if certain States include
"control measures to achieve the emission reductions required by the EPA's rule for
States in the OTAG region" in their SIPs they may be able to receive a new "transitional
status" which will relieve such states of "unnecessary local planning requirements" as
well as allow such States to benefit from revised new source review (NSR) and
conformity rules. Will emission reductions resulting from an emission trading program
beyond those required under a 0.12 ppm standard be required in order for an OTAG state
to receive "transitional status"? If EPA identifies "unnecessary local planning
requirements" will the Agency eliminate those requirements for all states? If not, why
not?

(6) How will EPA determine whether an area qualifies for "transitional" status and when
will that determination be made?

(7) What roles would air quality modeling and monitoring data play in the determination
of "transitional" areas? Why is agreement on a specific level of NOx emission reductions
required as a precondition to "transitional" status?

(8) Please clarify whether any states, or parts of states, included in the original 37-state
OTAG region will be excluded from EPA's September 1997 rulemaking. What is the
basis for excluding any state?

(9) Section 176A of the Clean Air Act contains mechanisms for mitigating ozone
transport in regions outside of the Northeast Ozone Transport Region through the
formation of regional ozone transport commissions. Under this section, the
Administrator may establish a transport region, on the Administrator's own motion,
where the Administrator "has reason to believe that the interstate transport of air
pollutants from one or more States contributes significantly to a violation of a national
ambient air quality standard in one or more States." Why hasn't this provision been
utilized to reduce the transport of ozone originating in states outside of the Northeast,
similar to efforts of the Northeast Ozone Transport Commission created by sections 176A
and 184 of the Clean Air Act? Why is EPA's proposed transport rulemaking necessary in
view of this existing -- albeit unused -- statutory authority? Has EPA considered

The Honorable Carol M. Browner
September 22, 1997
Page 4

whether states in the Northeast Ozone Transport Region must use the section 176A
procedures to seek NOx reductions in upwind states?

(10) In a March 17, 1997 letter from Deputy Assistant Administrator for Air Radiation,
Richard D. Wilson, to Mary Gade, Director of the Illinois Environmental Protection
Agency, a discussion of legal authority was included in a memorandum which addressed
the ability of the EPA to implement a cap and emissions trading program for NOx
sources. This memorandum indicated that the EPA relied on section 110(a)(2)(D) of the
Clean Air Act as the "principle legal provision" providing authority for EPA to require
that State SIPs include emission reductions based on their contribution to nonattainment
in other states. Although this memorandum indicated that EPA did not have legal
authority to require a "cap and trade" program, the memorandum further indicated that
such SIP revisions must be submitted to the EPA no later than 18 months after the SIP
call and that "EPA is considering developing an early FIP (Federal Implementation Plan)
proposal such that a final FIP could be promulgated shortly after the finding of non-
submittal."

Under the schedule outlined above, it would appear that EPA will be requiring states
to submit SIPs affecting NOx emissions in OTAG states by March, 2000. Furthermore, if
such states do not submit SIPs by the year 2000, such states will not be eligible for
"transitional" classification and presumably would be considered to be in
"nonattainment" if areas in such a state violated either the previous 0.12ppm or 0.08ppm
standard. Such states might also be subject to imposition of a FIP upon a finding of non-
submittal.

(a) Since state SIPs, under section 110(a)(2), must contain enforceable emissions
limitations and other control measures as well as a state program to provide for
enforcement, what is the basis for your statement before the House Agriculture
Committee on September 16, 1997, that "nobody is legally required to reduce
their pollution until the year 2009?" When would sources in a state, required to
submit a SIP affecting NOx emissions in an OTAG state by March, 2000, be
"legally required" to reduce their pollution if such a state submits an legally
acceptable SIP to the EPA? When would sources in a state be "legally required"
to reduce their pollution if EPA issues an "early FIP" for any state not meeting the
SIP submission deadline for OTAG states? When would sources in states that are
eligible for "transitional" classifications be "legally required" to reduce their
pollution?" Does EPA believe it could mandate a "cap and trade" program
through a FIP?

(b) Will the March, 2000, SIPs be required to demonstrate attainment of the 1
hour 0.12ppm standard? If so, what type of demonstration would be required?

The Honorable Carol M. Browner
September 22, 1997
Page 5

What would the applicable attainment date be for such SIPs?

(11) It is obvious that there have been significant delays in EPA's schedule regarding its planned SIP call. Under the current schedule, a final rulemaking will not be published until nearly a year following the plans outlined by EPA 10 months ago. Please provide an explanation of the factors which led to the delay of the OTAG SIP call and what elements in the OTAG SIP call have been changed or altered in any way during this delay.

Thank you for your attention to this request. As you know, the Committee has scheduled an October 1, 1997, hearing regarding implementation issues. Responses to the above questions are necessary for hearing preparation and should be provided no later than close of business September 26, 1997. If you have any questions, please do not hesitate to contact Mr. Robert Meyers, Committee counsel, at (202) 225-2927. Thank you for your prompt attention to this matter.

Sincerely,

Tom Bliley
Chairman

cc: The Honorable John D. Dingell

U.S. House of Representatives
Committee on Commerce
Room 2125, Rayburn House Office Building
Washington, DC 20515–6115

November 18, 1996

The Honorable Mary D. Nichols
Assistant Administrator for Air
 and Radiation
U.S. Environmental Protection Agency
401 M Street, S.W.
Washington D.C. 20460

Dear Administrator Nichols:

It has come to my attention that in a November 8, 1996, letter to Mary Gade, Chair of the Ozone Transport Assessment Group (OTAG), Assistant Administrator Mary D. Nichols indicated a new course of action regarding ozone control measures within the OTAG states.

Specifically, it was indicated that the Environmental Protection Agency (EPA) plans to issue SIP calls and by March, 1997, publish a Notice of Proposed Rulemaking regarding specific NOx and/or VOC emission reductions that each State will need to achieve. Moreover, this NPR will require SIP revisions to achieve the reductions which EPA decides are necessary for each State.

This action raises several issues which I would like you to address in a written response. First, it would appear that the March proposal date would preclude the EPA from fully considering either the final, and most critical, round of OTAG modeling and/or the OTAG states' recommendation to EPA on how to address regional ozone transport. As you noted in your letter, this modeling will be performed during January and February of 1997, yet the timetable established in your November 8th correspondence would appear to leave the Agency little or no time to assess this modeling.

When did the Agency first become aware that the final round of OTAG modeling would not be available until February 1997 and when did the Agency decide that delay was unacceptable due to "the continuing risk to public health"? Please provide specific dates.

The Honorable Mary D. Nichols
November 18, 1996
Page 2

Please also indicate on what scientific basis the EPA believes it can establish overall NOx and/or VOC emission reductions for each of the OTAG states by March of 1997; on what modeling such specific reductions will be based; whether such information or modeling has been subject to peer-review; and whether, and to what extent, the Agency will determine specific costs and health benefits which will occur from each states' NOx and VOC reduction requirements.

The Commerce Committee also notes that the Agency plans to issue final revised ozone National Ambient Air Quality Standards (NAAQS) by June 28, 1987. To what extent will the March, 1997, Notice of Proposed Rulemaking take into account public comments received regarding newly proposed ozone NAAQS? What specific ozone NAAQS will be used to establish the individual state NOx and VOC reductions indicated in your March, 1997 NPR? To what extent would specific reductions indicated in your March 1997 Notice be adjusted upward or downward based on the final ozone NAAQS?

Thank you for your prompt attention to these questions. I look forward to receiving your response by December 15th.

Sincerely,

Michael Bilirakis
Chairman, Subcommittee on
Health and Environment

TJB;rjm

UNITED STATES ENVIRONMENTAL PROTECTION AGENCY
WASHINGTON, D.C. 20460

DEC 17 1996

Honorable Michael Bilirakis
Chairman, Subcommittee on
   Health and Environment
House of Representatives
Washington, D.C. 20515

Dear Mr. Chairman:

This is in response to your letter of November 18, 1996 regarding my November 8, 1996 letter to Mary Gade, Chair of the Ozone Transport Assessment Group (OTAG). The purpose of my letter was to reiterate EPA's obligations under the Clean Air Act and to provide more detail on our schedule for issuing state implementation plan (SIP) calls to states affected by the OTAG recommendations. The letter is consistent with standing Environmental Protection Agency (EPA) policy as laid out in a memorandum of March 2, 1995 entitled "Ozone Attainment Demonstration" from me to EPA Regional Administrators.

EPA plans to issue SIP calls and to publish a Notice of Proposed Rulemaking by March 1997 regarding specific nitrogen oxides (NO_x) and or volatile organic compounds (VOC) emission reductions that each state will need to achieve. The Agency plans to publish a Notice of Proposed Rulemaking that will be based on all the technical information available from OTAG and other sources. This means that EPA will follow the OTAG results as closely as possible. If EPA does not have a recommendation from OTAG by the March 1997 timeframe, EPA will have to make some preliminary conclusions based on the work of the OTAG technical committees. However, it is our intention to publish a final rulemaking in the summer of 1997 which adheres to any OTAG recommendations.

Your letter raises several other points which are addressed in the enclosure. I appreciate this opportunity to be of service and trust that this information will be helpful to you.

Sincerely yours,

Mary D. Nichols
Assistant Administrator
   for Air and Radiation

Enclosure

<u>Response to Chairman Bilirakis</u>

1. **It would appear that the March proposal date would preclude the EPA from fully considering either the final, and most critical, round of OTAG modeling and/or the OTAG states' recommendation to EPA on how to address regional ozone transport.**

We are aware that the final round of modeling will occur in the January/February 1997 timeframe and that EPA will only have limited time to assess the results before publishing the Notice of Proposed Rulemaking. Again, we will base the proposal on all the information that is available from OTAG and other appropriate sources and will include the OTAG recommendations in the final rulemaking scheduled for summer 1997.

2. **When did the Agency first become aware that the final round of OTAG modeling would not be available until February 1997 and when did the Agency decide that delay was unacceptable due to "the continuing risk to public health"? Please provide specific dates.**

Please note that we did not indicate in the November 8, 1996 letter that the delay was unacceptable. At the September 26, 1996 OTAG Policy Group meeting, it was announced that the final round of OTAG modeling would be pursued and completed by February 1997. The Agency believes that OTAG needs to develop this information. However, as always with any delay in achieving the necessary emission reductions to achieve attainment of a national ambient air quality standard (NAAQS), we are concerned because of the continuing risk to public health.

3. **Please also indicate on what scientific basis the EPA believes it can establish overall NOx and/or VOC emission reductions for each of the OTAG states by March of 1997; on what modeling such specific reductions will be based; whether such information or modeling has been subject to peer-review; and whether, and to what extend, the Agency will determine specific costs and health benefits which will occur from each states' NOx and VOC reduction requirements.**

The Agency will use, as the scientific basis for the Notice of Proposed Rulemaking, a compilation of all available information which will include the extensive modeling runs completed within OTAG, Regional Oxidant Model results from work conducted with the Ozone Transport Commission, EPA work conducted for the Clean Air Power Initiative, and EPA monitoring information collected for the Title IV Acid Rain Program.

4. **The Commerce Committee also notes that the Agency plans to issue final revised ozone NAAQS by June 28, 1997. To what extent will the March 1997 Notice of Proposed Rulemaking take into account public comments received regarding newly proposed ozone NAAQS? What specific ozone NAAQS will be used to establish**

2

the individual state NOx and VOC reductions indicated in your March 1997 NPR? To what extent would specific reductions indicated in your March 1997 Notice be adjusted upward or downward based on the final ozone NAAQS?

The goal of OTAG is to develop a consensus recommendation of regional NOx and/or VOC controls that will provide for reductions in boundary conditions of ozone and ozone precursors. It is important to note that OTAG is not attempting to develop strategies that will provide for attainment in every nonattainment area. It is still the responsibility of each nonattainment area to develop an area-specific demonstration of attainment using a combination of regional and local controls. Expedited implementation of regional control strategies recommended by OTAG to facilitate attainment of the current standard would significantly reduce precursor emissions, which would also be beneficial in attaining any revised standard. In fact, it is likely that under a revised ozone standard regional reductions in ozone and ozone precursors would be even more critical. Such reductions could also minimize the number of areas designated nonattainment under a revised standard and/or lessen the severity of the nonattainment problem. The control strategies that are likely to result from OTAG will also provide progress toward attainment of a revised standard. In addition, information that is gathered by OTAG can be incorporated into the implementation of any revised standard.

Therefore, EPA does not believe that the Notice of Proposed Rulemaking on regional emission reductions should take into account public comments on the proposed NAAQS. In addition, EPA will use the current NAAQS to establish the individual state NOx and VOC reductions because we cannot prejudge the outcome of the ozone standard revision process. Furthermore, OTAG is using the current standard in its analyses. And, for the reasons cited above, the reductions indicated in the Notice of Proposed Rulemaking will not be adjusted based on a revised ozone NAAQS. However, they may be adjusted based on final OTAG recommendations or as a result of other public comments received.

U.S. House of Representatives
Committee on Commerce
Room 2125, Rayburn House Office Building
Washington, DC 20515–6115

September 22, 1997

The Honorable Janet Reno
Attorney General
Department of Justice
950 Pennsylvania Avenue, N.W.
Washington, D.C. 20530

Dear Attorney General Reno:

As you may know, the House Committee on Commerce has primary jurisdiction over the Clean Air Act. On September 16, 1997, in testimony before the House Agriculture Committee, Environmental Protection Agency ("EPA") Administrator Carol Browner indicated that the Department of Justice had been a participant in developing a "structured legal argument" with respect to the Administration's intent to provide "transitional" classifications to certain areas affected by the newly revised ozone National Ambient Air Quality Standard ("NAAQS"). Accordingly, pursuant to Rules X and XI of the United States House of Representatives, no later than close of business September 26, 1997, please provide the House Committee on Commerce written responses containing the following information and, where requested, please produce the relevant documents.

1.   Please provide the Committee with a detailed description of all work, including but not limited to legal analysis, performed by the Department of Justice on or before July 16, 1997, relating to the concept of "transitional" classifications for ozone nonattainment areas as referenced in the Memorandum for the Administrator of the Environmental Protection Agency, dated July 16, 1997, and the accompanying Implementation Plan for Revised Air Quality Standards, undated, both published in the July 18, 1997 issue of the <u>Federal Register</u>.

2.   Please provide the Committee with a detailed description of all work, including but not limited to legal analysis, performed by the Department of Justice on or after July 17, 1997, relating to the concept of "transitional" classifications for ozone nonattainment areas as referenced in the documents cited in (1) above.

The Honorable Janet Reno
Page 2

3. Please provide the Committee with copies of any legal opinions, analysis, option papers, draft guidance, draft regulations or writings of any kind produced by Department of Justice with respect to the concept of "transitional" classifications for ozone nonattainment areas as referenced in the documents cited in (1) above.

4. Please provide the Committee with a detailed description of all work, including but not limited to legal analysis, performed by the Department of Justice relating to any aspect of EPA's implementation of the recently revised ozone and particulate matter NAAQS that has not been described in your responses to (1) or (2) above. Please also provide the Committee with copies of any legal opinions, analysis, option papers, draft guidance, draft regulations or writings of any kind produced by the Department of Justice relating to any aspect of EPA's implementation of the recently revised ozone and particulate matter NAAQS that has not been provided in your response to (3) above.

Although the Committee is at this time limiting its request for records to the documents described in (3) and (4) above, please preserve all Department of Justice records relating to EPA's implementation of the recently revised ozone and particulate matter NAAQS (see Attachment for definition of the term "records").

If you have any questions, please do not hesitate to contact Mr. Joseph Stanko or Mr. Robert Meyers, Committee counsels, at (202) 225-2927. Thank you for your prompt attention to this matter.

Sincerely,

Tom Bliley
Chairman

cc: The Honorable John D. Dingell

153

ATTACHMENT

For the purposes of this request. the word "records" shall include but shall not be limited to any and all originals and identical copies of any item whether written. typed. printed. recorded. transcribed. punched. taped. filmed. graphically portrayed. video or audio taped. however produced or reproduced. and includes but is not limited to any writing. reproduction. transcription. photograph. or video or audio recording, produced or stored in any fashion. including any and all computer entries. memoranda. diaries. telephone logs. telephone message slips. tapes. notes. talking points. letters. journal entries. reports. studies. drawings, calendars. manuals, press releases. opinions. documents, analyses, messages, summaries. bulletins. e-mail. disks. briefing materials and notes. cover sheets or routing cover sheets or any other machine readable material of any sort whether prepared by current or former employees. agents. consultants or by any non-employee without limitation. "Records" shall also include redacted and unredacted versions of the same record. For purposes of this request. the terms "relating" or "relate" as to any given subject means anything that constitutes. contains. embodies. identifies. deals with. or is in any manner whatsoever pertinent to that subject. including but not limited to records concerning the preparation of other records.

UNITED STATES ENVIRONMENTAL PROTECTION AGENCY
WASHINGTON, D.C. 20460

September 29, 1997

The Honorable Thomas J. Bliley
Chairman, Committee on Commerce
U.S. House of Representatives
Washington, DC 20515

Dear Mr. Chairman:

I am providing an interim response to your two letters of last week, which were dated September 22 and received in this office on September 23 and 24, regarding EPA's updated clean air standards. As you know, the Administrator takes the oversight responsibility of the Commerce Committee very seriously, and the Agency strives to provide both timely and thorough responses to your oversight requests. However, it is simply not possible to respond to your latest inquiries in the limited time frame you requested.

Your first letter raises questions regarding one component of the implementation plan for the new standards -- the proposed call for State Implementation Plans ("SIPs"). We are still working through the interagency review process for the proposal, and as soon as the process is completed, we will provide you with an answer to the 11 questions you have posed.

Your second letter asks four questions regarding the legal bases for the implementation plan. While these questions raise issues that have been considered by the Agency, preparing a thorough response ██ and we expect to provide you with a response by October 6.

Please be assured that we are making a good faith effort to respond to your requests as thoroughly and promptly as possible. If you have further questions regarding the status of our replies, please feel free to call me.

Sincerely,

al

UNITED STATES ENVIRONMENTAL PROTECTION AGENCY
WASHINGTON, D.C. 20460

OCT 16 1997

The Honorable Thomas J. Bliley
Chairman, Committee on Commerce
U.S. House of Representatives
Washington, D.C. 20515-6115

Dear Mr. Chairman:

This is in response to your letter of September 22, 1997, in which you raised several questions regarding the legal basis of the Administration's implementation plan for the revised ozone air quality standard. The Administration's plan is designed to ensure that the new air quality standards for ozone and fine particulate matter are implemented in a way that will "maximize common sense, flexibility and cost-effectiveness." The President has directed the Administrator, "in consultation with all affected agencies and parties to undertake the steps appropriate under law to carry out the [implementation] plan and to complete all necessary guidance and rulemaking no later than December 31, 1998."

Attached is our analysis that responds to the questions in your letter. It addresses the legal basis for several key components of the implementation plan. The Agency will continue to develop and refine its legal and policy analysis through the process that the President has directed us to follow. This will include the opportunity for members of the public -- including States, industry, and other citizens -- to inform the Agency's policy and legal approaches. The Agency's final legal analysis will accompany the guidance and rulemakings which will be completed by December 1998.

I appreciate this opportunity to respond to your concerns and trust that this information will be helpful to you.

Sincerely,

Jonathan Z. Cannon
General Counsel

Enclosures

ATTACHMENT

Question 1: Title I of the Clean Air Act contains mandatory air pollution control requirements for all nonattainment areas. For example, section 172 requires that existing sources in nonattainment areas apply "reasonably available control technology" and that nonattainment areas achieve "reasonable further progress." Section 173 requires new ormodified sources, which includes new or expanding businesses, to obtain "offsetting" emission credits. Section 173 also requires that these same new or modified sources install the more stringent "lowest achievable emission rate" technology and requires that EPA review other sources of air pollution located at the site. Given these mandatory requirements, it is understandable that an EPA implementation fact sheet referred to the "stigma" associated with being a nonattainment area.

That same fact sheet stated that certain new, "transitional" nonattainment areas "will avoid burdensome new local planning requirements and restrictions on economic growth." While I applaud this general goal, I am concerned that the Clean Air Act's mandatory control requirements for nonattainment areas prevent EPA from granting any meaningful relief to the new transitional nonattainment areas. Accordingly, does EPA believe that is has the legal authority to exempt transitional nonattainment areas from mandatory Clean Air Act pollution control requirements? EPA's implementation plan states that Clean Air Act Section 172(a)(1) is the legal authority for transitional nonattainment areas. The relevant portion of Section 172(a)(1) reads:

> [T]he Administrator may classify the [nonattainment] area for the purpose
> of applying an attainment date... and for other purposes.

(emphasis added). It is well-established that exceptions to statutory law are not to be implied and cannot be created by artful construction. Nor can a general phrase such as "for other purposes" be read to invalidate mandatory Clean Air Act requirements. See Sutherland Stat. Const. Sections 47.11, 47.18-.19 (5th Ed.) Would it be EPA's interpretation that the general phrase "for other purposes" allows EPA to exempt transitional nonattainment areas from the statutory air pollution control requirements of the Clean Air Act? Please list the federal court opinions, if any, that interpret Section 172(a)(1) in this manner. Please also provide any writing or legal opinions developed by EPA or the Department of Justice that interpret Section 172(a)(1) or any other purported legal authority for EPA's implementation of the revised NAAQS for ozone and particulate matter.

Response: A primary goal of title I of the Clean Air Act is to ensure that areas come into attainment with the health-based national ambient air quality standards. Section 101(b)(1) provides that one of the purposes of title I is "to protect and enhance the quality of the Nation's air resources so as to promote the public health and welfare and the productive capacity of its population." Prior to the 1990 Amendments to the Clean Air Act, Congress provided substantial flexibility to States in preparing plans to attain ambient air quality standards. This flexibility was continued in the general nonattainment area planning provisions in subpart 1 of part D of title I, included in the 1990 Amendments. However, in the 1990 Amendments, Congress also provided more restrictive requirements for States to follow in developing attainment plans for areas designated nonattainment for the 1-hour 0.12 ppm ozone standard. See CAA title I, part D, subpart 2.[1]

As part of the statutory regime applicable to the implementation of the 1-hour ozone standard in subpart 2 of part D, Congress established a system of classifications for ozone nonattainment areas that differentiated among ozone nonattainment areas on the basis of the severity of their air quality problems. Under this classification system, nonattainment areas with more serious problems were required to undertake more steps to deal with their ozone problems but were also provided more time to attain the ozone standard.

EPA's Classification Authority under Section 172(a)

For areas not covered by the classification scheme of subpart 2, section 172(a) provides EPA with broad authority to establish classifications in connection with the implementation of new or revised national ambient air quality standards. Section 172(a)(1) states that "the Administrator may classify the area for the purpose of applying an attainment date pursuant to paragraph (2), and for other purposes. In determining the appropriate classification, if any, for a nonattainment area, the Administrator may consider such factors as the severity of nonattainment in such area and the availability and feasibility of the pollution control measures that the Administrator believes may be necessary to provide for attainment of such standard in such area." The language and structure of this provision indicate a clear intent to provide the Agency with broad authority to tailor requirements to the particular circumstances of nonattainment areas in a manner analogous to the Congressional scheme set forth in subpart 2.[2] Just as Congress specified in subpart 2 differing requirements for areas based on the severity of the nonattainment problem, so EPA may consider "the severity of nonattainment" and "the availability and feasibility of pollution control measures" in establishing classifications and in determining how the applicable control requirements of subpart 1 should be interpreted and applied with respect to different classifications. That Congress limited the applicability of the classification authority of section 172(a)(1) by providing that it "shall not apply with respect to

---

[1] Congress also provided more restrictive requirements for other existing air quality standards such as carbon monoxide (subpart 3) and PM-10 (subpart 4).

[2] As part of the ozone classification scheme established in subpart 2, Congress provided that certain otherwise applicable nonattainment area requirements of subpart 1 did not apply to marginal ozone nonattainment areas. See section 182(a) (stating that the section 172(c)(9) contingency measure requirement does not apply to marginal areas). Congress similarly established differing control requirements for different classifications of carbon monoxide and PM-10 nonattainment areas in subparts 3 and 4 of part D.

nonattainment areas for which classifications are specifically provided under other provisions of this part" confirms that Congress viewed EPA's authority under section 172(a)(1) to establish classifications to be at least as broad as, but independent of, the classifications Congress itself specified in subparts 2, 3 and 4 of part D.

EPA has previously exercised its discretion to interpret and apply the requirements of subpart 1 to classes of ozone nonattainment areas in a manner that recognizes the particular circumstances of those areas.[3] The classification scheme established by Congress under subpart 2 did not encompass all areas designated nonattainment for ozone at the time of the 1990 Amendments to the Clean Air Act. Thus, in 1991, EPA established classifications for certain other categories of ozone nonattainment areas (referred to as "submarginal" and "incomplete data" areas) with design values that did not fall within the range of design values specified in section 181(a)(1). See 56 Fed. Reg. 56694, 56697 (Nov. 6, 1991). These areas were not subject to the classification scheme provided in subpart 2 nor the control requirements linked to those classifications. In 1992, EPA interpreted how the requirements in section 172 of subpart 1 applied in these areas. See 57 Fed. Reg. 13498, 13524-27 (April 16, 1992). [4]

The Transitional Classification

For the revised 8-hour ozone standard, it is subpart 1, not subpart 2[5], that specifies the mechanisms for States to reach the goal of attainment. Compared to subpart 2 of part D, the provisions of section 172(c) in subpart 1 provide greater flexibility for States in reaching that goal. EPA believes that it may use the classification authority provided under section 172(a) in conjunction with the more flexible provisions of section 172(c) to establish a "transitional"

---

[3] Similarly, for different types of ozone nonattainment areas, EPA interpreted and applied the control requirements in section 172 prior to its revision by the 1990 Amendments. See State Implementation Plans; Approval of Post-1987 Ozone and Carbon Monoxide Plan Revisions for Areas not Attaining the National Ambient Air Quality Standards, 52 Fed. Reg. 45044 (Nov. 24, 1987).

[4] EPA took similar actions with respect to certain carbon monoxide nonattainment areas whose design values did not fit within the statutory classification regime established by subpart 3. See 56 Fed. Reg. 56694, 56697 (Nov. 6, 1991); 57 Fed. Reg. 13498, 13535 (April 16, 1992).

[5] EPA's interpretation that subpart 2 applies only for the 1-hour ozone standard is set forth in a February 10, 1997, letter to Congressman Barton. A copy of that letter is attached.

classification for areas meeting the eligibility criteria described in the President's July 16, 1997 memorandum.[6]

Several factors warrant a flexible approach for transitional areas. Transitional areas, by definition, will not be violating the 1-hour ozone standard. Moreover, a vast majority of these areas will be able to attain the new 8-hour standard due solely to the regional nitrogen oxide (NOx) reductions. In order to receive the transitional classification, areas will need to submit an air quality plan based on the regional strategy and, if necessary, include additional measures demonstrating how the standard will be attained, earlier than otherwise required; EPA believes that early adoption of a plan will lead to emission reductions and, therefore, health benefits, earlier than would otherwise occur.

We address below the specific requirements of subpart 1 raised in your letter.

Reasonable Further Progress

Section 172(c)(2) provides that State plans "shall require reasonable further progress," which is defined under section 171(1) as "annual incremental reductions in emissions of the relevant pollutant as are required by [part D] or may reasonably be required by the Administrator for the purpose of ensuring attainment ... ." (Emphasis added.) EPA believes that transitional areas that will attain the standard based on the regional NOx control strategy[7] will make sufficient "reasonable further progress" as a result of the phasing-in of regional NOx reductions. In addition, all areas that are classified as transitional will need to make sufficient progress in the early years in order to attain the standard within the statutory period provided for attainment. Therefore, EPA believes that areas that are classified as transitional will

---

[6] As provided in the attachment to the President's memorandum, an area will be eligible for the transitional classification if (1) it is attaining the 1-hour standard at the time EPA designates areas for the 8-hour standard (i.e., by the year 2000); (2) it is in a State that by 2000 submits an implementation plan that includes control measures to achieve emission reductions required by EPA's rule for regional NOx reductions; and (3) by 2000 (a date that is earlier than otherwise would apply for nonattainment areas), it provides an attainment demonstration and any necessary control requirements that may be needed to reach attainment if regional NOx reductions will not be sufficient to bring the area into attainment.

[7] Pursuant to the recommendations of a 37-State group known as the Ozone Transport Assessment Group, EPA plans to propose that many of these States participate in a regional strategy to reduce emissions of (NOx). NOx emissions are one of two precursors of ozone pollution. Both NOx and ozone are transported great distances; therefore, it is important for areas upwind from a nonattainment area to reduce emissions of NOx so that the air entering downwind nonattainment areas will be cleaner.

require implementation of any necessary regional and local measures on a schedule that will meet the reasonable further progress requirement.

Reasonably Available Control Technology

Section 172(c)(1) requires that State plans provide for reasonably available control measures (RACM), including reasonably available control technology (RACT). Under subpart 2, for purposes of applying RACT to sources that emit volatile organic compounds (VOCs), an ozone precursor, EPA has interpreted the RACT requirement to apply independently of what is needed to attain the ozone standard. However, with respect to particulate matter and other pollutants, EPA has interpreted the RACM/RACT requirement to mean that only those measures that are needed to demonstrate attainment by the applicable attainment date are "reasonably available" and thus required. EPA believes it has the authority under subpart 1, where appropriate, to apply the interpretation it has taken for pollutants other than ozone in determining whether additional stationary source controls are "reasonably available" for purposes of obtaining emission reductions for ozone precursors. In addition, EPA believes that a regional NOx control strategy may be sufficient to meet the RACT/RACM requirement.

New Source Review

Section 173 of the Act requires new sources locating in nonattainment areas to comply with the lowest achievable emission rate (LAER) and to offset emission increases. EPA is now developing new source review regulations for transitional areas, which the Agency expects to issue in a final rule in December 1998. These regulations will spell out how the offset and LAER requirements can be met in transitional areas in a manner consistent with the implementation plan. These regulations will take into account the fact that ozone air quality problems are often a result of the long-range transport of ozone precursors and that the Agency is taking steps to deal with this problem on a regional basis.

Offsets

In meeting the requirements for offsets for ozone precursors, EPA expects that sources will be able to rely upon a pool of offsets which a State may allocate from the larger set of emissions reductions generated by that State as part of the regional strategy. Hence, offsets will be required, but the burden on individual sources for finding such offsets will be eliminated. Some States have already used the alternative of pooled offsets under the one-hour NAAQS. For example, New York has used portions of a pool of emission reduction credits to provide offsets for major new source growth. The South Coast in California has also provided a pool of offsets for new source construction.

<u>Lowest Achievable Emission Rate</u>

Determinations of the technology to comply with LAER ultimately are made on a case-by-case basis by the State or EPA in acting on individual permit applications. EPA intends through rulemaking to develop a flexible interpretation of the LAER requirement for sources in transitional areas. Because of the circumstances unique to transitional areas, as discussed below, de minimis principles could support an approach that applies an enhanced "best available control technology" (BACT) analysis in selecting a control technology for a new source would meet the statutory LAER technology requirement.

There are a number of factors that, taken together, could provide a reasonable basis for concluding that any differences between this approach to selecting the technology requirement for new sources in transitional areas and the application of LAER under the Agency's current approach would be de minimis. First, an approach using an enhanced BACT analysis for transitional areas will, in many cases, result in an emission limit similar, if not identical, to what would be required through application of LAER under the Agency's current approach. Indeed, under any BACT determination the most stringent emission limit and associated technology must be considered in the analysis and, if not chosen, their rejection must be supported in the administrative record based on consideration of cost, environmental impacts, and other statutory factors. Moreover, in applying the analogous analytical concepts to transitional areas, the environmental impact of regional pollutants, such as NOx, would be required to be carefully considered. Second, EPA believes that the number of major new or modified sources in transitional areas that would be subject to a major new source program is likely to be very small. Thus, any differences between LAER under the Agency's current approach and the technology applied under the policy for transitional areas would not have any significant adverse effect on the achievement of the ambient air quality standard. Third, the transitional areas that will be eligible for this approach will already be in attainment of the 1-hour standard and will have begun their air quality planning and SIP submission activities for the 8-hour standard earlier than otherwise might be required. Finally, as in all cases, the ambient impact of a source subject to NSR ultimately is determined by whether its emissions are offset by reductions at the source or elsewhere, rather than by the applicable technology requirement. Since a source must offset any emissions remaining after application of controls, the source's ambient impact would be no greater if it applied a technology different than that which would have been selected in other areas. Beyond that, EPA is considering whether, consistent with de minimis principles, other elements, such as additional offsets, would be appropriate to further the purposes of the CAA and the goals of the implementation plan in transitional areas.

The use of de minimis principles in applying the LAER requirement in transitional areas will be further addressed in the proposed rule for new source review programs. Ultimate application of these principles will occur through SIP revisions and issuance of individual permits.

6

In providing common sense solutions to problems, EPA has previously applied de minimis principles where the strict application of the literal text of the Clean Air Act would produce trivial or non-existent environmental benefits. Moreover, the Agency has done so in circumstances precisely analogous to the present ones: where the Agency has adopted a classification scheme under section 172(a)(1) and believes that it is appropriate to provide a more flexible interpretation of the applicable requirements of section 172(c). For example, EPA believed it appropriate under de minimis principles to exempt the "submarginal" and "incomplete data" ozone nonattainment areas referred to above from the otherwise applicable requirement of section 172(c)(9) to submit SIPS providing for certain contingency measures. See 57 Fed. Reg. at 13525 (April 16, 1992).

Also, in its conformity rules, EPA established certain de minimis exceptions from the otherwise applicable requirements. These exceptions were upheld by the D.C. Circuit in Environmental Defense Fund, Inc. v. EPA, 82 F.3d 451, 467 (D.C. Cir. 1996), as being an "appropriate exercise" of EPA's de minimis authority. Thus, not only has EPA utilized de minimis principles in implementing the CAA, there is strong judicial support for the Agency's reasoned use of these principles. See also, Alabama Power Co. v. Costle, 636 F.2d 323 (D.C. Cir. 1979) (affirming that de minimis principles apply to new source rulemaking under the CAA). In contrast to the de minimis exception upheld in the EDF case, EPA is not here using a de minimis rationale to exempt categorically sources from any NSR statutory requirement; rather, EPA will enable the States in developing SIPs and permitting authorities in issuing permits to apply de minimis principles in making decisions regarding the technology that will meet the statutory LAER requirement.

EPA believes that the transitional classification for ozone nonattainment areas that do not meet the 8-hour standard but have met the 1-hour ozone standard is proper in light of the statutory authority Congress has provided under the Clean Air Act and general principles of administrative law and statutory construction. EPA has sought to provide flexibility for areas in the past and to interpret and apply the Clean Air Act pragmatically in a manner consistent with its objectives while avoiding the imposition of unnecessary burdens on States and sources. The transitional classification to be used in implementing the new ozone standard is consistent with those prior efforts and simply represents an application of those principles in a new context.

Question 2: State and local governments have expectations of certainty regarding Clean Air Act implementation schemes, as do the businesses that often must make operational decisions based on Clean Air Act requirements. I am concerned that EPA did not issue its ozone NAAQS implementation plan as a rule, although it proposed such a rule in November 1996. Moreover, the President's July 16, 1997, memorandum to you states that EPA's implementation plan is not judicially enforceable against the agency. Accordingly, it seems that there is no legal mechanism by which States could hold EPA to the specific terms of its July 19, 1997 implementation plan. Rather, EPA is free to change the plan at any time. If

EPA decided to change the terms of its implementation plan, what legal recourse is available to States and businesses that are regulated under the Act?

Response: In November 1996, Administrator Browner signed a proposed interim implementation policy which was published in the Federal Register on December 13, 1996. 61 Fed. Reg. 65752. The proposed interim implementation policy set forth EPA's proposed policy that would govern continued implementation during the period it will take to phase in the new standards. This document focused on continued implementation of the 1-hour ozone standard and pre-existing PM-10 standard, not implementation of the revised 8-hour standard and the new and revised particulate matter standards. Moreover, this document was a proposed policy and, as such, would not have had the binding effect of a rule.

The implementation plan for the revised ozone and new and revised particulate matter standards was issued contemporaneously with the standards. The President's memorandum to Administrator Browner, attaching the plan, directs the Administrator to consult "with all affected agencies and parties" and "to undertake the steps appropriate under law to carry out the ... plan." EPA plans to issue a series of guidance and rulemaking documents to assist States in developing plan submissions. In turn, the States will adopt rules that EPA will approve or disapprove. Such requirements on sources will be federally enforceable only upon approval into the SIP.

Although the President's directive is not a legally enforceable document, it describes the Executive branch's carefully considered strategy for implementation of the new standards and EPA will adhere to these guiding principles. As stated in the preambles to both the final particulate matter and ozone standards, see 62 Fed. Reg. at 38652 (July 18, 1997); 62 Fed. Reg. at 38856 (July 18, 1997), EPA is committed to carry out President Clinton's directive, and to propose rules, issue guidance, and take other actions consistent with the strategies set forth in that document. When EPA issues a proposed or final rule any interested party would be able to comment on any such proposed rule or file a petition for review in the appropriate court of appeals with respect to any such final rule. The bases for a court to overturn a final EPA rule are set forth in 5 U.S.C. 706 and section 307(d) of the Clean Air Act, where applicable.

EPA plans to work with affected stakeholders such as States, industry and public interest groups as it develops the policies that will guide implementation of the revised standards. Among other things, this could include stakeholder meetings or an opportunity for public comment on a draft of the policy. Since policy documents are not binding, States and sources also would have the flexibility to develop State plans that diverge from the terms of the final policy. EPA would then make a decision on whether such a plan met the requirements of the CAA despite its divergence from EPA's policy. If a party disagreed with EPA's ultimate approval or disapproval of such a State plan, the party could seek redress by filing a petition for review in the appropriate court of appeals. CAA section 307(b).

Question 3: I am also concerned that EPA's implementation plan will give courts an unnecessary opportunity to establish implementation policy. Section 304 of the Clean Air Act authorizes citizen suits to enforce certain mandatory provisions of the Act. For example, despite EPA's interpretation otherwise, a citizen group could conclude that Section 172(a)(1) does not authorize EPA to exempt transitional nonattainment areas from the more stringent "lowest achievable emission rate" technology required by Section 173. If a court sided with the citizen group in such a suit, couldn't the court require transitional nonattainment areas to use lowest achievable emission rate technology (or other mandatory controls for which EPA exempted transitional nonattainment areas)?

Response: Under long-standing principles of administrative law applied over a broad range of actions by EPA and other federal agencies, courts are required to defer to a reasonable interpretation proffered by the Agency that is responsible for administering the statute. The courts are not entitled in such cases to substitute their judgment for the Agency's. See Chevron USA Inc. v. EPA, 467 U.S. 837 (1984). Deference is also accorded to reasonable applications of de minimis principles by the agency responsible for administering a statute. See EDF v. EPA, 82 F.3d at 467 (stating that to "the extent that both Chevron and Alabama Power address agency power inherent in a statutory scheme, the same deference due to an agency's reasonable interpretation of an ambiguous statute may also be due to an agency's creation of a de minimis exemption"); State of Ohio v. EPA, 997 F.2d 1520, 1535 (D.C. Cir. 1993).

EPA believes that this implementation plan is properly based within the authority Congress has provided under the Clean Air Act and general principles of administrative law and statutory construction outlined above. In addition, EPA is prepared to defend this approach in any lawsuit brought against the Agency by citizens, States, or other interested parties. EPA believes that, consistent with established principles of administrative law, EPA's implementation plan and statutory interpretations are entitled to deference in any litigation under the Act. Hence, States and sources can rely on such interpretations.

Question 4: The PM2.5 NAAQS became effective on September 16, 1997. It is a new standard; there was not an existing NAAQS for PM2.5. Recently published reports indicate that PM2.5's new regulated status has immediate compliance effects under Titles I and V of the Clean Air Act despite EPA's claim that nothing will happen with regard to PM2.5 for at least five years. For example, these reports indicate that under Title V, regulated sources must obtain permits for emissions of all pollutants regulated under the Act, which now includes PM2.5. Additionally, Part C of Title I requires that new sources, including new or expanding businesses, located in attainment or "unclassifiable" areas prove that their operations will not impermissibly increase total emissions of regulated pollutants which now includes PM2.5. Both of these provisions arguably could require a source to control PM2.5. How does EPA plan to deal with these important and immediate Clean Air Act permitting issues? Does EPA plan to issue a rule addressing these matters? Because statutory requirements of the Clean Air Act are at issue here, couldn't a businesses' permit be challenged in a citizens suit under

Section 304 of the Act, even if EPA chooses not to enforce the Title I and V requirements discussed above?

Response: Title V of the CAA requires every major source to obtain an operating permit setting forth the "applicable requirements" under the various substantive CAA requirements that apply to that source. The new PM-2.5 standard has no immediate impact on the Title V operating permits requirement of the Act because the new standard does not at this time impose any "applicable requirements." PM-10 includes all particulate matter that is less than 10 micrometers and, therefore, includes all PM-2.5. Any source that emits 100 tons per year of PM-2.5 also emits, at a minimum, 100 tons per year of PM-10. Thus all such sources are currently subject to the operating permit requirements because of their PM-10 emissions. In addition, sulfur dioxide and NOx, which are PM-2.5 precursors, are also regulated pollutants and, therefore, sources that are major for sulfur dioxide or NOx emissions are already subject to the Title V operating permit requirements.

Regarding the Part C Prevention of Significant Deterioration of Air Quality (PSD) requirements, the EPA announced its position on December 13, 1996 (61 Fed. Reg. 65752, 65762) that PSD review for PM-2.5 would be deferred because of the lack of technical and analytical tools to predict PM-2.5 emissions levels and ambient impacts from proposed new and modified sources. The lack of these tools renders implementation administratively impracticable at this time. Instead, PSD review would continue to involve the review of PM-10 emissions only and that review would act as a surrogate for PM-2.5 PSD review until the PM-2.5 technical tools became available. The EPA believes that any citizen suit that seeks to enforce these requirements would be unsuccessful in light of the Agency's policy position and the very severe technical obstacles that would need to be overcome before a source could implement PSD for PM-2.5.

UNITED STATES ENVIRONMENTAL PROTECTION AGENCY
WASHINGTON, D.C. 20460

FEB 10 1997

OFFICE OF
AIR AND RADIATION

Honorable Joe Barton
Chairman, Oversight and Investigations
House Commerce Committee
House of Representatives
Washington, DC 20515-6115

Dear Mr. Chairman:

This is in response to your letter dated December 3, 1996 to Administrator Carol Browner. Your letter asked several questions regarding the Environmental Protection Agency's proposed revision to the national ambient air quality standards for ozone and particulate matter and its authority to revise specific requirements.

Enclosed is our response to these questions. I appreciate this opportunity to be of service and trust that this information will be helpful to you.

Sincerely yours,

Mary D. Nichols
Assistant Administrator
for Air and Radiation

Enclosure

cc: Honorable Ron Klink
Minority Ranking Member

ENCLOSURE

By letter dated December 3, 1996, you requested a legal
opinion as to the Environmental Protection Agency's (EPA)
authority to revise certain requirements in subparts 2 and 4 of
part D of Title I of the Clean Air Act (Act), based on a revision
to the national ambient air quality standard (NAAQS).
Specifically you asked about the authority to do the following:

a)   Revise timelines set out under 182(d)
b)   Revise categories (marginal, moderate, etc.), including
authority to eliminate categories
c)   Revise definition of "major sources"
d)   Revise, eliminate, or require new controls
e)   Revise various showings, including the so called 15 percent
plan and reasonable further progress
f)   Assuming that the EPA has the power to revise any of the
above (a-e) the EPA's authority to sanction in order to enforce
the revised classifications, controls, State implementation plans
(SIPs), etc.

For the reasons stated below, in the Background section, EPA
believes the requirements of subparts 2 and 4 will not carry
forward with respect to a revised ozone and new fine particulate
matter (PM-fine) NAAQS. Hence, EPA is not proposing to revise
the timelines and other requirements under subparts 2 and 4, but
rather adhering to the provisions of the Act that apply for a
revised NAAQS. The EPA, however, may engage in rulemaking under
sections 172 and 301 of the Act to specify requirements that
would apply in areas designated nonattainment under a new NAAQS.
More specific responses to the questions you have presented are
set forth following the Background section.

BACKGROUND

The EPA interprets the relevant portions of the Act to
provide that the general planning requirements of part A of title
I and the basic nonattainment planning requirements of subpart 1
of part D of title I govern the implementation of a new or
revised NAAQS. The detailed provisions of subparts 2 and 4 of
part D currently apply, respectively, to nonattainment planning
under the one-hour 120 ppb ozone NAAQS and the PM-10 NAAQS
(particles with an aerodynamic diameter less than or equal to a
nominal 10 micrometers). These provisions, however, would not
apply directly to the implementation of a new ozone NAAQS or a
new fine particle NAAQS. Because the provisions of subpart 4
are not linked to a specific PM-10 NAAQS (incontrast to subpart
2's linkage to one specific ozone NAAQS), the provisions of
subpart 4 would apply to the implementation of a revised PM-10
NAAQS.

2

The basis for this view lies in the language and structure of those subparts, which are clearly and explicitly tied to the 1-hour 120 ppb ozone NAAQS in the case of subpart 2, and to a PM-10 NAAQS in the case of subpart 4.[1] The provisions of subpart 1, however, apply to the implementation of any NAAQS, including revisions to NAAQS that were in effect at the time of the Clean Air Act Amendments of 1990 (1990 Amendments).

To understand this interpretation, it is helpful to begin with the process that the Act establishes for promulgating and implementing a revised NAAQS. Pursuant to section 109 of the Act, EPA must review the existing NAAQS every 5 years and, as appropriate, revise the standard. A revision to the NAAQS triggers the designation process set forth in section 107(d)(1).[2] Such designations would not be redesignations of existing areas, which are designated based on the current NAAQS, but rather would be initial designations for purposes of the new or revised NAAQS. Section 107(d)(1)(A) provides that no earlier than 120 days and no later than 1 year after promulgation of a _new or revised_ NAAQS, EPA shall require the Governor of each State to submit a designation for each area within the State. Moreover, EPA must designate such areas no later than 2 years after the new or revised NAAQS is promulgated, unless it has insufficient information to promulgate the designations, in which case, EPA has an additional year in which to act. [section 107(d)(1)(B).]

The designation process triggers the planning process under part D of title I of the Act. Under part D, the applicable requirements for purposes of a revised NAAQS are set forth in subpart 1, most notably section 172. The requirements specified in subparts 2 and 4 would not apply directly. The rationale for this interpretation is set forth in the following paragraphs that address the language and structure of subparts 2 and 4, as compared with section 172 of subpart 1.

---

[1] As you note in your letter, the classifications in subparts 2 and 4 of part D of title I of the Act are "based on the severity _as measured under the current standard_" and that these subparts establish "time lines, controls, and various showings _as determined by the area's classification_." (Emphasis added.)

[2] It is clear that the other designation provisions of section 107(d) do not apply to the situation where a new or revised NAAQS is promulgated. Section 107(d)(2) speaks to designations by operation of law at the time the 1990 Amendments were enacted. Section 107(d)(3) applies only to redesignations of areas designated for an existing NAAQS. Finally, section 107(d)(4) concerns only the initial designation process immediately following enactment of the 1990 Amendments.

3

Section 172(a) provides that EPA may classify any area designated nonattainment under section 107(d). Section 181(a) and (b) (located in subpart 2 of part D) also establishes a classification scheme for ozone nonattainment areas. Similarly, section 188(a) and (b) (located in subpart 4 of part D) establishes a classification scheme for PM-10 nonattainment areas.

Section 172(a) specifically provides the authority for establishing classifications for areas designated nonattainment for purposes of a revised standard, including any NAAQS in effect at the time of enactment of the 1990 Amendments. Section 172(a)(1) explicitly authorizes EPA to establish a new classification system with respect to a revision of a NAAQS. For purposes of the initial classifications under the Act, as amended in 1990, however, the more specific provisions in section 181(a), for ozone, and 188(a), for PM-10, applied rather than the general provisions of section 172(a)(1). Section 172(a)(1)(C) provides that section 172(a)(1) does not apply with respect to nonattainment areas for which classifications are specifically provided under other provisions of this part.

Section 181(a) provided the classification scheme for 1989 nonattainment areas -- i.e., those areas that were not attaining the ozone standard based on the 3-year period from 1987-1989.[3] This provision would not apply to classification of areas pursuant to a revised NAAQS for several reasons. First, it is clear from the heading that this section was meant to apply based on data from 1987-1989. Second, the design values listed in Table 1 that establish the classifications are explicitly linked to the 120 ppb 1-hour ozone standard. Third, the attainment dates in Table 1 and the SIP submission dates in section 182 are explicitly linked to the date of enactment of the 1990 Amendments. Because many of these dates have passed and because section 181(b), which allows modification of these dates, is inapplicable for purposes of a new or revised NAAQS (see below),

---

[3]Section 181(a) provides that the classification shall be based on "the interpretation methodology issued by the Administrator most recently before the date of the enactment of the Clean Air Act Amendments of 1990." An area is nonattainment for ozone if it has more than one exceedance per year, averaged over 3 years. A day is counted as an exceedance of the NAAQS if the measurement for at least 1-hour of the day is greater than 120 ppb. Once an area is designated as a nonattainment area, the table set forth in section 181(a) classifies areas based on the magnitude of each area's design value. The design value is the fourth highest daily maximum concentration, since the NAAQS allows an exceedance 1-day per year, on average, or 3 days over the 3-year compliance period.

4

it would be nonsensical to apply this provision for purposes of a
revised ozone NAAQS. Finally, if EPA revises the methodology for
calculating a violation of the ozone NAAQS, as the Agency has
proposed, the requirement in section 181(a)(1) that EPA rely on
"the interpretation methodology issued by the Administrator most
recently before the date of enactment of the [CAAA of 1990]" in
classifying areas would be absurd.

Section 181(b)(1) is the only provision for adjusting
attainment dates or SIP submission dates to take into account
ozone designations to nonattainment occurring after the initial
round of post-1990 designations. Section 181(b) addresses
classification of areas that were designated as attainment or
unclassifiable under section 107(d)(4), but are subsequently
redesignated to nonattainment in accordance with section
107(d)(3). However, section 107(d)(4) applies only to the
initial designations that occurred immediately following
enactment of 1990 Amendments. For the reasons stated earlier,
designation of areas pursuant to a revised NAAQS will occur in
accordance with section 107(d)(1), not section 107(d)(4).
Therefore, section 181(b)(1) is inapplicable to areas designated
nonattainment as to a new ozone NAAQS since it does not apply to
areas that are designated under section 107(d)(1). The fact that
this provision authorizes adjustments of attainment dates and SIP
submission dates only for areas redesignated to nonattainment of
the current 120 ppb ozone NAAQS, confirms the logical conclusion
that may be drawn from the explicit linkage of the provisions of
sections 181 and 182 to the 1990 time period and the ozone NAAQS
existing at that time -- which is that subpart 2 applies directly
only to that ozone NAAQS, not a revised one.

The situation with respect to PM-fine is also clear as
subpart 4 expressly applies only to PM-10. Section 188(a)
provides for classification of areas "designated nonattainment
for PM-10 pursuant to section 107(d)." (Emphasis added.)
Moreover, it explicitly references designations under section
107(d)(4), which is the designation provision that was explicitly
linked to the initial designations for ozone and PM-10 under the
Act, as amended in 1990. Section 188(b) applies to
reclassification of PM-10 nonattainment areas. The provisions
that address a PM-10 NAAQS simply woud not apply to nonattainment
planning for a new PM-fine NAAQS.

Because the control requirements in subparts 2 and 4 are
explicitly linked to the classification schemes set forth in
section 181, for ozone, and section 188, for PM-10, the control
requirements also would not apply directly to areas designated
nonattainment for a revised ozone and PM-fine NAAQS. As subparts
2 and 4 are limited in direct applicability to the 1-hour ozone
120 ppb NAAQS and PM-10 NAAQS, respectively, only subpart 1 of
part D directly applies to implementation of new or revised ozone

5

NAAQS or a fine particle NAAQS. Finally, another indicator that
subparts 2 and 4 do not apply of their own force and effect to
implementation of a new or revised ozone NAAQS or a fine particle
NAAQS is section 172(e), which requires EPA to establish control
requirements to prevent backsliding from subparts 2 and 4 in the
event of a relaxation of an existing NAAQS. Such a provision
would not have been necessary were those subparts to continue to
apply of their own force and effect.

    However, subpart 1 is not a blank slate. While certainly
not as detailed as subparts 2 and 4, section 172(c) establishes
specific control requirements and demonstrations that
nonattainment areas must make. As with the classification
provisions in section 172(a), the control requirements specified
in section 172(c) would apply to both ozone and PM-fine areas
designated as nonattainment for a revised ozone and PM-fine
NAAQS. These control requirements generally include reasonably
available control measures, including reasonably available
control technology; reasonable further progress; new source
review; an attainment demonstration; and contingency measures.
Section 172(a), the classification provision, appears to link the
level of control that may be applied in an area to the
classification of an area.

    In addition, section 172(e) establishes a no-backsliding
principle. Section 172(e) of the Act clearly provides that a no-
backsliding principle should apply upon a relaxation of an
existing NAAQS. It provides that EPA is to conduct a rulemaking
within 12 months of the promulgation of a relaxed NAAQS to
promulgate requirements applicable to areas not attaining the
existing standard that will provide for controls which are not
less stringent than the controls applicable to areas designated
nonattainment before such relaxation. The EPA believes that a
no-backsliding principle is even more important and by
implication was intended by the Act to be a governing principle
when an existing NAAQS is strengthened, as is the case with the
proposed ozone standard. Currently, EPA has proposed an interim
implementation policy based on a no-backsliding principle. See
Interim Implementation Policy on New or Revised Ozone and
Particulate Matter (PM) National Ambient Air Quality Standards
(NAAQS); Proposed Rule, 61 FR 65752, 65733 (Dec. 13, 1996)(IIP).
The EPA anticipates that it would adhere to a no-backsliding
principle under the planning requirements that would apply
subsequent to any nonattainment designations under the revised
NAAQS as well. Therefore, EPA may indirectly use the
requirements of subparts 2 and 4 to guide its interpretation of
the more general requirements of section 172(c) for specific
areas. The EPA has also proposed retaining the existing ozone
NAAQS for certain purposes until EPA determines that an area has
an approved SIP that provides for the achievement of the new
NAAQS. See, National Ambient Air Quality Standards for Ozone;

6

Proposed Decision, 61 FR 65716 (Dec. 13, 1996); IIP, 61 FR 65752, 65754. This would help ensure that no backsliding occurs and that the air quality benefits of the existing ozone NAAQS implementation program, which EPA believes are necessary to attain and maintain the potential new or revised ozone NAAQS, will be retained.

TIMELINES UNDER SECTION 182(d)

Section 182(d) of the Act, which is located in subpart 2 of part D of Title I, sets forth requirements that are applicable to ozone nonattainment areas that are classified as severe or extreme for the 120 ppb 1-hour ozone NAAQS. As explained above, the requirements of subpart 2 would not apply for purposes of a revised ozone standard. However, if the current 120 ppb 1-hour ozone NAAQS is retained for a period of time, certain of these requirements would continue to apply directly to areas designated as severe or extreme for the current ozone NAAQS, until such time that those areas are either redesignated to attainment for the current NAAQS or those areas have approved SIPs implementing the revised NAAQS. Moreover, as indicated above, EPA may rely on the provisions in section 182(d) in interpreting how the more general provisions of section 172(c) should apply to certain areas designated nonattainment for a new ozone NAAQS. (Analogous reasoning would apply to other ozone nonattainment areas, such as moderate areas, that are currently subject to other provisions of section 182.)

CLASSIFICATIONS

As explained above, designations and classifications for the new or revised ozone and PM-fine NAAQS will be independent of the current designation and classification scheme. In other words, areas will not be redesignated and reclassified; rather EPA will establish initial designations and possibly classifications for the new or revised NAAQS. The designations and classifications for purposes of a 120 ppb 1-hour ozone standard and for PM-10 will remain in place until such time as the areas are redesignated for the existing standards, or until the 120 ppb 1-hour ozone NAAQS or the PM-10 NAAQS is revoked.

The classification scheme set forth in section 181(a), for ozone, and 188, for PM-10, would not apply for purposes of a revised ozone NAAQS or a new PM-fine NAAQS because those classifications are explicitly tied to a 120 ppb 1-hour ozone NAAQS and a PM-10 NAAQS. If EPA retains the PM-10 NAAQS, as proposed, areas that are nonattainment for PM-10 will continue to remain classified in accordance with section 188. However, areas that are designated nonattainment for a new PM-fine NAAQS will be classified, if at all, in accordance with a classification scheme developed by EPA under section 172(a) of the Act. With respect to ozone, areas that are currently designated and classified for

7

purposes of the 120 ppb 1-hour ozone NAAQS, will retain those designations and classifications until such time as they meet the criteria for redesignation to attainment of that standard, or until such time as the 120 ppb 1-hour ozone NAAQS ceases to apply to an area.

DEFINITION OF MAJOR SOURCE

The definition of major stationary source set forth in section 302(j) would continue to apply with respect to a new or revised ozone and PM NAAQS. This definition defines a major source as one that "emits or has the potential to emit, 100 tons per year or more of any air pollutant." However, the specific major source limitations in section 182, for ozone, and in section 189(b)(3) for PM-10, would not apply directly for purposes of a revised ozone NAAQS or a new PM-fine NAAQS. The EPA believes that it has authority to develop lower major source thresholds through rulemaking for purposes of a new or revised NAAQS in accordance with the no-backsliding principle of section 172(e).

NEW CONTROLS, 15% PLAN, AND REASONABLE FURTHER PROGRESS

For purposes of a new or revised ozone and fine particulate NAAQS, the control requirements in subparts 2 and 4 of part D of Title I of the Act would not be directly applicable. The applicable control requirements for these areas are those set forth in subpart 1, including the requirements listed in section 172(c). In addition to subpart 1, certain mobile source requirements under Title II of the Act may also apply. With respect to section 172(c), the requirements are provided in a more general fashion than those provided under subparts 2 and 4. For example, under subpart 2, moderate and above ozone nonattainment areas were required to submit a SIP providing by 1996 for reductions of 15% of emissions from a 1990 baseline. In addition, for serious and above areas, the State was to provide for an additional 3% reduction in emissions per year from 1996 until such time as the area attained the standard. These requirements are generally referred to as "reasonable further progress." Section 172(c)(2), under subpart 1, provides that SIP submissions "shall require reasonable further progress," but provides no further elaboration. The EPA will need to determine what reasonable further progress means, either through a national rule or on a case-by-case basis, in acting on individual SIP submissions. The EPA could use its rulemaking authority under sections 172 and 301(a) to define this term and specify applicable requirements and timelines for their implementation. As noted earlier, in doing so EPA may rely on the no-backsliding principle of section 172(e) and the provisions of subparts 2 and 4 in interpreting the more general provisions of subpart 1.

8

SANCTIONS

Under EPA's interpretation of how the Act would apply to the revised ozone and PM-fine NAAQS, EPA's sanction authority, as set forth in sections 179(a) and (b) and 110(m), would remain in effect. Currently, EPA must impose either highway funding limitations or increased offset ratios for sources subject to new source review in areas for which EPA finds that: (1) a State failed to make a complete plan submission; (2) a State's planning submission was disapprovable; or (3) a State fails to implement an approved plan. One of these two sanctions is imposed 18 months after EPA makes such a finding, unless EPA determines in the interim the State has corrected the deficiency. The second sanction must be imposed 6 months after the first. Under section 110(m), EPA has discretion to impose either or both of these two sanctions earlier than the 18-month period in section 179(a) and also has discretion as to the geographic scope. In addition to the highway funding and offset sanctions, section 179(a) also provides EPA with discretion to withhold all or part of the grants that it may award under section 105 of the Act.

Under the Act, EPA does not have independent authority to enforce a classification. Rather, the classification results in the obligation to adopt specific control measures. The EPA's sanction authority, as described above, is tied to a State's failure to act with respect to these mandatory control obligations that apply because of an area's designation and classification. As with the control requirements in subparts 2 and 4, a State's failure to act with respect to the control requirements in subpart 1 (i.e., those in section 172(c)) would trigger mandatory sanctions. For a revised ozone and PM-fine NAAQS, where the applicable control requirements are set forth in section 172(c), there is no delineation of control requirements based on classification of areas. However, if EPA establishes through rulemaking a gradient of control requirements under section 172(c), based on the severity of air quality problems in an area, the sanctions would apply for a State's failure with respect to the control requirements established through regulation.

CONCLUSION

In conclusion, if new ozone and fine particulate matter NAAQS are promulgated, EPA would not revise the applicable classifications, controls, and timelines set forth in subparts 2 and 4 of part D of title I of the Act. Rather, EPA and the States would need to comply with the requirements set forth in subpart 1 of part D in lieu of the requirements set forth in subparts 2 and 4 for purposes of the new or revised NAAQS. While these controls and timelines are not established in the detail that requirements are established under subparts 2 and 4, EPA could use its rulemaking authority under sections 172 and 301(a) of the Act to further define these requirements and timelines and to further implement a no-backsliding principle. The EPA's sanction authority under section 179(a) and (b) and 110(m) would be unaffected by a revision to the NAAQS.

UNITED STATES ENVIRONMENTAL PROTECTION AGENCY
WASHINGTON, D.C. 20460

OCT 9 1997

OFFICE OF
AIR AND RADIATION

The Honorable Thomas J. Bliley, Chairman
Committee on Commerce
U.S. House of Representatives
Washington, DC 20515-6115

Dear Chairman Bliley:

This letter responds to your letter of September 22, 1997, generally regarding the calls for revisions of State implementation plans (SIPs) in the proposal that EPA is about to sign to reduce transport of nitrogen oxides (NOx) emissions to downwind ozone problem areas.

I am enclosing responses to your individual questions. I would also like to correct a possible misunderstanding. Your letter refers to a "disparity between representations made in December 1996 correspondence to Chairman Bilirakis and the present implementation plans being advanced by the Environmental Protection Agency." In the December 1996 correspondence, the Environmental Protection Agency (EPA) had not decided, and was not in a position to presume the outcome of the National Ambient Air Quality Standards (NAAQS) review for ozone, so therefore, planned at that time to develop the proposed NOx strategy rule based only on the then-current 1-hour 0.12 ppm standard. After EPA published the final 8-hour 0.08 ppm ozone standard and temporarily retained the 1-hour standard, EPA believed it appropriate to consider both the 1-hour and 8-hour standards in the NOx strategy rule.

I appreciate this opportunity to be of service and trust that this information will be helpful to you.

Sincerely,

Richard D. Wilson
Acting Assistant Administrator
for Air and Radiation

Enclosure

**RESPONSE TO LETTER OF SEPTEMBER 22, 1997 FROM
REPRESENTATIVE TOM BLILEY
REGARDING NOX SIP CALL AND TRANSITIONAL CLASSIFICATION**

Question 1: Please explain what is meant by the statement to Representative Dingell which indicates that the proposed rule regarding the OTAG SIP call will "consider" the reductions necessary for both the previous 0.12 ppm and new established 0.08 ppm ozone standard. Is it the intent of the EPA to require that SIPs in states in the OTAG region provide NOx reductions that may be in excess of the reductions necessary to achieve a 0.12 ppm ozone standard?

Response:
The regional NOx strategy is not designed primarily as an attainment strategy per se, but rather as a means of mitigating transport of ozone and NOx which contribute to downwind violations, or which interfere with attainment or maintenance of the standards. We believe mitigating regional transport is necessary to make attainment of the ozone standards a practical goal for some of the States in the eastern U.S. Nonattainment of the ozone standards is a consequence in many areas in the eastern U.S.--either in whole or in part--of ozone and/or its precursor, NOx, that are transported across state lines by weather patterns.

The OTAG process was developed to address ozone transport under the 1-hour ozone standard. Shortly after OTAG concluded and submitted their recommendations for a regional control strategy, EPA promulgated the new 8-hour ozone standard. Based on the OTAG modeling, EPA evaluated the regional reductions needed for both the 1-hour and 8-hour standards. EPA is proposing the same regional emissions reductions and needed to address transport for both the 1-hour and 8-hour standards. The emissions budgets that we are proposing are consistent with the OTAG recommendations.

Question 2: If the EPA intends to require or approve SIPs to achieve reductions beyond those required to meet a 0.12 standard in downwind states, on what specific modeling will a requirement for such NOx reductions be based? If EPA has made a decision to "adjust" or "consider" NOx reductions beyond those necessary to achieve a 0.12 ppm standard in its September 1997 proposed rule, what is the extent of the incremental tonnage "adjustment" or "consideration" that will be required in each of the states affected by the notice? What is the legal basis for including emission reductions for the purpose of attaining the new 0.08 ppm standard in the upcoming SIP calls?

Response: As noted above, EPA is proposing that the same level of emissions reduction is needed to address transport for both the 1-hour and 8-hour standards. EPA did not determine that any additional regional emissions reductions would be necessary as a result of the 8-hour standard. As noted in the response to question 1, while the SIP call addresses the new 8-hour standard, it is not a SIP call to address full attainment of the standard; rather it addresses transport of ozone and its major NOx precursors. Because areas are either currently not attaining the 8-hour standard or are projected to not attain the 8-hour standard in the future, even though they have not been designated nonattainment, EPA has the authority to require SIP revisions to mitigate or eliminate any significant contribution in downwind areas not attaining the standard resulting from upwind emissions. See CAA sections 110(k); 110(a)(2)(D).

Question 3: To what extent, if any, did OTAG modeling analyses address strategies for attaining an 8 hour ozone standard of 0.08 ppm as the average 4th highest value over three year averaging periods? To what extent did OTAG's assessment of the causes and nature of ozone transport distinguish between transport contributing to exceedences of the previous 0.12 ppm ozone standard and the newly promulgated 0.08 ppm standard?

Response: The OTAG modeling (the purpose of which was to address the transport of ozone and its precursors, not for attainment demonstrations for local areas) did not attempt to determine what levels of control would be needed to attain the NAAQS. The modeling did, however, examine effects of various levels of control on a variety of indicators, including predicted 1-hour and 8-hour daily maximum ozone concentrations using episodes taken from 3 separate years.

OTAG's assessment of the causes and nature of ozone transport did not distinguish between transport contributing to exceedences of the 0.12 ppm ozone standard and the newly promulgated 0.08 ppm standard. Rather the recommendations were intended to reduce levels of NOx emissions transported to downwind areas and thereby reduce overall levels of ozone.

Question 4: It is my understanding that a central recommendation of the OTAG process was that individual states should have the opportunity to conduct more refined subregional modeling before reaching judgments on what emission reductions might be sought from specific sources. Please explain what efforts EPA and OTAG states have made to date, in order to implement this recommendation and actually conduct more refined subregional modeling. In addition, please explain the relationship between state-specific emission reductions to be proposed by EPA in its September 1997 rulemaking and the results of such subregional modeling efforts.

Response:
We believe it is likely that some States will provide additional modeling results during the public comment period for a number of purposes, but we are not aware of any ongoing efforts or results which have been obtained to date. One purpose would be to determine how to meet the NOx emission budgets proposed in EPA's rulemaking, by modeling localized conditions to determine any ozone disbenefits from control on certain large sources. Another purpose could be to show the relative air quality benefits of alternative allocations of emissions, which might affect different mixes of sources. Such efforts are, of course, not required, but voluntary on the part of the State.

Because commenters may wish to submit technical information that may require additional time to develop -- including additional modeling that States wish to undertake -- EPA will accept additional pertinent information from States beyond the 120-day comment period EPA is establishing as part of the SIP call notice and will do whatever is possible to take the information into account for the final rulemaking.

Question 5: As you know, the July 16, 1997 Memorandum indicates that if certain States include "control measures to achieve the emission reductions required by the EPA's rule for the States in the OTAG region" in their SIPs they may be able to receive a new "transitional status" which will relieve such states of "unnecessary local planning requirements" as well as allow such States to benefit from revised new source review (NSR) and conformity rules. Will emission reductions resulting from an emission trading program beyond those required under a 0.12 ppm standard be required in order for an OTAG state to receive "transitional status"? If EPA identifies "unnecessary local planning requirements" will the Agency eliminate those requirements for all states? If not, why not?

Response: The emissions budgets EPA is proposing for purposes of mitigating interstate ozone transport are the same under the 1-hour and 8-hour standards. An evaluation of the reductions that would be achieved by the regional transport SIP call indicates that almost all of the areas that would become nonattainment for the new 8-hour standard would be brought into attainment by the regional strategy. The transitional status is available to areas that adopt and submit by the year 2000 measures to meet EPA's NOx SIP call and any additional measures, if needed, to meet the 8-hour standard. As indicated in the Presidential directive, EPA will be reexamining existing nonattainment area NSR requirements in light of EPA's approach to NSR for transitional areas.

Question 6: How will EPA determine whether an area qualifies for "transitional" status and when will that determination be made?

Response: The Presidential Directive of July 16, 1997 provides the overall criteria for eligibility for the transitional status. EPA is developing further policy and technical guidance for States that want to apply for transitional status concerning submittals and demonstrations. EPA anticipates developing preliminary guidance for those submittals in the next several months, and final guidance by the end of 1998. EPA anticipates making the final determination of which areas are transitional in 2000, but anticipates working closely in the interim with States in developing the transitional classification guidance. EPA will also work with individual States to help them prepare their submittals.

Question 7: What roles would air quality modeling and monitoring data play in the determination of "transitional" areas? Why is agreement on a specific level of NOx emission reductions required as a precondition to "transitional" status?

Response: Monitoring data will be needed to determine whether the 1-hour NAAQS is met. If the data indicate the 1-hour NAAQS is not met by 2000, an area cannot qualify as a transitional area. Monitoring data are also needed to determine whether the 8-hour NAAQS is met. One of the prerequisites for an area to qualify as a "transitional area" is to show that the 1-hour NAAQS is met, but the 8-hour NAAQS is not being met.

Additional modeling for a transitional area is only needed if the past OTAG analyses suggest that the NOx emission budgets in the EPA SIP call proposal will not suffice to meet the 8-hour NAAQS in the area by 2007. Additional modeling would also be needed to show attainment by the appropriate attainment date in locations wanting to qualify as transitional areas which are outside of the OTAG study area (e.g., in Western States).

Under the Presidential Directive of July 16, 1997, submission of a plan to implement controls by 2002 meeting the proposed NOx emission budgets is also a necessary precondition for consideration as a transitional area. (The NOx SIP call notice includes more specific proposals regarding the actual deadlines for compliance.) Based on the OTAG analyses, areas that can reach attainment through implementation of the regional transport strategy would not be required to adopt and implement additional local measures.

Question 8: Please clarify whether any states, or parts of states, included in the original 37 OTAG region will be excluded from EPA's September 1997 rulemaking. What is the basis for excluding any state?

Response: In its proposed NOx SIP call, EPA generally followed the recommendation of OTAG to require regionally applicable controls only in the States that were in the "fine grid" (those not near the edges) of the modeling domain that contribute significantly to nonattainment or prevent maintenance of the NAAQS. Of the original 37 OTAG region States, the following 15 are excluded from the proposed NOx SIP call: Arkansas, Florida, Iowa, Kansas, Louisiana, Maine, Minnesota, Mississippi, North Dakota, Nebraska, New Hampshire, Oklahoma, South Dakota, Texas, and Vermont. The basis for excluding States is generally that their emissions do not contribute significantly to nonattainment in, or interfere with maintenance by, any other State with respect to any national primary or secondary ambient air quality standard. The notice itself provides a more detailed explanation of the process EPA used to make the proposed determinations.

Question 9: Section 176A of the Clean Air Act contains mechanisms for mitigating ozone transport in regions outside of the Northeast Ozone Transport Region through the formation of regional ozone transport commissions. Under this section, the Administrator may establish a transport region, on the Administrator's own motion, where the Administrator "has reason to believe that the interstate transport of air pollutants from one or more States contributes significantly to a violation of a national ambient air quality standard in one or more States." Why hasn't this provision been utilized to reduce the transport of ozone originating in states outside the Northeast, similar to efforts of the Northeast Ozone Transport Commission created by sections 176A and 184 of the Clean Air Act? Why is EPA's proposed transport rulemaking necessary in view of this existing -- albeit unused -- statutory authority? Has EPA considered whether states in the Northeast Ozone Transport Region must use the section 176A procedures to seek NOx reductions in upwind states?

Response: EPA is well aware of the Section 176A mechanism. Several years ago, however, the Environmental Council of the States (ECOS) recommended formation of a regional workgroup to allow for a thoughtful assessment and development of consensus solutions to the problem of interstate transport of ozone and its precursors. The workgroup, the Ozone Transport Assessment Group, or OTAG, was a collaborative process conducted by representatives from 37 States, EPA, and interested members of the public, including environmental groups and industry, to evaluate the ozone transport problem and develop solutions. Because it brought together all significant stakeholder groups (not just States outside the Northeast) to formulate--and importantly, provide resources for--a thorough analytical assessment of the problem, EPA believed the OTAG process would be a more effective and expeditious process than that provided by the Section 176A mechanism. Given this decision several years ago, EPA's participation in the OTAG process, and the States' support of the OTAG recommendations, EPA believes it appropriate to use other Clean Air Act authority as indicated above at this time to implement measures similar to the OTAG recommendations.

EPA does not believe States in the Northeast Ozone Transport Region must rely solely on section 176A procedures to seek NOx reductions in upwind States. Section 126 of the Clean Air Act provides one mechanism for States to petition EPA to control emissions from sources in States that emit any air pollution in violation of the prohibition of Section 110(a)(2)(D)(I) of the Act. You are probably aware that 8 northeast states have already invoked authority under section 126 of the Clean Air Act to petition EPA to mandate NOx control from specific sources in upwind states; EPA is currently reviewing those petitions.

Question 10(a): Since State SIPs, under section 110(a)(2), must contain enforceable emissions limitations and other control measures as well as a state program to provide for enforcement, what is the basis for your statement before the House Agriculture Committee on September 16, 1997, that "nobody is legally required to reduce their pollution until the year 2009?" When would sources in a state, required to submit a SIP affecting NOx emissions in an OTAG state by March 2000, be "legally required" to reduce their pollution if such a state submits a legally acceptable SIP to the EPA? When would sources in a state be "legally required" to reduce their pollution if EPA issues an "early FIP" for any state not meeting the SIP submission deadline for OTAG states? When would sources in states that are eligible for "transitional" classifications be "legally required" to reduce their pollution? Does EPA believe it could mandate a "cap and trade" program through a FIP?

Response: The statement you quoted from testimony before the House Agriculture Committee on September 16, 1997, regarding the compliance date actually applied to implementation of the standard for PM-2.5, not ozone. The basis for the statement is that EPA plans to collect monitoring data for PM-2.5 starting 1998 - 2003. EPA plans to designate nonattainment areas using the data in the 2002 - 2005 timeframe. States would then be required to submit implementation plans no later than three years after designation (i.e., 2005-2008). Sources would then have to comply with the States' rules after their adoption by the States. With the possible availability of 2 one-year extensions, the attainment date could be up to 12 years following designation, or between 2012 and 2017. Of course, in the meantime, by 2002, EPA will have completed a 5 year scientific review of the particulate matter standard, so no one will have to comply with requirements under the PM-2.5 standard until after that review is complete.

EPA is proposing in the regional NOx SIP call rulemaking a requirement that States submit their NOx emission controls in response to that SIP call by September 1999. Failure of a State to submit an adequate response would subject the state to a number of consequences, including the inability to qualify for the transitional classification.

States affected by the NOx SIP call rulemaking will be required to adopt rules that reduce NOx emissions down to the levels needed to meet the emission budget. The EPA is proposing that sources affected would have to comply with the rules by September 2002. The EPA views seriously its responsibility to address the issue of regional transport of ozone and ozone precursor emissions. Decreases in NOx emissions are needed in the States named in the proposed rulemaking to enable the downwind States to achieve clean air for their citizens. Thus, although the Act allows EPA up to 2 years after the finding to promulgate a FIP, EPA intends to expedite that time frame to help assure that the downwind States realize the air quality benefits as soon as practicable. This is consistent with Congress's intent that attainment occur in these downwind nonattainment areas "as expeditiously as practicable" [sections 181(a), 172(a)]. EPA anticipates that compliance under a FIP would be on a timeframe similar to that in other States where EPA does not implement a FIP.

EPA anticipates that States with transitional classifications that are subject to the NOx SIP call rulemaking would be required to ensure compliance with the necessary emission limitations by the 2002 timeframe specified above. In order to minimize the burden on the electric generating units and other large boilers, EPA would mandate in the FIP an interstate emissions cap and trading program for the affected large stationary sources. The FIP trading program would be designed to be compatible with the emissions trading program anticipated for States that comply with the provisions of the NOx SIP call and therefore do not need a FIP. Development of such emissions trading programs would be a joint EPA/State effort with appropriate stakeholder input, as recommended by OTAG.

Question 10(b): Will the March 2000 SIPs be required to demonstrate attainment of the 1-hour 0.12 ppm standard? If so, what type of demonstration would be required? What would the applicable attainment date be for such SIPs?

Response: The SIP revisions required under the NOx SIP call would be due one year after EPA's final SIP call. These revisions are intended to address transported ozone and NOx precursors and thus only need to address requirements for implementing the statewide NOx budgets. (Note that there will probably not be a "March 2000" SIP as you have assumed in your letter). EPA anticipates requiring no attainment demonstration in that submission, either for the 1-hour or 8-hour standard. EPA's forthcoming final policy on implementing the 1-hour standard is expected to specify a submission date of April 1998 for the attainment demonstrations for the 1-hour standard. The attainment dates for these SIPs are those specified under Title I, Part D, Subpart 2 of the Clean Air Act (e.g., 1999 for serious areas, 2005 or 2007 for severe areas, etc.).

Question 11: It is obvious that there have been significant delays in EPA's schedule regarding its planned SIP call. Under the current schedule, a final rulemaking will not be published until nearly a year following the plans outlined by EPA 10 months ago. Please provide an explanation of the factors which led to the delay of the OTAG SIP call and what elements in the OTAG SIP call have been changed or altered in any way during this delay.

Response: As you may be aware, OTAG was originally scheduled to make its recommendations in September 1996, but those recommendations were delayed until June 1997. The SIP call has been delayed at the request of the OTAG States to accommodate the delay of the OTAG recommendations. EPA's thinking in developing the SIP call has been influenced by the OTAG recommendations. In fact, the SIP call has been developed based on the OTAG underlying analyses and recommendations adopted by the OTAG States. As noted above, EPA's proposed SIP call would require controls similar to those recommended by OTAG; previous EPA thinking on the SIP call did not have the benefit of those OTAG recommendations.

U. S. Department of Justice

Office of Legislative Affairs

Office of the Assistant Attorney General   *Washington, D.C. 20530*

October 21, 1997

The Honorable Thomas J. Bliley, Jr.
Chairman, Committee on Commerce
U.S. House of Representatives
Washington, DC   20515-6115

Dear Mr. Chairman:

This responds to your letter of September 22, 1997 to the
Attorney General concerning the newly revised ozone National
Ambient Air Quality Standard (NAAQS).

The Department was involved in interagency meetings
regarding the issue of implementation of NAAQS.   The interagency
meetings were part of the Office of Management and Budget's (OMB)
review of the standards.   The Environment and Natural Resources
Division (ENRD), as the Department's representative, was one of
many agencies that participated in these meetings.

ENRD generally does not provide formal advisory opinions to
client agencies on pending rulemakings, and did not in this
instance provide a legal opinion to the Environmental Protection
Agency (EPA).   ENRD staff did informally discuss with EPA the
"transitional" classification.   They also informally discussed
with EPA the issue of which areas will fall into nonattainment
status as a result of the revised standards.   More recently, ENRD
reviewed EPA's legal analysis of the implementation plan.   EPA's
analysis was attached to an October 6, 1997, letter from EPA
General Counsel Jonathan Cannon to you.   As we have advised EPA
and Committee staff, it is the Department's position that the
implementation plan set forth in the attachment to the October 6
letter is legally defensible.

In response to your request for documents, we have provided
the Committee with a memorandum which we understand had already
been disclosed to you by the White House.   Also, on October 14,
additional materials were made available for Committee staff
review and Department representatives answered staff questions
about them.   The Department has strong confidentiality interests
in these documents, which include an internal memorandum about
preliminary issues relating to nonattainment areas and Department
attorneys' notes of meetings.   These documents were made
available for staff review in order to accommodate the
Committee's interests, consistent with the Department's long-

The Honorable Thomas J. Bliley, Jr.
Page 2

standing policy regarding such materials. We remain available to staff if there are additional questions regarding these documents. We believe that in light of EPA's recent letter to the Committee, this arrangement should address any remaining questions regarding this matter.

Please do not hesitate to contact me if you would like additional information about this or any other matter.

Sincerely,

Andrew Fois
Assistant Attorney General

cc: The Honorable John D. Dingell
      Ranking Minority Member

UNITED STATES ENVIRONMENTAL PROTECTION AGENCY
WASHINGTON, D.C. 20460

OCT 2 8 1997

The Honorable Michael Bilirakis, Chairman
Subcommittee on Oversight and
  Investigations
Committee on Commerce
U.S. House of Representatives
Washington, D.C. 20515

The Honorable Joe Barton, Chairman
Subcommittee on Health and
  Environment
Committee on Commerce
U.S. House of Representatives
Washington, D.C. 20515

Dear Messrs. Chairmen:

The purpose of this letter is to provide the Environmental Protection Agency's (EPA's) corrections to the transcript of the joint Subcommittees on Health and Environment and Oversight and Investigations hearing on October 1, 1997 concerning the implementation of the new standards for ozone and particulate matter under the Clean Air Act. Also attached for inclusion in the hearing record are copies of documents and information provided in response to the requests that various Members of the two Subcommittees made during the hearing. Please note that in two cases, more time is needed to provide the responses for the reasons noted below. Please also note that the responses listed below were prepared by EPA's Office of Air and Radiation.

The responses to the specific requests included with this letter are as follows:

Request 1:
Provide any legal opinions, memorandums, or writings of any nature whatsoever, regarding the NAAQS implementation plan.

Response:
Documents responsive to this request were made available to the Commerce Committee on October 10th and 16th.

Request 2:
Provide responses to outstanding Committee oversight requests by October 6. Any associated document production should occur as soon as possible.

Response:
Responses to outstanding Commerce Committee oversight requests have been provided or are being addressed in this response. Please see Request 1 for information concerning the status of the document production requests.

Request 3:
Submit for the record the legal citations found in the CAA that provide the authority for the assurances that EPA provided to Representative Kucinich in a letter dated May 16, 1997.

Response:
The answer to this specific request will be provided in EPA's response to a request dated October 9, 1997 from the Honorable John D. Dingell and the Honorable Ron Klink for information concerning RACT for ozone nonattainment areas.

Request 4:
Submit for the record correspondence between EPA and the United Steel Workers of America concerning the NAAQS.

Response:
A copy of the letter is enclosed with this response.

Request 5:
Submit for the record the recent PSE&G study on anticipated cost savings from electricity restructuring and the anticipated cost increases that will accrue through NAAQS implementation.

Response:
A copy of the report is enclosed with this response.

Request 6:
If EPA has drafted contingency plans to address options that might be pursued in the event that the NAAQS were successfully challenged in court, submit such plans for the record.

Response:
There are no plans that are responsive to this request.

Request 7:
Information regarding gross-emitting vehicles from the Mexican border.

Response:
Please see the letter dated August 7, 1997 from Mary D. Nichols to the Honorable Brian Bilbray, which was introduced into the hearing record by Congressman Bilbray. For your convenience, an additional copy of the letter is enclosed.

3

Request 8:

Information regarding the identity of any attorneys from EPA's Office of General Counsel who may have asked environmental groups whether they intended to sue under the Clean Air Act to challenge the implementation strategy for the new regulations.

Response:

Attorneys from the Office of General Counsel discussed with environmental attorneys David Hawkins and Robert Yuhnke whether environmental groups they represent intended to bring suit to challenge elements of the President's implementation strategy that were reflected in the final rules promulgating the revised ambient air quality standards for ozone and particulate matter. The elements in question related to continued implementation of programs that had been implemented under the pre-existing standards. These conversations involved Richard Ossias, Kevin McLean, and Michael Prosper of the Office of General Counsel.

Request 9:

Characterize any oral conversations between the Administrator and her lawyers, or between EPA and the Department of Justice, on this issue. Did the Department of Justice inform EPA that the earliest any citizen suit could be brought involving these standards would be in three years?

Response:

In response to requests from Chairman Bliley, EPA has already provided the Committee all Agency documents that might contain written opinions or correspondence relating to the requested subject matter. When you refer to "citizen suits," we take that term to include not only citizen suits under Section 304 of the Clean Air Act to compel, among other things, performance of a nondiscretionary duty, but also any petition under Section 307(b) of the Act to review a final action of the Administrator in connection with implementation of the NAAQS. EPA attorneys had several conversations with the Department of Justice, among others, at which the possibility of litigation against elements of the implementation strategy was discussed. Those discussions involved the form in which the implementation plan would be adopted. The participants concluded that the implementation plan, which was developed through an interagency process and not as part of the Agency's rulemaking, should not be included in EPA's final rule. Instead, the plan would be issued as a Presidential directive. As such, it would not be subject to immediate judicial review, but its elements would be reviewed by the courts as they were implemented, and after the completion of necessary rulemaking or other procedures. Judicial review of certain elements of the plan could take place as early as next year. The Administrator also had discussions with lawyers at EPA regarding the reviewability of the implementation plan, but our review of documents has disclosed no records of these discussions. We are unaware of any advice from EPA or Department of Justice attorneys to the effect that litigation could not take place for three years.

-4-

In addition to the foregoing requests, a request was made concerning the legal authority for the measures contained in the attachment to the President's Memorandum. Please note that on October 6, 1997 EPA's General Counsel, Jon Cannon, provided a written response to Chairman Bliley's request concerning the legal bases for several key components of the implementation plan attached to the President's Memorandum. EPA's Office of General Counsel is preparing a supplemental response to provide additional information, a copy of which will be provided to you shortly.

I appreciate the opportunity to be of service, and trust that this information will be helpful to you.

Sincerely yours,

Joseph R. Crapa
Associate Administrator
Office of Congressional and Intergovernmental Relations

UNITED STATES ENVIRONMENTAL PROTECTION AGENCY
WASHINGTON, D.C. 20460

OCT 2 9 1997

OFFICE OF
GENERAL COUNSEL

The Honorable Thomas J. Bliley
Chairman, Committee on Commerce
U.S. House of Representatives
Washington, D.C. 20515-6115

Dear Mr. Chairman:

At the October 1, 1997, hearing of the Commerce Committee on implementation of the new ozone and particulate matter standards, EPA Administrator Browner agreed to provide the Committee a writeup citing the sections of the Clean Air Act on which the Agency relies for the interagency implementation plan attached to the July 16, 1997, Presidential memorandum. As you know, I subsequently wrote to you on October 6, 1997, responding to several specific questions related to EPA's legal authority that you had raised in your September 22, 1997, letter. That response sets forth the specific statutory authority on which EPA relies for many aspects of the implementation plan, including authority for the "transitional classification" and several associated requirements.

The attachment to this letter describes the statutory authorities that support remaining portions of the implementation plan related to the revised standards. As with the October 6, 1997 letter, this response generally explains the legal flexibility provided under the Act as relevant to the implementation plan. Based on consultation with the Office of Air and Radiation, the attachment also reflects how current information could be expected to inform the Agency's exercise of its discretion under the Act in carrying out the implementation plan. Where legal authorities described in this and the October 6 letters will involve future rulemakings or other determinations by the Agency, these rulemakings or determinations will be based on the record before the Agency at the time it takes such action.

I appreciate the opportunity to respond to the concerns of the Committee and trust that this information will be helpful to you.

Sincerely,

Jonathan Z. Cannon
General Counsel

Attachments

ATTACHMENT

<u>Dates for Ozone Designations, State Implementation Plan (SIP) Submittals, and Attainment</u>:
The implementation plan (62 Fed. Reg. at 38424)[1] states that "areas would be designated as
nonattainment for the 8-hour standard by 2000 and would submit their nonattainment SIPs by
2003. The Act allows up to 10 years plus two 1-year extensions from the date of designation
for areas to attain the revised NAAQS." The plan (62 Fed. Reg. At 38426) also notes that
areas not qualifying for the transitional classification "will need to submit an implementation
plan within 3 years of designation as nonattainment for the new standard for achieving the 8-
hour standard."

On September 9, 1997, Acting EPA Assistant Administrator Richard Wilson
responded to your request of August 27, 1997, regarding, among other things, the legal
authority for designations under the implementation plan. That response, a copy of which is
attached to this letter, sets forth a detailed description of the specific sections of the Clean
Air Act governing the designation process. In short, under section 107(d)(1)(A), the
Administrator must specify a date, no later than 1 year from promulgation of a revised
standard, for the Governors to submit their recommendations on how each portion of their
States should be designated in relation to the new standard. Section 107(d)(1)(B) then
requires the Administrator to promulgate designations as expeditiously as practicable, but no
later than 2 years from the date of promulgation of the standards, except that period may be
extended for up to one year (i.e., up to 3 years from the date of promulgation of the
standards -- or July 2000) "in the event the Administrator has insufficient information to
promulgate the designations" at the 2-year mark.

EPA will be issuing initial guidance by the end of 1997 to aid States in making the
submissions required pursuant to section 107(d)(1)(A) of the Clean Air Act. Throughout
1998, EPA and States will be analyzing the effects of regional and local transport on existing
or potential nonattainment areas within their borders and EPA will be moving to finalize its
regional NOx rule.[2] States will need time to assess the effects of transport on individual
areas and determine which locations "contribute" to "nearby" monitored violations, as those
terms are used in the definition of "nonattainment area" in section 107(d)(1)(A)(i). For
example, if an area that is monitoring a violation is expected to attain the standard through
implementation of regional NOx measures, it may not be reasonable to include all of the
surrounding suburban counties in the nonattainment area if those counties' contribution to

---

[1] Citations are to President Clinton's Memorandum regarding "Implementation of
Revised Air Quality Standards for Ozone and Particulate Matter," 62 Fed. Reg. 38421 (July
18, 1997).

[2] The regional NOx rule is more fully discussed in the President's implementation
plan; <u>see</u> 62 Fed. Reg. 38425.

nonattainment can be shown to be insignificant in light of the controls to be installed in upwind areas.

EPA expects that the need to await the outcome of the regional NOx rulemaking and other analyses of regional and local transport will result in the Agency's determining in 1999 that it has insufficient information to decide how area boundaries should be drawn for the revised standard. EPA anticipates that upon completion of its regional NOx rulemaking and other regional and local transport analyses, Governors will wish to submit during 1999 revised designation recommendations to account for the results of the transport analyses. EPA also intends to issue additional guidance to States upon conclusion of its regional NOx rule to aid States in drawing nonattainment area boundaries. After Governors submit revised recommendations in 1999, EPA could be expected to need until 2000 to determine finally how the area boundaries should be drawn, to notify the Governors of any expected modifications to their recommendations, and to promulgate the designations.

Section 172(b) states that, at the time the Administrator promulgates a nonattainment designation for an area, she must establish a schedule for the required submission of a SIP for the area "meeting the applicable requirements of [section 172(c)] and section 110(a)(2). Such schedule shall . . . include a date or dates, extending no later than 3 years from the date of the nonattainment designation, for the submission" of that plan. Thus, assuming that EPA promulgates nonattainment designations for the 8-hour ozone standard in 2000, the SIPs for those areas will be due by a date no later than 2003. At present, EPA does not expect that it will call for submission of these SIPs earlier than that 3-year date of 2003.

Section 110(a)(1), read with section 110(a)(2), calls for States to submit SIPs implementing a revised standard meeting section 110(a)(2) within 3 years of <u>promulgation of such standard</u> -- which in this case would be 2000. However, as just described, section 172(b) requires States to submit SIPs meeting section 110(a)(2) no later than 3 years from <u>designation</u>. EPA believes that the best way to reconcile these provisions is to require that the section 110 SIP due 3 years from promulgation of a standard include only those SIP elements of section 110(a)(1) and (2) that are not integrally related to the selection of local controls for an area identified as nonattainment. Under this interpretation, only more general requirements, such as any transport-related SIP that EPA calls for under section 110(a)(2)(D) and any necessary monitoring SIP under section 110(a)(2)(B), would be due by July 2000. The SIPs incorporating any necessary local nonattainment-area controls would not be due until as late as July 2003.

Section 172(a)(2)(A) states that the attainment date for an area designated nonattainment for a standard shall be as expeditiously as practicable, but no later than 5 years from the nonattainment designation, except that the Administrator may extend the date for up to 5 additional years if she deems appropriate, considering the severity of nonattainment and the availability and feasibility of pollution control measures. That provision is the basis for the implementation plan's statement that attainment must be achieved "up to 10 years" from the nonattainment designation. Based on current information, EPA expects that many areas that do not qualify for the transitional classification may need the full 10 years to attain the new standard.

Section 172(a)(2)(C) states that the Administrator may provide "[n]o more than 2 one-year extensions" of the attainment date, consistent with the requirements of that subparagraph. That provides the basis for the implementation plan's reference to "two 1-year extensions" beyond the 10-year maximum attainment date otherwise provided.

<u>Dates for PM-2.5 Designations, SIP Submittals, and Attainment</u>: The implementation plan (62 Fed. Reg. 38428) states:

> Three years of data will be available from the earliest [PM-2.5] monitors in the spring of 2001, and 3 years of data will be available from all monitors in 2004. Following this monitoring schedule and allowing time for data analysis, Governors and the EPA will not be able to make the first determinations as to which areas should be designated nonattainment until at least 2002, 5 years from now. The Clean Air Act, however, requires that the EPA make designation determinations (i.e., attainment, nonattainment, or unclassifiable) within 2 to 3 years of revising a NAAQS. To fulfill this requirement, in 1999 the EPA will issue "unclassifiable" designations for PM-2.5. . . . When the EPA designates PM-2.5 nonattainment areas pursuant to the Governors' recommendations beginning in 2002, areas will be allowed 3 years to develop and submit to the EPA pollution control plans showing how they will meet the new standards. Areas will then have up to 10 years from their redesignation to nonattainment to attain the PM-2.5 standards with the possibility of two 1-year extensions.

As described above in relation to ozone designations, section 107(d)(1)(B) requires EPA to promulgate designations for the new PM-2.5 standards as expeditiously as practicable, but no later than 2 years from promulgation of the new standards, except that up to an additional year may be taken if the Administrator has insufficient information to promulgate the designations at the 2-year mark. As provided in the implementation plan, the three years of Federal reference method monitoring data necessary to determine whether areas are attaining the standard will not be available for any areas until 2002, at the earliest. In the absence of such monitoring data, EPA will not have sufficient data or information on which to base nonattainment designations for PM-2.5 at the 2-year mark (July 1999), and does not expect to have the information it needs within the additional year. Consequently, the Agency believes it would be appropriate to proceed with promulgating designations of "unclassifiable" by July 1999. The "unclassifiable" designation is appropriate in the absence of adequate Federal reference method monitoring data because that designation applies whenever the area "cannot be classified on the basis of available information as meeting or not meeting the [standard]." CAA section 107(d)(1)(A)(iii).

Section 107(d)(3) sets forth a process under which Governors may (on their own initiative or at EPA's request) submit recommendations to revise the designation of an area previously designated. Under that process, whether or not Governors submit recommendations, EPA is empowered to promulgate redesignations of areas from unclassifiable to nonattainment for PM-2.5 once monitoring data reveal that areas are violating a PM-2.5 standard. That process provides the basis for the implementation plan's reference to designating PM-2.5 nonattainment areas pursuant to the Governors' recommendations beginning in 2002. As described above, sections 172(a) and (b) provide the basis for the plan's statements that the resulting PM-2.5 nonattainment areas will be allowed 3 years (from the nonattainment designation) to submit attainment SIPs and up to 10 years (with the possibility of two 1-year extensions) to attain.

OTAG NOx SIP Call: The implementation plan (62 Fed. Reg. 38425) states EPA's intent to issue by September 1998 a rule "requiring States in the OTAG region that are significantly contributing to nonattainment or interfering with maintenance of attainment in downwind States to submit SIPs to reduce their interstate pollution." (A proposed rule to this effect was announced on October 10, 1997.) The plan then refers to this "regional strategy" as the basis for certain aspects of the transitional classification that EPA intends to create for some types of areas.

On October 9, 1997, Acting Assistant Administrator Wilson responded to your September 22, 1997, request regarding the OTAG NOx SIP call. That response touches on several issues, but a more expansive treatment of the legal issues is appropriate here. Section 110(a)(2)(D) requires SIPs to prohibit emissions activities that "contribute significantly to nonattainment [of a national ambient air quality standard] in, or interfere with maintenance [of such a standard] by, any other State . . . ." The proposed rule that EPA announced on October 10 would require 22 States and the District of Columbia to submit to EPA SIP revisions containing controls to reduce NOx emissions which have been proven to cross State boundaries and which are contributing significantly to downwind violations of the ozone standard. EPA is authorized to require these SIP revisions to implement the 1-hour ozone standard, under section 110(k)(5), which may be termed the "SIP call" provision. Through this SIP call, as proposed, EPA would find current SIPs for the 1-hour ozone standard to be inadequate to address the NOx contribution their sources are making, and EPA would mandate additional NOx reductions. In addition, EPA has proposed that the same States and the District of Columbia address section 110(a)(2)(D) for the new 8-hour ozone standard by meeting the same NOx budgets as those proposed for the 1-hour standard. EPA has the authority under sections 110(a)(1) and 301 to issue a rule (as proposed on October 10, 1997) at this time requiring these types of SIP revisions under the new ozone standard. While EPA would be requiring these States to achieve a specified amount of reductions in NOx emissions, EPA would mandate neither the type of sources that the State must control nor the emissions reductions program that the State must use to provide for these reductions.

Section 110(k)(5) provides for States to submit SIPs satisfying a SIP call by a reasonable date established by the Administrator, not exceeding 18 months from the SIP call. Section 110(a)(1) requires States to submit at least the early SIP elements needed to implement a new standard (as described above) by the date 3 years from promulgation of the standard, "or such shorter period as the Administrator may prescribe." Pursuant to both of these provisions, EPA has proposed to require the submission of these section 110(a)(2)(D) NOx-reduction SIPs no later than 1 year from the issuance of the final rule calling for the SIPs.

EPA has proposed that sources affected would have to comply with the resulting SIPs by September 2002 (although EPA is soliciting comment on a range that extends from then until September 2004). The proposed schedule is consistent with section 172(a)'s requirement that attainment of a new or revised standard occur as expeditiously as practicable.

<u>Areas Not Eligible for the Transitional Classification</u>: The implementation plan (62 Fed. Reg. 38426) states that, "[w]hile the additional local reductions that [these areas] will need to achieve the 8-hour standard must occur prior to their 8-hour attainment date (e.g., 2010), for virtually all areas the additional reductions needed to achieve the 8-hour standard can occur after the 1-hour attainment date. This approach allows them to make continued progress toward attaining the 8-hour standard until the 1-hour standard is attained. These areas, however, will need to submit an implementation plan within 3 years of designation as nonattainment for the new standard for achieving the 8-hour standard. Such a plan can rely in large part on measures needed to attain the 1-hour standard."

Section 172(c)(2) provides that nonattainment areas should make continued progress toward achieving emission reductions in the years preceding the attainment date by providing that SIPs "shall require reasonable further progress." Section 171(1) defines "reasonable further progress" as "such annual incremental reductions in emissions of the relevant air pollutant as are required by this part or may reasonably be required by the Administrator for the purpose of ensuring attainment of the applicable national ambient air quality standard by the applicable date."

EPA believes that, for all areas except the Los Angeles area, the emissions reductions that non-transitional areas would achieve by implementing the regional NOx strategy and the currently applicable provisions of subpart 2 of part D of title I of the Act (including reasonably available control technology, new source review, and the numerical rate-of-progress requirements) will provide enough annual incremental emissions reductions for the purpose of ensuring attainment of the 8-hour standard by the applicable date. Thus, EPA expects that any additional local controls that such areas will need to attain the 8-hour standard will not need to be applied prior to an area's attainment date for the 1-hour standard in order to meet the requirement for reasonable further progress; it will be sufficient for those controls to be applied in the remaining years between the 1-hour standard attainment date and the attainment date for the 8-hour standard. In this way, the SIPs can be anticipated to provide for both reasonable further progress and timely attainment.

<u>Conformity</u>: The implementation plan (62 Fed. Reg. 38425) states that EPA will revise its rules for conformity so that States will be able to comply with only minor revisions to their existing programs in areas classified as transitional.

As stated in an August 1, 1997, letter from Mary Nichols to Representatives Dingell and Klink, and in accordance with section 176(c)(5), the conformity requirements in section 176(c) apply to all areas designated nonattainment, including transitional areas. EPA believes it has the authority, in developing a classification scheme, to interpret and apply the provisions of subpart 1, including the conformity requirement, in a way that recognizes the particular circumstances of areas within each classification.

Furthermore, EPA believes that the substantive requirements of section 176(c) provide flexibility for the Agency to provide simplified, streamlined conformity procedures for transitional areas. For example, EPA is exploring whether it might provide the technical basis -- including projections that the emissions reductions provided by the regional NOx strategy would be sufficient to accommodate all emissions growth expected to result from planned transportation programs -- to allow local and federal agencies to make conformity determinations without significant additional technical work.

EPA has not yet selected the procedures that will apply for these areas, but will be doing so in revisions to the existing conformity rules to be published by December 1998. In that rulemaking, EPA will ensure that the alternative conformity procedures for transitional areas comply with the statutory requirements of section 176(c), as applied in light of applicable principles of statutory construction.

(197)

Identification of studies used by the U.S. Environmental Protection Agency in the promulgation of a National Ambient Air Quality Standard for PM$_{2.5}$

REFERENCES

1. Abbey, D. E.; Mills, P. K.; Petersen, F. F.; Beeson, W. L. (1991a) Long-term ambient concentrations of total suspended particulates and oxidants as related to incidence of chronic disease in California Seventh-Day Adventists. Environ. Health Perspect. 94: 43-50.

2. Abbey, D. E.; Lebowitz, M. D.; Mills, P. K.; Petersen, F. F.; Beeson, W. L.; Burchette, R. J. (1995a) Long-term ambient concentrations of particulates and oxidants and development of chronic disease in a cohort of nonsmoking California residents. In: Phalen, R. F.; Bates, D. V., eds. Proceedings of the colloquium on particulate air pollution and human mortality and morbidity; January 1994; Irvine, CA. Inhalation Toxicol. 7: 19-34.

3. Abbey, D. E.; Ostro, B. E.; Petersen, F.; Burchette, R. J. (1995b) Chronic respiratory symptoms associated with estimated long-term ambient concentrations of fine particulates less than 2.5 microns in aerodynamic diameter (PM2.5) and other air pollutants. J. Exp. Anal. Environ. Epidemiol. 5: 137-159.

4. Abbey, D. E.; Ostro, B. E.; Fraser, G.; Vancuren, T.; Burchette, R. J. (1995c) Estimating fine particulates less that 2.5 microns in aerodynamic diameter (PM2.5) from airport visibility data in California. J. Exp. Anal. Environ. Epidemiol. 5: 161-180.

5. Ackermann-Liebrich, U.; Leuenberger, P.; Schwartz, J.; Schindler, C.; Monn, C.; Bolognini, B.; Bongard, J. P.; Brändli, O.; Domenighetti, G.; Elsasser, S.; Grize, L.; Karrer, W.; Keller, R.; Keller-Wossidlo, H.; Künzli, N.; Martin, B. W.; Medici, T. G.; Perruchoud, A. P.; Schöni, M. H.; Tschopp, J. M.; Villiger, B.; Wüthrich, B.; Zellweger, J. P.; Zemp, E. (1996) Lung function and long term exposure to air pollutants in Switzerland. Am. J. Respir. Crit. Care Med: submitted

6. Braun-Fahrländer, C.; Ackermann-Liebrich, U.; Schwartz, J.; Gnehm, H. P.; Rutishauser, M.; Wanner, H. U. (1992) Air pollution and respiratory symptoms in preschool children. Am. Rev. Respir. Dis. 145: 42-47.

7. Burnett, R. T.; Dales, R. E.; Raizenne, M. E.; Krewski, D.; Summers, P. W.; Roberts, G. R.; Raad-Young, M.; Dann, T.; Brook, J. (1994) Effects of low ambient levels of ozone and sulfates on the frequency of respiratory admissions to Ontario hospitals. Environ. Res. 65: 172-194.

8. Burnett, R. T.; Dales, R.; Krewski, D.; Vincent, R.; Dann, T.; Brook, J. R. (1995) Associations between ambient particulate sulfate and admissions to Ontario hospitals for cardiac and respiratory diseases. Am. J. Epidemiol. 142: 15-22.

9. Cifuentes, L.; Lave, L. B. (1996) Association of daily mortality and air pollution in Philadelphia, 1983-1988. J. Air Waste Manage. Assoc.: in press.

10. Dassen, W.; Brunekreef, B.; Hoek, G.; Hofschreuder, P.; Staatsen, B.; De Groot, H.; Schouten, E.; Biersteker, K. (1986) Decline in children's pulmonary function during an air pollution episode. J. Air Pollut. Control Assoc. 36: 1223-1227.

11. Derriennic, F.; Richardson, S.; Mollie, A.; Lellouch, J. (1989) Short-term effects of sulphur dioxide pollution on mortality in two French cities. Int. J. Epidemiol. 18: 186-197

12. Dockery, D. W.; Ware, J. H.; Ferris, B. G., Jr.; Speizer, F. E.; Cook, N. R.; Herman, S. M. (1982) Change in pulmonary function in children associated with air pollution episodes. J. Air Pollut. Control Assoc. 32: 937-942.

13. Dockery, D. W.; Speizer, F. E.; Stram, D. O.; Ware, J. H.; Spengler, J. D.; Ferris, B. G., Jr. (1989) Effects of inhalable particles on respiratory health of children. Am. Rev. Respir. Dis. 139: 587-594.

14. Dockery, D. W.; Schwartz, J.; Spengler, J. D. (1992) Air pollution and daily mortality: associations with particulates and acid aerosols. Environ. Res. 59: 362-373.

15. Dockery, D. W.; Pope, C. A., III; Xu, X., Spengler, J. D.; Ware, J. H.; Fay, M. E.; Ferris, B. G., Jr.; Speizer, F. E. (1993) An association between air pollution and mortality in six U.S. cities. N. Engl. J. Med. 329: 1753-1759.

16. Dockery, D. W.; Cunningham, J.; Damokosh, A. I.; Neas, L. M.; Spengler, J. D.; Koutrakis, P.; Ware, J. H.; Raizenne, M.; Speizer, F. E. (1996) Health effects of acid aerosols on North American children: respiratory symptoms. Environ. Health Perspect: in press.

17. Dusseldorp, A.; Kruize, H.; Brunekreef, B.; Hofschreuder, P.; de Meer, G., van Oudvorst, A. B. (1994) Associations of PM10 and airborne iron with respiratory health of adults living near a steel factory. Am. J. Respir. Crit. Care Med. 152: 1932-1939.

18. Fairley, D. (1990) The relationship of daily mortality to suspended particulates in Santa Clara county, 1980-86. Environ. Health Perspect. 89: 159-168.

19. Gordian, M. E.; Özkaynak, H.; Xue, J.; Morris, S. S.; Spengler, J. D. (1996) Particulate air pollution and respiratory disease in Anchorage, Alaska. Environ. Health Perspect. 104: 209-297.

20. Hefflin, B. J.; Jalaludin, B.; McClure, E.; Cobb, N.; Johnson, C. A.; Jecha, L.; Etzel, R. A. (1994) Surveillance for dust storms and respiratory diseases in Washington State, 1991. Arch. Environ. Health 49: 170-174.

21. Hoek, G.; Brunekreef, B. (1993) Acute effects of a winter air pollution episode on pulmonary function and respiratory symptoms of children. Arch. Environ. Health 48: 328-335.

22. Hoek, G.; Brunekreef, B. (1994) Effects of low-level winter air pollution concentrations on respiratory health of Dutch children. Environ. Res. 64: 136-150.

23. Hoek, G.; Brunekreef, B. (1995) Effect of photochemical air pollution on acute respiratory symptoms in children. Am. J. Respir. Crit. Care Med. 151. 27-32.

24. Ito, K.; Thurston, G. D. (1996) Daily PM10/mortality associations: an investigation of at-risk sub-populations. J. Exposure Anal. Environ. Epidemiol.: in press.

25. Ito, K.; Thurston, G. D.; Hayes, C.; Lippmann, M. (1993) Associations of London, England, daily mortality with particulate matter, sulfur dioxide, and acidic aerosol pollution. Arch. Environ. Health 48: 213-220.

26. Ito, K., Kinney, P.; Thurston, G. D. (1995) Variations in PM-10 concentrations within two metropolitan areas and their implications for health effects analyses. In: Phalen, R. F.; Bates, D. V., eds. Proceedings of the colloquium on particulate air pollution and human mortality and morbidity, part II; January 1994; Irvine, CA. Inhalation Toxicol. 7: 735-745.

27. Katsouyanni, K.; Karakatsani, A.; Messari, I.; Touloumi, G.; Hatzakis, A.; Kalandidi, A.; Trichopoulos, D. (1990a) Air pollution and cause specific mortality in Athens. J. Epidemiol. Commun. Health 44: 321-324.

28.   Katsouyanni, K.; Hatzakis, A.; Kalandidi, A.; Trichopoulos, D. (1990b) Short-term effects of atmospheric pollution on mortality in Athens. Arch. Hellen. Med. 7: 126-132.

29.   Katsouyanni, K.; Pantazopoulou, A.; Touloumi, G.; Tselepidaki, I.; Moustris, K.; Asimakopoulos, D.; Poulopoulou, G.; Trichopoulos, D. (1993) Evidence for interaction between air pollution and high temperature in the causation of excess mortality. Arch. Environ. Health 48: 235-242.

30.   Kinney, P. L.; Özkaynak, H. (1991) Associations of daily mortality and air pollution in Los Angeles County. Environ. Res. 54: 99-120.

31.   Kinney, P. L.; Ito, K.; Thurston, G. D. (1995) A sensitivity analysis of mortality/PM_{10} associations in Los Angeles. In: Phalen, R. F.; Bates, D. V., eds. Proceedings of the colloquium on particulate air pollution and human mortality and morbidity; January 1994; Irvine, CA. Inhalation Toxicol. 7: 59-69

32.   Koenig, J. Q.; Larson, T. V.; Hanley, Q. S.; Rebolledo, V.; Dumler, K.; Checkoway, H.; Wang, S.-Z.; Lin, D.; Pierson, W. E. (1993) Pulmonary function changes in children associated with fine particulate matter. Environ. Res. 63: 26-38.

33.   Li, Y.; Roth, H. D. (1995) Daily mortality analysis by using different regression models in Philadelphia County, 1973-1990. In: Phalen, R. F.; Bates, D. V., eds. Proceedings of the colloquium on particulate air pollution and human mortality and morbidity; January 1994; Irvine, CA. Inhalation Toxicol. 7: 45-58.

34.   Lyon, J. L.; Mori, M.; Gao, R. (1995) Is there a causal association between excess mortality and exposure to PM-10 air pollution? Additional analyses by location, year, season and cause of death. In: Phalen, R. F.; Bates, D. V., eds. Proceedings of the colloquium on particulate air pollution and human mortality and morbidity, part II, January 1994; Irvine, CA. Inhalation Toxicol. 7: 603-614.

35.   Moolgavkar, S. H.; Luebeck, E. G.; Hall, T. A.; Anderson, E. L. (1995a) Particulate air pollution, sulfur dioxide, and daily mortality: a reanalysis of the Steubenville data. In: Phalen, R. F.; Bates, D. V., eds. Proceedings of the colloquium on particulate air pollution and human mortality and morbidity; January 1994; Irvine, CA. Inhalation Toxicol. 7: 35-44

36.   Moolgavkar, S. H.; Luebeck, E. G.; Hall, T. A.; Anderson, E. L. (1995b) Air pollution and daily mortality in Philadelphia. Epidemiology 6: 476-484.

37.   Neas, L. M.; Dockery, D. W.; Ware, J. H.; Spengler, J. D.; Ferris, B. G., Jr.; Speizer, F. E. (1994) Concentration of indoor particulate matter as a determinant of respiratory health in children. Am. J. Epidemiol. 139: 1088-1099.

38.   Neas, L. M.; Dockery, D. W.; Koutrakis, P.; Tollerud, D. J.; Speizer, F. E. (1995) The association of ambient air pollution with twice daily peak expiratory flow rate measurements in children. Am. J. Epidemiol. 141: 111-122.

39.   Ostro, B. D.; Lipsett, M. J.; Wiener, M. B.; Selner, J. C. (1991) Asthmatic responses to airborne acid aerosols. Am. J. Public Health 81: 694-702.

40.   Ostro, B. D.; Lipsett, M. J.; Mann, J. K.; Krupnick, A.; Harrington, W. (1993) Air pollution and respiratory morbidity among adults in Southern California. Am. J. Epidemiol. 137: 691-700.

41.   Ostro, B. D.; Lipsett, M. J.; Mann, J. K.; Braxton-Owens, H.; White, M. C. (1995) Air pollution and asthma exacerbations among African-American children in Los Angeles. In: Phalen, R. F.; Bates, D. V., eds. Proceedings of the colloquium on particulate air pollution and human mortality and morbidity, part II; January 1994; Irvine, CA. Inhalation Toxicol. 7: 711-722

42   Ostro, B : Sanchez, J. M.; Aranda, C.; Eskeland, G. S. (1996) Air pollution and mortality:
      results from a study of Santiago, Chile In: Lippmann, M., ed. Papers from the ISEA-ISEE
      annual meeting; September 1994; Research Triangle Park, NC. J. Exposure Anal. Environ.
      Epidemiol. in press.

43.   Özkaynak, H.; Xue, J.; Severance, P.; Burnett, R.; Raizenne, M. (1994) Associations
      between daily mortality, ozone, and particulate air pollution in Toronto, Canada.
      Presented at: Colloquium on particulate air pollution and human mortality and morbidity:
      program and abstracts; January; Irvine, CA. Irvine, CA: University of California Irvine,
      Air Pollution Health Effects Laboratory; p. P1.13; report no. 94-02.

44   Pope, C. A., III. (1991) Respiratory hospital admissions associated with PM_{10} pollution in
      Utah, Salt Lake, and Cache Valleys. Arch. Environ. Health 46: 90-97.

45   Pope, C. A., III; Dockery, D. W. (1992) Acute health effects of PM_{10} pollution on
      symptomatic and asymptomatic children. Am. Rev. Respir. Dis. 145: 1123-1128.

46.   Pope, C. A., III; Kalkstein, L. S. (1996) Synoptic weather modeling and estimates of the
      exposure-response relationship between daily mortality and particulate air pollution.
      Environ. Health Perspect. 104: in press.

47   Pope, C. A., III; Kanner, R. E. (1993) Acute effects of PM_{10} pollution on pulmonary
      function of smokers with mild to moderate chronic obstructive pulmonary disease. Am.
      Rev. Respir. Dis. 147: 1336-1340.

48   Pope, C. A., III; Schwartz, J.; Ransom, M. R. (1992) Daily mortality and PM_{10} pollution
      in Utah valley. Arch. Environ. Health 47: 211-217

49   Pope, C. A., III; Thun, M. J., Namboodiri, M. M.; Dockery, D. W.; Evans, J. S.; Speizer,
      F. E.; Heath, C. W., Jr. (1995b) Particulate air pollution as a predictor of mortality in a
      prospective study of U.S. adults. Am. J. Respir. Crit. Care Med. 151: 669-674.

50.   Quackenboss, J. J.; Krzyzanowski, M.; Lebowitz, M. D. (1991) Exposure assessment
      approaches to evaluate respiratory health effects of particulate matter and nitrogen
      dioxide. J. Exposure Anal. Environ. Epidemiol. 1: 83-107.

51.   Raizenne, M.; Neas, L. M.; Damokosh, A. I.; Dockery, D. W.; Spengler, J. D.; Koutrakis,
      P.; Ware, J. H.; Speizer, F. E. (1996) Health effects of acid aerosols on North American
      children: pulmonary function. Environ. Health Perspect.: accepted.

52.   Roemer, W., Hoek, G.; Brunekreef, B. (1993) Effect of ambient winter air pollution on
      respiratory health of children with chronic respiratory symptoms Am. Rev. Respir. Dis.
      147: 118-124.

53   Saldiva, P. H. N.; Lichtenfels, A. J. F. C.; Paiva, P. S. O.; Barone, I. A.; Martins, M. A.;
      Massad, E.; Pereira, J. C. R.; Xavier, V. P., Singer, J. M.; Böhm, G. M. (1994)
      Association between air pollution and mortality due to respiratory diseases in children in
      São Paulo, Brazil: a preliminary report. Environ. Res. 65: 218-225.

54.   Saldiva, P. H. N.; Pope, C. A., III; Schwartz, J.; Dockery, D. W.; Lichtenfels, A. J.; Salge,
      J. M.; Barone, I; Bohm, G. M. (1995) Air pollution and mortality in elderly people: a
      time-series study in São Paulo, Brazil. Arch. Environ. Health 50: 159-163.

55.   Samet, J. M.; Zeger, S. L.; Berhane, K. (1995) The association of mortality and
      particulate air pollution. In: Particulate air pollution and daily mortality: replication and
      validation of selected studies, the phase I report of the particle epidemiology evaluation
      project [preprint]. Cambridge, MA: Health Effects Institute; pp. 1-104

56.   Samet, J. M.; Zeger, S. L.; Kelsall, J. E.; Xu, J. (1996a) Air pollution and mortality in
      Philadelphia, 1974-1988, report to the Health Effects Institute on phase IB: Particle
      Epidemiology Evaluation Project. Cambridge, MA: Health Effects Institute; accepted

57.   Samet, J. M.; Zeger, S. L.; Kelsall, J. E., Xu, J.; Kalkstein, L. S. (1996b) Weather, air pollution and mortality in Philadelphia, 1973-1980, report to the Health Effects Institute on phase IB, Particle Epidemiology Evaluation Project. Cambridge, MA: Health Effects Institute; review draft.

58.   Schwartz, J. (1991a) Particulate air pollution and daily mortality in Detroit. Environ. Res. 56: 204-213.

59.   Schwartz, J. (1993a) Air pollution and daily mortality in Birmingham, Alabama. Am. J. Epidemiol. 137: 1136-1147.

60   Schwartz, J. (1994a) Total suspended particulate matter and daily mortality in Cincinnati, Ohio. Environ. Health Perspect. 102: 186-189

61.   Schwartz, J. (1994d) Air pollution and hospital admissions for the elderly in Detroit, Michigan. Am. J. Respir. Crit. Care Med. 150: 648-655.

62.   Schwartz, J. (1994e) Air pollution and hospital admissions for the elderly in Birmingham, Alabama. Am. J. Epidemiol. 139: 589-598.

63   Schwartz, J. (1994f) PM_{10}, ozone, and hospital admissions for the elderly in Minneapolis, MN. Arch. Environ. Health 49: 366-374.

64   Schwartz, J. (1994g) Nonparametric smoothing in the analysis of air pollution and respiratory illness. Can. J. Stat. 22: 1-17.

65.   Schwartz, J. (1995a) Short term fluctuations in air pollution and hospital admissions of the elderly for respiratory disease. Thorax 50: 531-538.

66.   Schwartz, J. (1996) Air pollution and hospital admissions for respiratory disease. Epidemiology 7: 20-28.

67.   Schwartz, J.; Dockery, D. W. (1992a) Increased mortality in Philadelphia associated with daily air pollution concentrations. Am. Rev. Respir. Dis. 145: 600-604.

68.   Schwartz, J.; Dockery, D. W. (1992b) Particulate air pollution and daily mortality in Steubenville, Ohio. Am. J. Epidemiol. 135: 12-19

69.   Schwartz, J.; Morris, R. (1995) Air pollution and hospital admissions for cardiovascular disease in Detroit, Michigan. Am. J. Epidemiol. 142: 23-35.

70.   Schwartz, J.; Spix, C.; Wichmann, H. E.; Malin, E. (1991a) Air pollution and acute respiratory illness in five German communities. Environ. Res. 56: 1-14.

71.   Schwartz, J.; Slater, D.; Larson, T. V.; Pierson, W. E.; Koenig, J. Q. (1993) Particulate air pollution and hospital emergency room visits for asthma in Seattle. Am. Rev. Respir. Dis. 147: 826-831.

72.   Schwartz, J.; Dockery, D. W.; Neas, L. M.; Wypij, D.; Ware, J. H.; Spengler, J. D.; Koutrakis, P.; Speizer, F. E.; Ferris, B. G., Jr. (1994) Acute effects of summer air pollution on respiratory symptom reporting in children. Am. J. Respir. Crit. Care Med. 150: 1234-1242.

73.   Schwartz, J.; Dockery, D. W.; Neas, L. M. (1996a) Is daily mortality associated specifically with fine particles? J. Air Waste Manage. Assoc.: accepted.

74.   Shumway, R. H.; Azari, A. S.; Pawitan, Y. (1988) Modeling mortality fluctuations in Los Angeles as functions of pollution and weather effects. Environ. Res. 45: 224-241.

75. Spektor, D. M.; Hofmeister, V. A.; Artaxo, P., Brague, J. A. P.; Echelar, F.; Nogueira, D. P.; Hayes, C.; Thurston, G. D.; Lippmann, M. (1991) Effects of heavy industrial pollution on respiratory function in the children of Cubatao, Brazil: a preliminary report. Environ. Health Perspect. 94: 51-54

76. Spix, C.; Heinrich, J.; Dockery, D.; Schwartz, J.; Völkach, G.; Schwinkowski, K.; Cöllen, C.; Wichmann, H. E. (1993) Air pollution and daily mortality in Erfurt, East Germany, 1980-1989. Environ. Health Perspect. 101: 518-526.

77. Studnicka, M. J.; Frischer, T.; Meinert, R.; Studnicka-Benke, A.; Hajek, K.; Spengler, J. D.; Neumann, M. G. (1995) Acidic particles and lung function in children: a summer camp study in the Austrian Alps. Am. J. Respir. Crit. Care Med. 151: 423-430.

78. Styer, P.; McMillan, N.; Gao, F.; Davis, J., Sacks, J. (1995) The effect of airborne particulate matter on daily death counts. Environ. Health Perspect. 103: 490-497.

79. Sunyer, J.; Sáez, M.; Murillo, C.; Castellsague, J.; Martinez, F.; Antó, J. M. (1993) Air pollution and emergency room admissions for chronic obstructive pulmonary-disease: a 5-year study. Am. J. Epidemiol. 137: 701-705.

80. Thurston, G. D.; Ito, K.; Lippmann, M.; Hayes, C. (1989) Reexamination of London, England, mortality in relation to exposure to acidic aerosols during 1963-1972 winters. In: Symposium on the health effects of acid aerosols; October 1987; Research Triangle Park, NC. Environ. Health Perspect. 79. 73-82.

81. Thurston, G. D.; Ito, K.; Kinney, P. L.; Lippmann, M. (1992) A multi-year study of air pollution and respiratory hospital admissions in three New York State metropolitan areas: results for 1988 and 1989 summers. J. Exposure Anal. Environ. Epidemiol. 2: 429-450

82. Thurston, G. D.; Gorczynski, J. E., Jr.; Currie, J. H.; He, D.; Ito, K.; Hipfner, J.; Waldman, J.; Lioy, P. J., Lippmann, M. (1994a) The nature and origins of acid summer haze air pollution in metropolitan Toronto, Ontario. Environ. Res. 65: 254-270.

83. Thurston, G. D.; Ito, K.; Hayes, C. G.; Bates, D. V.; Lippmann, M. (1994b) Respiratory hospital admissions and summertime haze air pollution in Toronto, Ontario: consideration of the role of acid aerosols. Environ. Res. 65: 271-290.

84. Touloumi, G.; Pocock, S. J.; Katsouyanni, K.; Trichopoulos, D. (1994) Short-term effects of air pollution on daily mortality in Athens: a time-series analysis. Int. J. Epidemiol. 23: 957-967.

85. Ware, J. H.; Ferris, B. G., Jr.; Dockery, D. W.; Spengler, J. D.; Stram, D. O.; Speizer, F. E. (1986) Effects of ambient sulfur oxides and suspended particles on respiratory health of preadolescent children. Am. Rev. Respir. Dis. 133: 834-842.

86. Wyzga, R. E.; Lipfert, F. W. (1995b) Temperature-pollution interactions with daily mortality in Philadelphia. In: Particulate matter: health and regulatory issues: proceedings of an international specialty conference; April; Pittsburgh, PA. Pittsburgh, PA: Air & Waste Management Association; pp. 3-42. (A&WMA publication VIP-49).

87. Xu, X.; Gao, J.; Dockery, D. W.; Chen, Y. (1994) Air pollution and daily mortality in residential areas of Beijing, China. Arch. Environ. Health 49: 216-222.

Ambient Particles and Health:
Lines That Divide

Dr. Sverre Vedal, MD, MSc
Professor of Medicine
University of British Columbia

INTRODUCTION

Increases in ambient particle concentrations are associated with an array of adverse health outcomes. These range from the least adverse, such as increases in symptoms of respiratory irritation and small decreases in level of lung function, to the most adverse: mortality. Since most of the data supporting this association comes from observational studies, there has been legitimate concern that the observed association may not reflect physical or chemical causes. Arguments against the association being causal include: (1) use of inappropriate statistical methodology, (2) inability to account for other factors such as meteorology and co-pollutants, and (3) lack of biological plausibility. Arguments have supported effects from particle features such as size, acidity, and emission source. Since much of the observational data supporting the effects of these particle features on health come from studies in the eastern United States, the application of these observations to other settings is also a reasonable concern.

The 1997 Critical Review focuses on the lines of division that characterize much of the discussion on particle health effects. It addresses differences of opinion about the interpretation of health studies and the methods that should be used to analyze health indicators and ambient air quality data. It describes attempts to divide particles into those with lesser or greater pathogenicity based on particle size or composition. This includes geographical divisions that separate "eastern" from "western" in North America, and that separate North America from Europe and the rest of the world. Particle emission sources, meteorology, terrain, and consequently particle size and composition show substantial differences for different regions. Finally, the review evaluates the concept of a "line," or air quality standard, defined by particle concentration that divides concentrations to which people are exposed into those thought to cause adverse effects from those that do not. The Critical Review intends to provide insight, especially for those who are not epidemiologists, into the role of epidemiology, the concept of causation, the roles of professional and geographic perspective, and alternative approaches to expressing air quality standards or objectives.

HOW WE KNOW WHAT WE KNOW

The philosophy behind risk assessment is to accept uncertainty as a given and to make use of relevant scientific data to calculate estimates of risk for the purpose of guiding public health policy. Epidemiological data play a central role in guiding health policy. This has required the use of judgment and generation of consensus in interpreting the epidemiological data. Some (taking the determinist's perspective) have bemoaned the relative lack of rigor used in arguing for causation when epidemiology serves the interests of public

> *"The study of the health effects of ambient particles is plagued by a weak biological foundation."*

health policy. Further, this lack of rigor has, it is argued, critically weakened the more fitting role of epidemiology in contributing to our knowledge of disease.

When serving policy interests, the validity of each individual study does not become critical in making a case for causation. Instead, arguments are advanced that involve judgment and seek consistency, coherence and plausibility for the entire body of research in an area. Reference to the sheer bulk

<u>of studies</u> in support of such criteria therefore has value in this setting. Well-performed studies in which no associations are observed, if they happen to be relatively uncommon, can no longer successfully refute arguments of causality. The different modes of causal reasoning have been contrasted by describing the more traditional mode as attempting to build a <u>chain</u> of well-validated causes as opposed to the more modern mode of building a <u>mosaic</u> of evidence, no single element of which needs be indisputable.

The study of the health effects of ambient particles is plagued by a weak biological foundation. This makes it tempting to base what understanding we have on the results of statistical mod-

Critical Review Published in Journal and Presented at Annual Meeting

The complete 1997 Critical Review is published in the May 1997 *Journal of the Air & Waste Management Association*. Dr. Sverre Vedal will present the Critical Review in its entirety at the Association's Annual Meeting on Wednesday, June 11, in Room 107 Convention Centre in Toronto, Ontario, Canada.

As part of the Critical Review process, various responses to Dr. Vedal's discussion will be presented including comments from the floor. All final responses will be published in the September 1997 issue of the *Journal of the Air & Waste Management Association*. Dr. John Watson, Desert Research Institute, is the chairman of the Association's Critical Review Subcommittee.

for which the available data are not adequate.

An observational study is one in which the investigator does not control assignment of the exposure (or treatment) in the study subjects. In an experimental study, in contrast, the investigator is free to determine how exposure is distributed among the study subjects. Random assignment of the exposure (randomization) is a common way of distributing exposure in an experimental study since, if successful, it counters potential biases resulting from subjects with varying exposures harboring important differences in characteristics other than the differences in exposure. The strength of randomizing exposure, however, is not merely in preventing bias due to factors of which the investigator is aware and can measure (such as cigarette smoking), but in preventing bias due to factors of which the investigator is either unaware or cannot measure. These "hidden" confounders cannot be accounted for in an observational study, whereas analytic techniques are available for countering the effects introduced by the known and measured confounders in observational studies. When successful, randomization equalizes the distribution of <u>all</u> confounding factors across exposure levels, whether these are known or "hidden."

In the type of time series epidemiological study often applied in air pollution health research, frequently measured pollution concentrations are compared to frequent and regular measurements of an adverse health outcome. A typical example consists of one series of daily, 24-hour particle concentration measurements and another series of daily counts of deaths. The goal of the analysis of such a time series study is to evaluate the association between the particle series and the mortality series while attempting to control for effects of other time-varying factors that might result in a spurious association between the two time series.

In a cohort epidemiological study, a sample population is identified, exposure is measured and estimated for the members of the sample, and the sample is then observed over a period of time for the occurrence of an adverse health outcome in the individual sample subjects. For example, a sample population from geographic regions experiencing different particle concentrations is followed for a period of time to identify the individuals who die during the follow-up as well as the point in time that deaths occur. The mortality rate observed during the follow-up period is then compared across the different regions characterized by different air pollution concentrations.

The methods and results of more than 100 epidemiological studies of various designs conducted throughout the world are tabulated in the complete Critical Review which is published in the May 1997 *Journal of the Air & Waste Management Association*. This tabulation shows that: (1) most often particle pollution is but one of several components of air pollution present in any area, with most studies also making use of serial measurements of these other pollution components; (2) the duration of the time series in the different studies ranges from less than one year to over one decade; and (3) a variety of ways of measuring particle concentrations have been used. Most of the U.S. studies, although not all, have reported an association between short-term increases in ambient particle concentrations and daily mortality; most were statistically significant associations with the respective 95 percent confidence intervals not including an effect estimate of no effect.

The U.S. studies as a group have been used to make some compelling arguments in support of a causal link between short-term increases in particle pollution and increases in daily mortality. Because this series of studies has covered a wide spectrum of settings characterized by substantial variability in the types and concentrations of co-pollutants and by variability in meteorology, it has been argued that the one common factor that is associated with mortality in this series of studies, as measured by concentrations of particles, is in fact the particle concentration itself. Recent European studies assessing the association between short term increases in particle concentrations and increases in respiratory hospitalizations have not shown effects as consistently, nor have they been able to attribute the effects specifically to the particle component of the pollution mix, as those from the United States and Canada. The reasons for these differences are not known.

Most human experimental studies investigated the effects of acid aerosol in mild asthmatics, and results have been variable. Few studies show significant effects at typical ambient concentrations, though several show effect at high concentrations (e.g., $100\,\mu g/m^3$ of sulfuric acid aerosol). Non-acidic particle experimental exposure studies are essentially lacking. Ferric sulfate challenges have been carried out in normal and asthmatic subjects, with no significant effects on level of lung function being produced in either group. The human experimental data have not provided convincing evidence for adverse effects of particle exposures under realistic exposure conditions. The relevance of the human experimental data will be greatly enhanced as techniques for generating the complex mix of particles present in ambient air are applied in these experimental settings.

LINES THAT DIVIDE

The <u>first line</u> represents the division between health researchers who believe that the health studies provide strong evidence in favor of a causal relationship between ambient particle concentration increases and adverse health effects and those who do not. The <u>second line</u> represents the division of particles into more and less pathogenic particles based on differences in particle size or chemical composition, differences that also produce divisions based on geographical lines. The <u>third line</u> represents the particle concentration cut-point defined by an ambient air quality standard.

Many criticisms of the methods used in the studies reviewed have been put forward by air pollution health researchers, including: (1) misclassification of the exposure, (2) inadequate control for confounding factors, and (3) lack of plausibility.

Misclassification of exposure is a problem in studies where the exposure of a population over a large urban area is estimated from only a few, and sometimes only one, monitoring sites. Particle concentrations vary spatially over an urban area, with use of only a few sites not adequately reflecting that variability. But more seriously, concentrations measured at fixed, central monitoring sites do not adequately represent exposures of a population that spends the large majority of time indoors.

The two main approaches that have been used in dealing with confounding by time-varying covariates is either to attempt to remove effects of these factors from the health outcome and the pollution time series before analyzing the exposure-outcome association, or to attempt to control for their effects by incorporating the time-varying covariates into the same models in which the association is analyzed. The choice of which approach is taken seems to be determined more by familiarity of the investigator with the respective analytical techniques than by their relative merits. One point in favor of incorporating the covariates into the same models is to allow estimation of the covariate effects, which for some covariates such as those defining meteorology, may also be of interest.

The most basic issue concerns that of biological plausibility and the nature of the association between the pollution concentration and health outcomes increases. If one accepts that the observational studies present a consistent and coherent picture of particle-induced health effects, then the observational findings should allow a stronger case for causality if they were complemented by

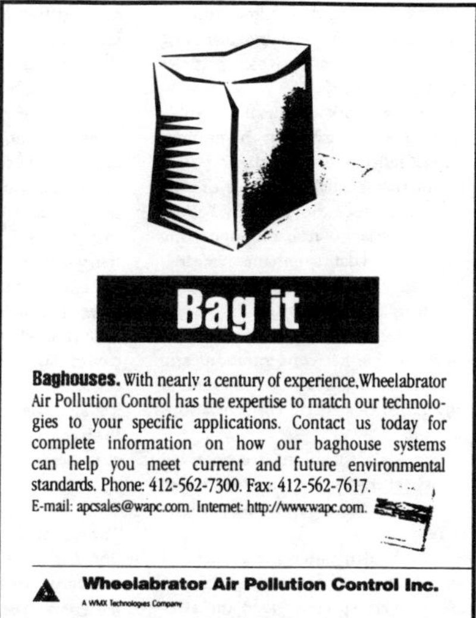

relevant experimental animal and human data, and by some understanding of the mechanisms by which particles exert their effects.

One could ask whether it is possible to make a good case that the findings do not reflect a cause and effect association that is equally plausible, or even more than the case for causality. Besides the relative lack of relevant, supporting experimental data to enhance plausibility, a case against causality can be based on the following observations:

Similar associations between increases in particle concentrations and the adverse health effects are found at <u>any</u> range of particle concentrations studied.

Similar associations are observed across settings where the source and chemical composition of the particles differ.

The health outcomes associated with particle concentration increases are not very specific, given the associations observed for cardiovascular and other outcomes.

There must be significant misclassification of exposure that results from using central monitors of ambient particle concentrations for populations that spend the vast majority of time indoors.

Mortality effects are observed across settings where particle concentrations differ by an order of magnitude. In many studies, when the analyses are stratified by season, short-term variations in ambient particle concentrations, at both relatively high and low concentrations, they are often observed to have similar effects. While it is conceivable that variation in particle concentrations at low levels results in similar adverse effects as variation at high levels, it seems equally plausible, if not more plausible, that the phenomenon of short-term variability in particle concentrations is reflecting temporal variability in another factor or process.

Observing similar associations in set-tings with different sources of particle emissions has been used to argue that the effects are related to particles generated by combustion processes, regardless of the specific composition differences of these particles. Particle mixes may be composed predominantly of particles from automobile combustion, industrial emissions, wood burning or coal burning, as well as transported particles. It is surprising that the effects of such heterogeneous particles are similar. It seems equally plausible that whatever contributes to the day-to-day variability in particle concentrations in these different settings, if it also affects health, is confounding the particle-health effects association in every setting.

The recent U.S. Environmental Protection Agency (EPA) proposal to include standards for fine inhalable particles ($PM_{2.5}$) was based on an extensive review in which a strong case was make for considering the fine particle fraction to be more pathogenic than the coarse fraction of the inhalable particle fraction. Although associations of $PM_{2.5}$ concentration with several health outcomes have been evaluated in many observational studies, most have not been able to evaluate in parallel the association with the coarse inhalable particle fraction (CP). These studies are therefore unable to directly assess the relative impacts of the fine and coarse fractions. In a few studies the investigators had access to measurements of both inhalable and coarse fraction concentrations, and were able to evaluate their relative impacts. Although the findings of these are not entirely consistent, a good case can be made that the fine fraction is more pathogenic than the coarse fraction.

A case has been made that the <u>ultrafine</u> particle fraction (particles smaller than 0.1 µm) within the fine particle fraction is the most pathogenic and is largely responsible for the particle-associated findings reported in the observational studies. The combined particle surface area for a given mass concentration increases exponentially as the diameter of the particles decreases. If

> *"The evidence favoring a role for particle composition in determining the health effects of ambient particles is not as convincing as the evidence for a role for particle size."*

particle surface area is the primary determinant of particle pathogenicity, then the smaller the particles, for a given mass concentration, the greater the pathogenic effect. It is not clear that combined surface area is a critical determinant of pathogenicity. Further, the fact that surface area increases given the above assumptions may not be relevant if significant exposure to ultrafine particles does not occur. It is observed that the mass concentration of ambient ultrafine particles is much lower than the mass concentration of larger particles. Therefore, the combined ultrafine particle surface area is much smaller than would be predicted for equivalent mass concentrations. Even the count concentration of the ultrafine fraction tends to decrease as particle size decreases below 0.1 µm. Given the propensity of particles within the ultrafine fraction to rapidly agglomerate into larger particles, significant numbers of ultrafine particles may not reach the portions of the lung that are susceptible to ultrafine particle effects. Recent work on human autopsy lungs has shown that only low concentrations of ultrafine particles are resident in human lungs, implying either that ultrafine particles do not reach those portions of the lung, or if they do, that they disappear either because they are soluble or are transported.

The evidence favoring a role for particle composition in determining the health effects of ambient particles is not as convincing as the evidence for a role for particle size. Although particles emitted from combustion processes seem to be more consistently pathogenic than crustal particles, it is not clear that this difference is not merely due to the fact that combustion particles are predominantly in the fine inhalable fraction, whereas crustal particles predominate in the coarse fraction; that is, the difference in the observed effects of combustion and crustal particles may relate to differences in particle size rather than particle composition. Nevertheless, it seems reasonable to suspect that chemical composition is one feature of particles that determines pathogenicity. Neither the experimental studies nor the epidemiological studies have been convincing in demonstrating a critical role for particle acidity. Initial human chamber exposure studies showed that exposure to acid aerosol caused a small reduction in level of lung function in adolescent asthmatics. However, more recent work has failed to reproduce this finding even in adolescent asthmatics. Subjects without asthma do not appear to experience adverse effects on level of lung function from acid aerosol exposure.

Of the 70 areas designated in 1990 as nonattainment areas, at least 50 (71 percent) of the areas were in western states. In 1993, five areas, all in western states, were classified as "serious nonat-tainment" areas. Based on 1993 to 1995 monitoring data, of the 41 regions not meeting the current EPA standards, 31 (76 percent) were regions in the western states, and another five were in mid-western states. Only five (12 percent) were in the eastern United States. Of the 167 sites estimated to not meet the proposed EPA $PM_{2.5}$ standard, only 44 (26 percent) are in western states. Emission sources for airborne particles are typically quite different in non-western and western states, with fugitive dust which consists of a relatively large proportion of particle mass in the CP fraction, being a significant source of PM_{10} in western states. If particles in the CP range are less pathogenic than smaller particles, then the newly proposed standards appropriately lessen the burden of the western states whose largest contributor to PM_{10} is fugitive dust.

The particle characteristics that distinguish non-western and western states in the United States also distinguish the non-western and western provinces in Canada. A recent report described a study in Canada in which 19 sites across Canada underwent serial particle concentration measurements with simultaneous measurement of TSP, PM_{10}, $PM_{2.5}$, CP and sulfate concentrations. Prairie sites in Calgary, Edmonton (comparable to typical western states) and Winnipeg (north of the mid-western states) had the lowest $PM_{2.5}$ contribution to PM_{10} and had, apart from one site in Montreal sited in a heavy traffic area, the highest CP concentrations. The crustal source of these particles was reflected in the relatively high silica content of the particles. Interestingly, the sites on the west coast of Canada (Vancouver and

Scrubbers. Wheelabrator Air Pollution Control offers wet FGD, dry scrubbing, dry injection and NO_x control technologies to match your most demanding requirements. Contact us today for complete information on how our scrubbers can help you meet current and future environmental standards. Phone: 412-562-7300. Fax: 412-562-7617. E-mail: apcsales@wapc.com. Internet: http://www.wapc.com.

Wheelabrator Air Pollution Control Inc.
A WMX Technologies Company

CIRCLE 52 ON READER SERVICE CARD

Victoria in British Columbia) had relatively high $PM_{2.5}$ concentrations, and therefore large ratios of $PM_{2.5}$ to PM_{10}. Sulfate concentrations were markedly lower in the prairie provinces and the west coast reflecting the relative absence of SO_2 emissions in those regions.

As noted previously, the findings of the recent observational studies on particle pollution health effects performed in Europe did not consistently show associations between increases in particle concentrations and relevant health outcomes. Effect estimates were also generally smaller than those estimated in the U.S. studies. In addition, these findings did not strongly implicate the particle component of the pollutant mix as being specifically responsible for the health effects. Possible explanations for these differences include different susceptibility of the populations to pollution effects, differences in pollutant mix or particle composition or differences in study methodology.

The EPA has recently proposed changes to the U.S. National Ambient Air Quality Standards (NAAQS) for both particulate matter and ozone. For particulate matter these proposed standards include the addition of standards for fine particles ($PM_{2.5}$): an annual average concentration of 15 µg/m³ and a 24-hour average concentration of 50 µg/m³. The proposed changes to the NAAQS carry on two traditions in setting particle pollution standards: (1) definition of a concentration cut-point and (2) definition of both an annual and a 24-hour standard.

A concentration cut-point implies that the chosen concentration has some meaning, either in terms of the significance of the health effects that occur above that concentration, or in terms of the certainty with which particle pollution effects have been demonstrated above it. The first, in turn, implies the existence of a threshold concentration below which no, or less significant, health effects are caused by particle exposure. However, the large majority of relevant observational health studies provides no evidence that such a threshold can be identified. Many exposure-response plots for several of the health outcomes have been presented in the literature, sometimes plotting effects by quantile of particle concentration as estimated using regression indicator variables for the quantiles or using smoothed nonparametric estimates. Most of these plots indicate a relatively linear relationship with no obvious threshold particle concentration below which no relationship is apparent.

Shortcomings of a particle standard that focuses on a single cut-point are:

There is no straightforward way in which the amount or degree of exceedance above that cut-point can be incorporated.

The frequency with which such exceedances of varying degree occur is not easily incorporated.

There is an inordinate emphasis on the specific cut-point chosen, which is difficult to justify given the findings of the relevant health studies.

A concentration cut-point is selected to allow a "margin of safety" between that concentration and the point where adverse effects will be experienced. However, it seems difficult to justify the selection of any cut-point, unless it were very low indeed, that would provide such a margin of safety for every member of the population.

An alternative to the cut-point type of standard incorporates a mechanism for incrementally accumulating the number and extent of exceedances above a selected particle concentration. The assumptions required for such an approach are: (1) a short-term particle concentration can be identified above which health effects occur; (2) the size of short-term concentration <u>increase</u> that causes an acute increase in adverse health effect can be identified; and (3) the exposure-response relationship is linear, at least above a concentration where effects can be identified to occur. Several methods of applying this approach are included in the complete Critical Review.

The advantages of an approach that is based on the <u>degree and frequency</u> of increases in PM_{10} concentrations are that it: (1) is a direct application of the form in which exposure-response relationships are expressed in the relevant epidemiological study reports; (2) allows flexibility in setting standards and objectives based on estimated health impacts; and (3) readily allows modifications in defining the particle concentration "increments" based on new evidence, both with respect to the concentration where counting begins and the size of concentration increase that defines the increase in count.

CONCLUSIONS

There is a basic disagreement on the interpretation of the epidemiological data on health effects associated with increases in particle air pollution. Although a strong case has been presented that this association reflects a causal relationship, plausible alternative explanations for the association can be justified. In particular, changes in meteorology also correspond with changes in pollution and health indicators, thereby potentially confounding statistical associations. Concerns about confounding by such a factor are aggravated by the small size of the estimated par-

> *"Issues of confounding and misclassification of exposure add to concerns about the biological plausibility of the observed associations between particle pollution and adverse health effects."*

ticle effects.

Future observational studies should explore novel approaches to controlling for all potentially confounding meteorological factors. Additional studies in settings with very low ambient particle concentrations would be valuable in determining whether the exposure-response relationships extend to low concentrations.

Measurement error in the particle exposure of individuals results in misclassification of exposure, which can in turn bias the effects estimated from the epidemiological studies in unpredictable ways. It is not certain whether the impact of this misclassification on the particle-associated health findings has been large or small.

Although sometimes it is justifiable to maintain that an association is causal in the absence of supporting experimental data, or data on the biological mechanisms involved, in the case of particle-associated health effects, such data will be critical in supporting the case for causality.

Issues of confounding and misclassification of exposure add to concerns about the biological plausibility of the observed associations between particle pollution and adverse health effects. Although sometimes it is justifiable to maintain that an association is causal in the absence of supporting experimental data, or data on the biological mechanisms involved, in the case of particle-associated health effects, such data will be critical in supporting the case for causality.

Findings from recent European epidemiological studies do not as consistently show associations between increases in particle concentrations and increased mortality or increased hospitalizations as the American studies. The reasons for these less consistent findings are not known, although differences in particle composition is a possibility.

The EPA presents a strong case for regulating the fine inhalable particle fraction ($PM_{2.5}$). Given the different emission and particle size profiles of "eastern" and "western" ambient particles, the impact of the proposed standard, will significantly alter the regional distribution of nonattainment areas in the United States.

The evidence that acid aerosol in inhalable particles is important in determining particle pathogenicity is not consistent. The pathogenicity of ambient particles in western regions of the United States and Canada, given the very low ambient acid concentrations measured in those regions, is clearly not related to the acid aerosol component.

Although particles in the ultrafine particle fraction are pathogenic, as demonstrated in experimental studies, it is not clear that the concentrations of ultrafine particles in ambient air contribute significantly to the adverse effects associated with the increases in inhalable particle concentration demonstrated by the epidemiological data.

Expressing ambient particle standards in terms of concentration cutpoints, a tradition carried on in the proposed EPA standards, is inconsistent with epidemiological studies in which relatively linear increases in health risks are observed in association with increases in particle concentrations. One alternative is to express standards in terms of incremental increases in particle concentration.

The proposed EPA standards also carry on the tradition of maintaining both a long-term (annual) standard and a short-term (24-hour) standard. There is no evidence that long-term, low level exposure to particles, apart from repeated, short-term increases in particle concentrations, is associated with adverse health effects. Even the evidence supporting development of chronic illness following long-term particle exposure, either from repeated exposures to short-term particle concentration increases or from chronic exposures to relatively low concentrations, is weak.

AIR & WASTE MANAGEMENT ASSOCIATION
•
SINCE 1907

BNL–64310
Informal Report

TECHNICAL COMMENTS ON EPA'S PROPOSED REVISIONS TO THE NATIONAL AMBIENT AIR QUALITY STANDARD FOR PARTICULATE MATTER

Frederick W. Lipfert

March 1997

Prepared for
Office of Fossil Energy
United States Department of Energy

BIOMEDICAL AND ENVIRONMENTAL ASSESSMENT GROUP
ANALYTICAL SCIENCES DIVISION
DEPARTMENT OF APPLIED SCIENCE
BROOKHAVEN NATIONAL LABORATORY
ASSOCIATED UNIVERSITIES, INC.

Under Contract No. DE-AC02-76CH00016 with the
U. S. Department of Energy

Technical Comments on EPA's Proposed Revisions to the
National Ambient Air Quality Standard for Particulate Matter

Executive Summary

Background

The U.S. Environmental Protection Agency (EPA) has proposed new ambient air quality standards specifically for fine particulate matter, regulating concentrations of particles with median aerodynamic diameters less than 2.5 um ($PM_{2.5}$). Two new standards have been proposed: a maximum 24-hr concentration (50 ug/m^3) that is intended to protect against acute health effects, and an annual concentration limit (15 ug/m^3) that is intended to protect against longer-term (chronic) health effects. EPA has also proposed a slight relaxation of the 24-hr standard for inhalable particles (PM_{10}), by allowing additional exceedances each year (150 ug/m^3). These proposals were published in the Federal Register (FR 61:65638, Dec. 13, 1996) and are the result of a lawsuit that mandated EPA action under a court-defined schedule.

Fine particles are currently being indirectly controlled by means of regulations for PM_{10} and TSP, under the Clean Air Act of 1970 and subsequent amendments. Although routine monitoring of $PM_{2.5}$ is rare and data are sparse, the available data indicate that ambient concentrations have been declining at about 6% per year under existing regulations.

The proposed NAAQS revisions are not entirely consistent with the recommendations of EPA's Clean Air Scientific Advisory Committee (CASAC). Only two of the 21 CASAC members recommended an annual $PM_{2.5}$ standard as low as the 15 ug/m^3 level that EPA proposed and eight of them felt that no annual standard was appropriate at any level. In addition, the proposed NAAQS revisions may be unique in the history of the Clean Air Act, in that they are based almost solely on epidemiology, none of the epidemiological studies is capable of identifying a "safe" exposure level, and no plausible biological mechanisms have been identified that would support the epidemiological findings.

Technical Issues Pertaining to Studies of Both Short- and Long-Term Health Effects

1. <u>Health effects specific to fine particles</u>. At the heart of the EPA proposals for new NAAQS for $PM_{2.5}$ is their contention that health responses have been determined that are specific to this class of particles, from which the public would not be protected by means of controls on PM_{10}. EPA cites seven epidemiological studies in support of this position, three for acute effects and four for long-term effects. These studies are listed and described in Appendix A.

First, there is no health end point that is unique to fine particles, to any type of particle, or to air pollution in general; all of the health effects are multifactoral, responding to a variety of environmental and non-environmental stimuli. Second, based on the studies that EPA has selected, in no case does the mean health response associated with $PM_{2.5}$ significantly exceed that associated with PM_{10}; in most cases, PM_{10} effects are greater and the effects attributed to sulfate (SO_4^{2-} is an important component of both PM_{10} and $PM_{2.5}$) are a distant third. Finally, $PM_{2.5}$ consists of a multitude of different types of particles that would be expected to exhibit a range of responses in the lung. Some particles are solid, some are liquid, some are acidic, some are chemically active, some are soluble, some are insoluble, and some have complex chemical composition. Thus, *a priori* we might expect that some, but not all, types of particles could be associated with adverse health effects. The proposed new standards do not distinguish by particle characteristics, only by size, and thus assume that all particles in a given size range will have the same effects. This means that the proposed regulations will be inefficient and may impose controls having no benefits.

For these reasons, there is presently no justification for a separate $PM_{2.5}$ air quality standard based on health effects.

2. <u>Pollutant collinearity.</u> Mixtures of air pollutants are found in all urban areas, because of common sources. In addition, many pollutants respond in concert to changes in weather. In epidemiological analysis, failure to consider all relevant exposures to air pollution and to weather extremes will result in overstating the responses to those agents that are considered. For example, during hot weather, elevated PM, ozone (O_3), and temperature levels tend to occur together. Thus, inadequate consideration of heat-wave effects will tend to overstate the contributions of air pollution. Many of the EPA-selected studies considered only a few pollutants; carbon monoxide (CO) has been a common omission, for example. To the extent that CO and particulate matter (PM) are correlated (and the PM from vehicles tends to be highly correlated with CO), the health effects of PM are likely to have been exaggerated or mis-stated. Recent studies of hospital admissions have shown that O_3 is more important than PM for respiratory diseases such as asthma and that CO is more important than PM for certain cardiac diagnoses. Another recent EPA-funded mortality study showed that inclusion of relative humidity in the regression model eliminated PM as a significant risk factor.

3. <u>Uncertainties in the air pollution levels to which the putative victims of air pollution may have actually been exposed.</u> All epidemiological studies use air quality data as obtained from one or more centrally located air monitoring stations in each city or metropolitan area, as opposed to direct measurements of individual exposures. The limited studies of individual personal exposures show poor correlations with the routine monitoring data, in part because of variations in outdoor pollution levels across a metropolitan area and in part because people spend 85% or more of their time indoors where the levels and types of pollutants are quite different from outdoor air. These spatial variations are most pronounced for coarse particles and tend to be smaller for fine particles and sulfates, because the smaller particles settle out of the atmosphere more slowly. As a result, the actual exposure levels are poorly known, the contributions of outdoor air to these exposures can only be estimated, and important statistical difficulties result with respect to the epidemiology. These difficulties tend to obscure the true shapes of the dose-response functions, including the presence of thresholds, and they have not been adequately addressed by EPA or by the authors of the studies that EPA has selected.

As an example, air monitors in EPA's key study of acute mortality by particle size were located up to 60 km away from the population centers in three of the six cities involved. This makes it highly unlikely to find any effects attributed to coarse particles and other reactive (O_3) or local (CO) pollutants, since the measurements were made so far away from the bulk of the population and since concentration of these pollutants tend to vary greatly with distance from their sources.

4. <u>Lack of plausible biological mechanisms.</u> The problems of unknown exposures and of collinear pollutants make it difficult to reliably link specific pollutants with specific diseases, which is a prerequisite to formulating hypotheses for biological mechanisms. In contrast to PM, such information exists for some of the gaseous pollutants, notably CO (heart disease) and O_3 (slightly elevated hospital admissions for asthma). Implausible statistical combinations that have been reported include CO and respiratory disease, O_3 and heart disease, and sulfates and cancer. When the combination of PM and CO is linked with heart disease and a biological mechanism is known for CO but not for PM, it is difficult to accept the PM contribution as real, given the problems of determining actual exposures discussed above.

Technical Issues Pertaining to Long-Term Studies

1. <u>Regional confounding by lifestyle, ethnic, and socioeconomic factors.</u> Long-term air pollution epidemiological studies depend on cross-sectional (spatial) contrasts, in which dose-response relationships are developed by comparing spatial patterns in health indicators with those for air quality. If either of the health or the air quality patterns were randomly distributed across the country, this procedure might provide useful results, but unfortunately, this is not the usual case. Regional air pollution patterns that have been reasonably stable over time include very high O_3 and PM levels in Southern California, moderate ozone levels across almost the entire region east of the Mississippi, and higher $PM_{2.5}$ and sulfur oxide levels in Appalachia and parts of the Northeast. Ischemic heart disease tends to be higher in the

northeastern states, strokes are more common across a wide region of the South, respiratory disease mortality seems to be more common in certain mountain states, and people generally tend to live longer in Sunbelt cities and in the West. Further, some of these patterns have been changing over time, presumably due to diffusion of new medical procedures and changes in diet and lifestyle. There are many reasons for these regional disease patterns that have nothing to do with air pollution, but, unless these exogenous factors are considered, a naive analysis will inappropriately assign some of the regional differences in health indicators to air pollution.

For example, levels of medical care are higher on both coasts of the U.S. than in Appalachia, where studies show that people also tend to be less physically active. Thus, health-effects models that relate spatial differences in air pollution to mortality rates may find a statistical association between some measure of air pollution and health if they fail to account for intervening variables such as lifestyle and/or access to medical care. Other examples of non-pollution intervening factors that should be considered include diet, selective migration, income or employment status, and ethnicity. The EPA studies did not consider these factors and regional differences in physical activity could have accounted for almost all of the mortality gradients that one of these studies assigned to air pollution, including fine particles and sulfates.

2. <u>Inadequate consideration of all pollutants</u>. One of EPA's key mortality studies considered only two pollutants, $PM_{2.5}$ and SO_4^{2-} and thus provides no information on the role of particle size. The other mortality study had data on only six locations, so that the confidence limits on the relative effects of different pollutants were unacceptably wide. Neither study considered peak O_3 levels as a risk factor (note that O_3, SO_4^{2-}, and $PM_{2.5}$ tend to be collinear during hot weather and that there were major heat waves during the periods of study). None of the long-term studies have considered CO as a potential risk factor.

3. <u>Latency periods between exposure and development of chronic disease.</u> EPA intends its annual standards to protect against chronic effects (they could also protect against the annual sum of acute effects, but in this interpretation, 24-hr standards would not be needed). Chronic effects, by definition, have their origins well before their existence is actually observed; typical latency periods run from 10-40 years. Neither of the long-term epidemiological studies selected by EPA considered prior exposures, which were considerably higher in the most polluted cities. Had they done so, the resulting dose-response functions would have more closely resembled those developed from acute studies. Thus, even if these two long-term studies are taken at face value (which is problematic for other reasons, discussed above and below), they should not be interpreted as specifically supporting the existence of chronic responses to air pollution. This alone calls into question the need for a separate annual standard for PM.

4. <u>Investigation of pollutant thresholds.</u> None of the long-term studies considered by EPA made an attempt to distinguish health effects in cities that violated the NAAQS from those in compliance. Two of the six cities in one mortality study exceeded the NAAQS for TSP during the study. Almost half of the cities in the other long-term mortality studies exceeded the TSP/PM_{10} NAAQS for the period of study; in earlier periods, even more cities were in violation. Almost all major U.S. cities violated the O_3 NAAQS during these studies (and many still do). Given the problems in estimating true exposures and thresholds and the collinearity among pollutants, the need for stricter air standards cannot be inferred from these studies.

Technical Issues Pertaining to Short-Term Studies

1. <u>Misinterpretation of the apparent consistency of findings.</u> EPA regards the apparent consistency of the many (over 30) studies of short-term PM effects throughout the U.S. and in other countries as a major element in its causal hypothesis. However, for each data set that has been reanalyzed by different investigators, agents (pollutants and/or weather factors) other than PM have been found to be equally plausible risk factors. In addition, the consistency noted by EPA extends to all air pollutants (gases and particles) and to virtually all subsets of PM, in terms of mean effects on all-cause mortality. Thus, the apparent consistency should be regarded, not as support for causality, but instead as a warning that the similarity in findings

may be derived from the common structure of the studies, rather than from physiological considerations per se. First, it is difficult to rationalize how such a wide range of air pollutants could all exhibit the same types of responses, let alone the same magnitudes, given the differences in personal characteristics and exposures to be expected in this wide variety of situations. Second, it is well known that PM has very different characteristics in different parts of the country, even when classified by particle size. EPA's claims of consistency across studies implies the same responses to complex wood smoke, diesel exhaust, and soluble sulfate particles; such similarity in response is incompatible with the hypothesis of pathological health effects.

2. <u>Inadequate controls for seasonal and weather effects.</u> The short-term epidemiological studies use temporal relationships between daily values of health indices and of air quality to postulate dose-response relationships. This requires accounting for all non-pollution factors that affect short-term health. Such factors include season (respiratory disease and mortality peak in the winter in all climates), weather extremes (hot and cold), days of the week (more people die on Mondays; O_3 is lower but primary pollutants are higher on weekends), and holidays (mortality is higher on Christmas). Since the putative effects of air pollution are so weak (1-4%), it is necessary to account for even very small effects in the regression models. The studies in the literature vary considerably in their attention to these details; if days of the week are not accounted for, for example, (as sometimes has been the case) spurious relationships will result. Inadequate consideration of heat wave effects can result in overstating the effects of both O_3 and PM. Another example of weather confounding was seen recently in an EPA-funded mortality reanalysis study, which showed that including relative humidity in the regression model eliminated PM as a significant risk factor, as had originally been claimed.

3. <u>Improper consideration of lag effects.</u> The only identified plausible mechanism for mortality occurring at the low air pollution concentrations we now enjoy is that of loss of homeostasis in severely compromised individuals (however, EPA has not advanced this hypothesis). In such cases, virtually any small environmental disturbance, including weather and pollution changes, might be sufficient to advance death by a few days. In such cases, a corresponding dip in the daily mortality counts should be seen on following days, and this has been in fact reported in several studies. In some cases, the net excess mortality over a week or so thus approaches zero. However, other authors have not reported this phenomenon but instead report only the combination of mortality delay (i.e., "lag") and pollutant averaging time that maximizes the apparent positive response. This is one of the reasons that reanalysis of a given data set by different investigators almost always produces results at odds with the original conclusions.

A further consideration here is the physiological plausibility of either mortality or hospitalization occurring on the same day as an air pollution excursion. Clinicians and others have expressed doubt that such effects can happen, especially when one considers that most air pollutants are not uniform over the course of a day but exhibit well-defined diurnal peaks (O_3 in the afternoon or CO in the morning, for example). Heart attacks are most common in the morning, so that there may be a real probability of the response preceding the cause in those studies that purport to find a PM response on the same day.

Overall Conclusion

This review finds a lack of statistical and physiological support for the specific ambient air quality standards that EPA proposes for fine particles. While the epidemiological studies do not exclude the existence of associations between fine particles and health responses, they fail to show responses that are specific to this class of particles; hence the need for a separate standard has not been established. Finally, EPA has relied on only a portion of the epidemiological literature and has neglected to consider several important factors bearing on the likelihood of causality. Given the fact that ambient air quality (including fine particles) will likely continue to improve under current regulations, EPA could better serve the public interest by devoting additional resources to a comprehensive targeted program of research and monitoring.

**Technical Comments* on EPA's Proposed Revisions to the
National Ambient Air Quality Standard for Particulate Matter**

1. Background

EPA's proposed revisions to the National Ambient Air Quality Standards (NAAQS) for particulate matter (PM) are based almost entirely on selected epidemiological studies that have been published in recent years. The important reference documents for EPA's proposal include: the revised criteria document for particulate matter (EPA/600/P-95/001cF, April 1996); the revised EPA Staff Paper (EPA-452\R-96-013, July 1996); the proposed regulations (Federal Register 61:65638, Dec. 13, 1996), and the draft regulatory impact analysis (no document number, dated December 1996). In these comments, the self-consistency of this body of information, key technical issues that arise and have been incompletely addressed, and new information that has since appeared in the scientific literature are examined. EPA proposed to retain the present standards for "inhalable" particles, i.e., those with aerodynamic diameters nominally less than 10 um, i.e., PM_{10}, and slightly relaxed the 24-hr standard by allowing more exceedances per year. However, EPA proposed strict new standards for "fine" particles (those with aerodynamic diameters nominally less than 2.5 um, i.e., $PM_{2.5}$) and this review emphasizes the smaller particles. The review begins with some introductory concepts that are needed to understand the scientific issues; reviews the studies that EPA selected as supporting their proposal, new studies that have since appeared, and existing studies that EPA should have considered. Then, some basic scientific issues that remain unresolved are discussed and the conclusions and recommendations of the review are presented.

1.1 Types of Particles in the Atmosphere

In general, fine particles are distinguished from larger particles by their sources and chemistry as well as by size. The first distinction is that particles formed in the atmosphere from gaseous precursors are initially quite small, mainly less than 0.1 um, and then grow in size due to agglomeration and absorption of atmospheric water. These water soluble particles are often mixtures of sulfate, nitrate, hydrogen (acid) and ammonium compounds and may be slightly acidic if they have not been in contact with atmospheric ammonia (gas). These particles can also be quite hygroscopic and tend to grow in size at high relative humidity, and thus tend to contain substantial particle-bound water. Organic particles may be formed as part of the photochemistry that produces ozone. Particles are formed from combustion processes that include volatilization of material at high temperatures followed by condensation in the atmosphere. These (insoluble) particles include soot and metal oxides. Health effects of some of these particles have been demonstrated in animal testing, at concentrations much higher than found in the ambient. Wood smoke is among those for which a biological mechanism has been demonstrated (*Science*, 1995); breathing coal smoke from domestic cooking has also been shown to have health effects (Xu and Wang, 1993). Transient health effects from acidic particles have only been demonstrated at concentrations about 2 orders of magnitude than those found in the ambient (EPA, 1996).

Particles larger than 2.5 um include pollen, spores, fibers, other biological aerosols, tire erosion products, and small fragments of material from the earth's crust. These particles tend to be insoluble. Sources of crustal particles include road dust, material handling, mining, cement manufacturing, agriculture. Some of these (silica, asbestos, and biological materials) are also known to produce health effects. Shapes of these particles are also important (e.g., asbestos); soot is typically found in large irregular agglomerations with high surface-to-volume ratios, such that pollutant gases may readily be adsorbed. Soot may be found in both size fractions.

----

* Note that these comments incorporate material that was originally prepared for other purposes, some of which has been published in various forms. These publications include Section 12.4 of the PM Criteria Document and various publications by Lipfert and colleagues.

The arbitrary size cut at about 2.5 um does not provide a complete separation of the two types of particles; the lower tail of the biological and crustal particle distributions will often extend into the "fine" particle collection, and some secondary water-soluble particles (notably nitrates) may exceed 2.5 um.

It is important to realize that all air pollutants are defined for regulatory purposes by their officially-designated EPA "reference" measurement methods. For pollutant gases, these methods are generally quite specific to the actual chemical compounds (e.g., CO, SO_2, O_3, NO_2), but this is not the case for particles. The reference method for PM_{10}, for example, defines this quantity by the mass of material collected on a filter using a sampler meeting EPA specifications, without regard to the chemical compounds involved. This mass determination may be augmented or diminished by the processes of collection and analysis; volatile materials such as nitrates and organics may be lost and reactive gases may form compounds on the filters that were not present in the ambient. This latter process is most important for nitrate and sulfate aerosols as collected on the glass fiber filters used in high-volume (TSP) samplers. Analysis has shown that this "artifact" sulfate is about 5 ug/m^3 in the Eastern U.S., which is about half of the total SO_4^{2-} mass reported (Lipfert, 1994a). A large part of the typical difference between TSP and PM_{10} is suspected to be artifactual, since mass is is gained on the glass fiber TSP filters (SO_4^{2-} and NO_3^-) and lost from the Teflon PM_{10} filters (organics and NO_3^-) (Lipfert, 1994a). Chemical-specific PM measurement methods include SO_4^{2-} and NO_3^- ions and carbonaceous particles. The latter are typically related to the darkness of filter paper stained by passing the air sample through; these determinations are referred to in various contexts as "smoke", "British smoke", coefficient of haze (COH), or (in California) "KM". The samplers used for these determinations are thought to be selective to particles less than about 5-6 ug/m^3 in diameter, although the typically irregular shapes of soot particles may result in different lung deposition patterns than more spherical particles.

Finally, not all air pollutants are measured with current routine air sampling; potentially important omissions include aerosol acidity, elemental carbon, metals, aldehydes, and benzo(a)pyrene, which has recently been identified as a known human carcinogen. Sulfates and nitrates are measured routinely by only a few states.

1.2 Types of Epidemiological Studies Used to Support EPA's Proposal

All of the information on health effects associated with PM at current ambient levels comes from epidemiological studies. However, robust, specific pollutant-disease relationships have largely been lacking, in part because most of the studies have considered PM in general rather than specific types of particles. Further, no plausible biological mechanisms have been advanced in support of these various statistical associations. Like almost all environmental epidemiology studies, these studies are opportunistic in that they take advantage of data that were originally collected for other purposes. The dependent variables include daily counts of mortality and hospital admissions, annual mortality rates, diary studies on symptoms, and measures of lung function. The independent variables include data on individual and population characteristics and data from centrally located air pollution and weather instruments, but no information on individual exposures to pollution or to weather extremes. Also, only one study has considered the actual detailed chemical characteristics of the particles collected (Dockery et al., 1992). Further, the particle collection systems can create both positive and negative artifacts with respect to actual ambient conditions. Positive artifacts result from conversion of pollutant gases (such as SO_2) to particles (such as SO_4^{2-}) on the filter; negative artifacts result from loss of volatile material (such as nitrates and organics) after collection (Lipfert, 1994a). These uncertainties in exposure lead to serious difficulties in the statistical analyses that have been used to support the proposed standards (Lipfert and Wyzga, 1995a). These problems are discussed below in Section 1.3.

There are two types of observational epidemiological (statistical) studies in common use for studying air pollution. One type uses long-term contrasts among locations to derive statistical relationships between the air quality in different places and various health characteristics of the local population. This type is essentially a study of differences among places and is called a "cross-sectional" study. Studies of this type have been used to support annual air quality standards.

The second basic type of statistical study involves similarities in the timing of acute events, in which daily variations in air quality and weather are contrasted with daily variations in various health indicators. These time-series studies essentially contrast population health statistics on the "bad" days with those on the "good" days, and the worst days usually have adverse weather conditions as well as poor air quality. Studies of this type have been used to support shorter-term ambient air quality standards such as 1-, 3-, or 24-hour standards.

Since failure to achieve either short- or long-term standards can have similar consequences in requiring reductions in emissions from specified sources, there are some interactions among the standards. However, large isolated sources with tall stacks (power plants, smelters) will tend to be more important for short-term concentrations, and a dense distribution of low-level area sources (vehicles, space heating) will tend to have the most effect on annual air quality standards (per unit of pollutant emission). These rules-of-thumb apply to emissions of primary pollutants; in contrast, secondary pollutants that are formed in the atmosphere from chemical reactions of secondary pollutants tend to affect larger areas relatively far from the primary sources and may be important at all time scales.

1.3 Exposure Error Issues Common to All Air-Pollution Epidemiological Studies

However, there have been several separate studies of personal exposures for a few individuals for periods up to 2 weeks (Lipfert and Wyzga, 1996)). Instead, data from outdoor centrally located air monitoring stations that are intended for routine environmental management and enforcement are used as surrogates for exposure to the complex mixtures that individuals receive in various indoor and outdoor situations. These data rarely (if ever) include speciation of particulate mass by compound or element, which makes it all the more difficult to postulate specific pollutant-disease relationships or to hyopthesize biological mechanisms.

It must be emphasized that epidemiological studies in general cannot be used to "prove" cause and effect; their best use is in generating hypotheses to be tested in controlled settings. A major problem with observational epidemiology studies is the lack of data on individual exposures to air pollution. The use of surrogate exposure measures (such as ambient air quality data from central monitors) results in biased "dose-response" functions, such as a slope or regression coefficient (Pickles, 1982). If the error induced by using a surrogate is normally distributed with zero mean (in other words, when it creates random noise with no effect on the mean value), the slope is biased towards a shallower relationship, since the scale of exposures has been expanded due to the variability added from using a surrogate (Armstrong, 1990). This also biases the x-intercept (which might be regarded as a threshold) to the left (Lipfert and Wyzga, 1996). If the true dose-response function is non-linear (i.e., has curvature), this measurement noise will obscure this fact and make it appear to be linear (Yoshimura, 1990). Thus, in general, the types of exposure data that EPA has relied upon is setting NAAQS will tend to underestimate the risks of high exposures and overestimate the risks of low exposures. Note that people have respiratory defenses against most air pollutants, and adverse health effects can occur in an individual only when these defenses have been overcome. This is why individual (rather than group-average) exposures are important and why it is reasonable to expect to find pollutant thresholds, even though the levels may vary among individuals.

Although detailed data on personal exposures are scarce for most pollutants, some generalizations can be made. Smaller particles settle out of the atmosphere more slowly and thus tend to have longer residence times and smoother spatial distributions. However, there are many sources of fine particles indoors that tend to obscure the relationships between actual personal exposures and the ambient monitoring data used in epidemiological studies. An important exception is sulfate aerosol (Dockery and Spengler, 1981), which is the only pollutant for which personal exposures correlate well with ambient measurements (probably because of the absence of indoor sources). Pollutants with especially poor relationships between personal exposures and ambient data include SO_2, CO, O_3, TSP. This is a result of the combined effects of indoor sources and sinks and of outdoor spatial variability due to local emission sources (Lipfert and Wyzga, 1996).

None of the "new" statistical studies have been able to firmly identify pollutant thresholds that might serve to support ambient standards, although only a few have made serious attempts. However, since all of them are based on surrogate exposure measures, this does not constitute evidence that such thresholds do not exist. A part of the failure to find thresholds results from the fact that the highest pollution days tend to not correspond with the worst health days; instead, such correspondence occurs more often at lower pollution values, nearer the mean. This could be interpreted either as a measurement error problem (perhaps people are not actually exposed to the highest pollution levels) or as a problem in the statistical analysis, but in any event this mismatching sounds a warning that a closer look should be taken at both possibilities. Knowledge of biological mechanisms are needed to aid in such interpretation.

Finally, when correlated pollutants have different exposure measurement errors, these differences will tend to control their relative performance in joint regression models that are intended to partition their effects (Lipfert and Wyzga, 1995a). The pollutant with the least error will tend to dominate, assuming equivalent underlying toxicity. This means that the results of studies involving joint regressions of correlated pollutants with different estimated exposure reliabilities (which includes almost all of them) cannot necessarily be taken at face value (Lipfert and Wyzga, 1995a).

1.4 Bases for Comparing Epidemiological Studies

A review of this type will inevitably involve comparisons of associations involving different pollutants within a given study and between studies. Different authors use different paradigms for this purpose that are not always well defined. Common expressions in the literature include "consistent" or "stronger" associations, or even "more significant." In this review, statistical significance is considered a threshold of credibility that must be overcome, taken here as a 5% probability level ($p < 0.05$, which requires a regression coefficient about twice its standard error). Significance depends on sample size and measurement error and should not necessarily be considered as a figure of merit; a study with $p < 0.001$ does not necessarily outrank one with $p < 0.01$, for example. "Mean effect" or elasticity at the mean is used as a quantitative, continuous figure of merit relating the endpoint to a risk factor; this measure is obtained by multiplying the regression coefficient by the mean value of the risk factor. If Poisson or log-linear models are used, the resulting mean effect is expressed as a fraction of the endpoint. If ordinary least squares regression is used, the mean effect is expressed as a fraction by dividing this product by the mean value of the endpoint variable. The relative risk at the mean pollution levels is obtained by adding 1.0 to the fractional mean effect.

Many authors express their results as relative risks, a measure that is probably most appropriate for specific individual health risks such as smoking or being a diabetic. It is less appropriate for air pollution, since we have no knowledge that the putative victims were actually exposed to the levels on which these figures are based. Further, authors differ with respect to the bases on which they express relative risks. Some authors express relative risks in terms of some arbitrary concentration level, such as 50 or 100 ug/m^3. This makes the risks due to small particles seem higher, since 50 ug/m^3 of $PM_{2.5}$ is a much rarer situation than 50 ug/m^3 of PM_{10}, for example. Other studies have used the entire range (maximum - minimum) of the pollutant variable as a basis for risk comparison, or the "interquartile range" (IQR = 75th percentile - 25th percentile). This practice tends to make the pollutant with the most disperse distribution appear to be the most harmful. For example, Thurston et al. (1994) reported that aerosol acidity posed the largest relative risk (1.5) for respiratory hospital admissions in Toronto, based on its range from minimum to maximum concentration. Ozone was stated to pose lower risks on this basis (1.34). However, the **mean** effect of ozone was about five times that of acidity, because high concentrations of acidity occur so much less frequently than high concentrations of ozone.

Based on the results of Thurston et al. (1994), ozone should thus be selected as the pollutant whose control should yield the most benefit (assuming no confounding due to temperature, etc.).

Also, the qualitative appearance of confidence limits for risk ratios depends on the risk levels selected for illustration; for example, a relative risk of 1.02 with confidence limits of 1.005 to 1.035 seems to be less impressive than one given as 1.2 with confidence limits of 1.05 to 1.35, yet these represent exactly the same situation, differing only in the exposure level selected to illustrate the risk. Use of mean effects has none of these problems.*

2. Key Epidemiological Studies in the EPA Proposal

2.1 Studies Intended to Support the Proposed Short-term (24-hr) NAAQS for $PM_{2.5}$

2.1.1 <u>Mortality Studies</u>. Although costs and benefits are not formally part of the consideration for revising the NAAQS, EPA has emphasized that their estimates of health benefits widely exceed their cost estimates. The major part of their estimated economic benefits comes from changes in mortality, in part because each life that is projected to be extended (note that no lives are actually "saved"[as reported by EPA]) is valued at several million dollars, regardless of the length of such life extensions. Epidemiological studies with mortality endpoints are thus very important in the analysis. However, only two studies specifically considered the association between daily mortality and fine particle concentrations, measured as $PM_{2.5}$, and their results are somewhat conflicting. Other studies have used TSP (particles up to 50 um), PM_{10}, COH, and KM. TSP and PM_{10} are the most common PM metrics used in recent mortality studies.

Dockery et al. (1992) considered metropolitan St. Louis and a large area in Eastern Tennessee in the first attempts at time-series studies of particles classified by size and chemistry. These are two of the "Harvard Six Cities" that are discussed more fully below. Unfortunately, this effort was compromised by the short periods of record used and the large geographic areas that were required to obtain sufficient daily death counts for analysis. The St. Louis SMSA comprises eight counties, plus the city, for a total of almost 13,000 km^2. For eastern Tennessee, the geographic area was extended to include 11 counties (total area of about 12,000 km^2).

A single monitoring site was used in each area, which recorded daily values for PM_{10}, $PM_{2.5}$, $SO_4^=$, H^+, SO_2, ozone, and NO_2. Carbon monoxide (CO) measurements were not included. Elemental composition data were also obtained from the particulate catch. The St. Louis site was located on the south side of the city in a residential area (Carondelet). The TSP data collected for use in a 1980 cross-sectional study (Lipfert, 1993) offer a means of comparing mean particulate values across the SMSA; in 1980, the SMSA-wide mean of 27 air monitoring stations was 84.2 ug/m^3, with a standard deviation of 17.2. The difference between the mean TSP values for the city and for the SMSA was statistically significant. Based on TSP data recorded at Carondelet in 1980 (80 ug/m^3 [Dockery et al., 1989]), this site did not seem out of line with the rest of the SMSA on the basis of mean values; however, the question of interest here is the temporal correlation, which was not available.

Dockery et al. determined PM_{10} by adding the fine and coarse particle mass determinations, thus capturing the measurement errors of both. In spite of this handicap, PM_{10} was the only pollutant of the seven species to achieve statistical significance in St. Louis, although $PM_{2.5}$ was close as was the coarse particle fraction. Thus, a major difference in mortality association by

------

* The selection of the most appropriate measure of health response also relates to the existence of thresholds. Using the mean effect is tantamount to treating all concentration excursion the same, regardless of their absolute magnitudes. This "no threshold" paradigm is consistent with the fact that none of the published epidemiological studies has been able to establish the existence of such thresholds (measurement error may be partly responsible for this outcome). If, on the other hand, thresholds do exist, then it would be appropriate to consider risks specific to the higher concentrations, as was done, for example, by Abbey et al. (1991).

particle size was not indicated. One- and two-day lags were about equally effective. Neither $SO_4^=$ nor H^+ were even close to significance, nor were any of the gaseous species ($SO_4^=$ is a major component of the fine particle fraction). Elements normally associated with the coarse fraction were reported to have positive associations with mortality. The mean effects for PM_{10} and $PM_{2.5}$ were 0.041 and 0.030, respectively. The linearity plot showed that the first three PM_{10} quintiles all had about the same risk (1.0 - 1.01), with mean values from about 12 to 27 ug/m^3. The highest two quintiles both had an average risk of about 1.03 (note that these results could also be interpreted as a "hockey stick" dose-response function, with a threshold at about 25-30 ug/m^3 (24-hr average PM_{10})). The authors pointed out that the highest PM_{10} value in this data set was only 97 ug/m^3, well below the 24-hr NAAQS level of 150 ug/m^3. However, we have no assurance that levels are this low throughout the SMSA, nor do we know what TSP levels the decedents are actually exposed to. The estimated average TSP level for St. Louis for this time period was 64 ug/m^3, which is about 27% higher than the level implied by the PM_{10} data used in the study. Obviously, if the average PM level is 27% higher, there must be other locations in the metropolitan area with still higher PM levels, some of which could have been in violation of the NAAQS.

The portion of the study dealing with Eastern Tennessee was structured around a monitoring data set for Harriman, TN, which was similar to the St. Louis data. There was more topographical relief in the Tennessee study area, and almost half of the population lived in one county (Knox), which was not specifically represented by air quality monitoring data. None of the air pollutants was even close to statistical significance, although the regression coefficients were generally similar to those found in St. Louis. The mean effects were 0.048 for both PM_{10} and $PM_{2.5}$ and 0.07 for $SO_4^=$. The linearity plot showed that the second highest PM_{10} quintile had a risk of about 1.0, relative to the lowest quintile, while the remaining quintiles formed an approximate linear relationship. The estimated average TSP level for Knox County (the major population center of the area under study) for this time period was about 55 ug/m^3. The lack of significance shown for Eastern Tennessee may have resulted from the smaller average number of daily deaths (15.5 vs. 56 for the St. Louis area). If the algorithm of Ito et al. (1996) is used to adjust the standard errors to an equivalent daily death rate, both PM_{10} and $PM_{2.5}$ would have achieved significance, with about the same regression coefficients as in St. Louis. However, $SO_4^=$, H^+, SO_2, NO_2, and O_3 would have not have achieved statistical significance. This implies that the non-sulfate particles must have been more important.

Another important contribution of Dockery et al. is the testing of chemical constituents of PM for associations with daily mortality. Positive associations were reported in both metropolitan areas for Al, Ca, Cr, Fe, and SiO_2, all of which are commonly found in particles associated with the earth's crust (rather than with combustion products, per se). Crustal particles were found to be associated with respiratory hospital admissions in Anchorage, Alaska, for example (Gordian et al., 1996 (discussed in more detail below)).

This study was followed by an extended analysis of all six cities, covering the period 1976 to 1987 (Schwartz et al., 1996). Although the authors assigned the entire acute mortality effect to "combustion-related" fine particles (even though such particles had not been specifically addressed), this outcome was largely preordained by the design of the study. As in the previous study, the extended analysis was based on data from a single dichotomous sampler located in each of the six areas (and not always in the population center), while the mortality data being fit to these data were taken from surrounding areas sometimes comprising as many as 9 counties and areas of up to 150 km in extent. This design provides much more reliable estimates of exposure for fine particles than for coarse particles, by virtue of the more precise and accurate measurements of $PM_{2.5}$ from the dichotomous samplers used* and the much more uniform spatial distributions of fine particles across each of the six metropolitan areas. Since random

----

* The samplers used allocated only 10% of the sampled air stream to the coarse particle catch, which greatly reduces the amount of mass available for analysis. In addition, there was a problem with loss of coarse particles during shipment from the field sites to the analytical laboratory (Spengler et al., 1986; Dzubay and Barbour, 1983). These two factors result in a measurement precision at least twice as good for fine particles, compared to coarse particles.

errors in exposure tend to bias the regression slope towards the null and towards nonsignificance, the Six City study design contains a strong bias in favor of fine particles. This bias is discussed in detail by Lipfert and Wyzga (1997). The resulting increased measurement error for coarse particles will have biased their regression result towards non-significance, and this fact alone could explain the differences reported by Schwartz et al. Further, it is necessary to consider gaseous pollutants (such as CO or O_3) simultaneously with particles to provide robust estimates of the relative responsibilities for the implied daily mortality gradients.

Finally, the model specifications used by Schwartz et al. may be sensitive to differences in pollutant frequency distribution characteristics according to particle size. For example, during most of their study, size-classified particles were collected only every other day. Schwartz et al. elected to use a 2-day average of PM concentrations in their regression analysis, which amounts to assuming a 2-day persistence for air quality. Since smaller particles tend to have longer residence times in the atmosphere, this assumption is more realistic for fine than for coarse particles and thus creates an additional statistical bias in favor of $PM_{2.5}$.

The authors' conclusions in this paper are thus not supported by their data and analyses. As concluded by Lipfert and Wyzga (1997), a more appropriate conclusion would have been that it was incapable of determining differences in association with mortality by particle size. In addition, their failure to consider gaseous pollutants raises questions as to whether the mortality associations can be realistically assigned to any PM constituent.

2.1.2 <u>Hospitalization Studies</u>. Thurston et al. (1994) examined associations between daily admissions to 22 acute care hospitals in metropolitan Toronto and various air pollution concentrations monitored daily at a central site for 6 weeks during July and August for 1986-88. The end points were admissions for respiratory causes (all respiratory and asthma) and for selected "control" causes. These data were apparently for all ages. Air quality data included specially collected size-classified particle concentrations and routine data on TSP, O_3, SO_2, and NO_2. PM_{10} was obtained by summing the fine and coarse fractions of the dichotomous samplers and is thus affected by errors in both measurements. Thurston et al. noted that 1988 had a very hot summer and higher pollution levels. Ozone was highly correlated with temperature, as were SO_4^{2-}, NO_2, PM_{10}, and the coarse particle fraction of PM_{10} (but not $PM_{2.5}$); these correlations were based on data from which seasonal trends and day-of-week effects had been removed. The regressions model used trigonometric functions to control for seasonal variability; this has the disadvantage of treating all years the same. Since 1988 was clearly different, this model may have not controlled adequately for weather effects. Humidity was not considered as a potential confounder and no special treatment was given to days of peak temperature and humidity. Such days have been shown to be risk factors for mortality in the Northeastern U.S., for example (Schwartz and Dockery, 1992).

Thurston et al. presented their regression results in terms of p values and regression coefficients; for the purposes of this review, the latter were converted into mean effects, as discussed above. All of their models included daily maximum temperature and either one or two air pollutants. In the models that included only one air pollutant, the mean effects on total respiratory admissions were, in decreasing order: O_3, PM_{10}, coarse fraction of PM_{10}, TSP, $PM_{2.5}$, SO_4^{2-}, TSP - PM_{10}, and H^+. However, Thurston et al. chose to rank in terms of statistical significance, which places acidity first, instead of ozone. Similar results were obtained for asthma admissions. Ranking by mean effect ranking strongly suggests that fine particles are not the most important pollutants in Toronto in terms of their effect on respiratory hospital admissions. For example, Thurston et al. (1994) reported that aerosol acidity posed the largest relative risk (1.5) for respiratory hospital admissions in Toronto, based on its range from minimum to maximum concentration. Ozone was stated to pose lower risks on this basis (1.34). However, the mean effect of ozone was about five times that of acidity, because high concentrations of acidity occur so much less frequently than high concentrations of ozone. On this basis, we conclude that ozone is the pollutant of most concern in this study of Toronto hospital admissions.

In two-pollutant models, the combined mean effects were about the same for all pollutants (as has been noted by Lipfert and Wyzga (1995b) for mortality studies). In models including O_3 with $PM_{2.5}$ or SO_4^{2-}, O_3 always had the larger mean effect. However, ozone could have been confounded by hot day effects. Detailed results were not presented for the "control" admissions, but none of the cross-correlations between control admissions and the air pollution variables were significant.

2.1.3 <u>Other Morbidity Effects</u>. No other acute morbidity studies specifically involving fine particles were cited in the Decision Document.

2.1.4 <u>Summary of Acute Health Effects by Type of Particle</u>. Two related acute mortality studies considered size-classified particles. Both of them were compromised by substantially increased measurement errors for the coarse particles that are included in PM_{10}. In spite of this handicap, PM_{10} had about the same or larger effects on daily mortality than either $PM_{2.5}$ or SO_4^{2-}. For daily respiratory hospital admissions, PM_{10} had a larger mean effect than $PM_{2.5}$ in Toronto, and in a study not specifically considered by EPA (Gordian et al., 1996), crustal particles (PM_{10}) were specifically shown to predict daily respiratory outpatient visits. Sulfate had lower mean effects on respiratory hospital admissions than non-specific particles and was not significant in one of the acute mortality studies.

2.1.5 <u>Conclusions on Short-term Supporting Studies</u>. The studies used by EPA in support of their proposed 24-hr NAAQS for $PM_{2.5}$ do not find effects that can clearly be attributed to $PM_{2.5}$. Instead, associations with acute health effects do not seem to be specific to any single type of particle or gas (Lipfert and Wyzga, 1995b). There appears to be as much evidence implicating PM_{10} or the coarse fraction of PM_{10} as there is for $PM_{2.5}$. There is no evidence specific to $SO_4^=$ suggesting that those particles are any more toxic than other PM constituents; instead, given its statistical advantage resulting from its much smaller exposure error, the evidence suggests that $SO_4^=$ must be less toxic than other constituents of PM.

2.2 Studies Intended to Support the Proposed Annual NAAQS for $PM_{2.5}$

Although several cross-sectional mortality studies involving long-term mortality were discussed in detail in the Criteria Document (Section 12.4), EPA focused on only two of them in their Decision Document. These were recently published prospective cohort studies, which have the advantage of dealing with individuals rather than community-wide averages, but are still limited by the use of community based air quality monitoring data, rather than individual exposures.

<u>2.2.1 Long-Term Mortality Studies</u>. EPA only considered two long-term mortality studies, which are critiqued below. Studies that were not considered by EPA include the prospective study by Abbey et al. (1991) of California Seventh-Day Adventists that failed to find mortality effects due to any measure of PM, and the population based study of Lipfert (1993) that found equivocal effects due to TSP (but no significant associations with PM_{15}, $PM_{2.5}$, or SO_4^{2-}). These omitted studies are considered further below (Section 3).

2.2.1.1 <u>The Harvard Six Cities Study</u>. Dockery et al. (1993) analyzed survival probabilities among 8111 adults who were first recruited in the mid-1970s in six cities in the eastern portion of the United States. The cities are: Portage, WI, a small town north of Madison; Topeka, KS; a geographically-defined section of St. Louis, MO; Steubenville, OH, an industrial community near the West Virginia-Pennsylvania border; Watertown, MA, a suburb of Boston; and Kingston-Harriman, TN, two small towns southwest of Knoxville. This selection of locations thus comprises a transect across the Northeastern and Northcentral United States, from suburban Boston, through Appalachia, and into the upper Midwest and Great Plains.

The adults were white and aged 25 to 74 at enrollment. Follow-up periods ranged from 14 to 16 years, during which from 13 to 22% of the enrollees died. Of the 1430 death certificates, 98% were located, including those for persons who had moved away and died elsewhere. However, no information was given in the paper about the actual locations of death. The bulk of the analysis was based on all-cause mortality; no mention was made of subtracting external causes of death. The Cox proportional hazards model was used to estimate coefficients for the individual risk factors after stratifying by sex and age (5-year groups).

Air monitoring data were obtained from routine sampling stations and from special instruments set up by the research team. The air pollution exposure data used in this analysis were based on data from a central station in each community rather than on individual exposures. Steubenville, Kingston-Harriman, and St. Louis were the most polluted (and also had the oldest and least educated cohorts and the heaviest rates of smoking) among the six cities.

Most of the air quality measures were averaged over the period of study, in an effort to study long-term (chronic) responses; the specific averaging periods varied by pollutant. However, the size-classified particulate data began in 1980 while TSP data began in 1974; from 1974 to 1980 there were large reductions in TSP (and probably in the size-classified particles as well), so that it appears that the size-classified data may be less representative of cumulative exposures than TSP. Sulfate appeared to be intermediate in this regard. Exposures prior to the study (which were much higher in the dirtiest cities and often violated the NAAQS) were not considered. In this sense, there is a mismatch in time between the air quality data, which were obtained after the study began, and the descriptive data on individuals, which pertain to the period before entry into the study.

The effects of air pollution were evaluated in two ways: by evaluating the risks of residence in each city relative to Portage (the city with the lowest pollution levels for most indices), and by including air quality levels directly in the regression models. Since only six different values were available for each pollutant, the effective degrees of freedom are greatly reduced by this procedure, relative to the other variables and to what could have been done had personal exposure data been available.

Individual characteristics of the members (and thus of the decedents) considered included smoking habits, an index of occupational exposure, body mass index, and completion of a high-school education. The index of smoking rate used here was pack-years, defined as the average number of packs of cigarettes smoked per day times the number of years of smoking. This metric is also a function of age. Current and former smokers were treated separately. The index of socioeconomic status used was having less than a high school education; Rogot et al. (1992) show that this index is a good measure of mortality differences due to differences in education for white men, but not for white women. For women, there is a more continuous relationship between (reduced) relative mortality risk and increased levels of educational attainment.

The index of occupational exposure to air pollution (dusts or fumes) did not take into account the nature of the agents involved or the details of exposure. Occupational exposure to dusts or fumes was not found to be a significant risk factor; this outcome may have resulted from the lack of specificity of the index used. The average percentages of subjects having occupational exposure were high, ranging from 28 to 53%, with an average across all cities of 45%.

The index of physiology used was the body mass index (BMI), defined as weight divided by height squared (kg/m^2), treated as a linear relationship. The relative risk (RR) of increased body mass was similar to that found by Sandvik et al. (1993) (where it was not statistically significant), but other investigators have found that the relationship is U-shaped rather than linear and that it may interact with other risk factors, especially smoking (Gronbaek et al., 1994). The nonlinear relationships imply increased mortality risk for those at either end of the scale, either obese or very thin.

In multi-factor regressions, the effect of inadequate representation of a confounder is to inflate the importance of other variables. In this analysis, for example, the excess mortality risk in Steubenville was reduced from 67% to 26% by adjusting for age, smoking status, education, and body-mass index. It is likely that the adjustments would have further reduced the "unexplained" risk if more detailed attention had been given to each of these factors. In addition, a number of potential confounders were not considered.

No consideration was given to possible independent effects of occupation classification, other personal lifestyle variables such as diet or physical activity, migration, or income. Presumably, each subject was characterized by his status at entry to the study; follow-up data on possible changes in risk factors over time (such as smoking status, occupational exposure, or body mass) were not mentioned. Figure 1 (from Lipfert, 1995) shows how differences in the fraction of the elderly population with sedentary lifestyles in each area (these data were not available for Topeka) can account for most of the mortality gradient, and how the slope of that relationship matches that found in an independent study in California (Breslow and Enstrom, 1980). The figure shows that differences in lifestyle among the six areas could explain most of the mortality risk that was attributed to air pollution.

It should also be noted that new independent studies (Mesink et al. 1996) continue to identify lack of physical activity as an important risk factor for heart disease and premature mortality. Also, new studies (Brody, 1996) have found that improper maternal diet can predispose an infant towards heart disease later in life; it is not unlikely that the Appalachian region of the U.S. that is now high in $SO_4^=$ aerosols (and $PM_{2.5}$) could have been affected by poor nutrition during the depression years when today's adults were *in utero*.

Statewide mortality data indicate substantial differences in survival rates across this transect of the Northeastern U.S that are similar to those that were observed among the six cities; this suggests that the responsible factors constitute regional trends. Dockery et al. report a long-term average mortality rate of 16.2 deaths per 1000 person-years in Steubenville and a rate of 9.7 in Topeka, yielding a range in average (crude) relative risk of 67% among the six cities. After individual adjustment for age, smoking status, education, and body-mass index, the range in average relative risk was reduced to 26%. In comparing the most and least polluted cities, Dockery *et al.* also report elevated risks for cardiopulmonary (ICD-9 401-440, 460-519) causes (1.37, [1.11-1.68]) and lung cancer (ICD-9 162) (1.37, [0.81-2.31], not significant). The relative risk for all other causes of death was 1.01 (0.79-1.30).

Dockery et al. report that "mortality was more strongly associated with the levels of fine, inhalable, and sulfate particles" than with the other pollutants, which they attributed primarily to factors of particle size. They provided relative risk estimates and confidence limits based on the differences between air quality in Steubenville and in Portage for these three pollutants. However, it is a relatively simple matter to independently estimate these coefficients from the adjusted risks and pollutant levels reported for each of the six communities; these estimates are given in Table 1 (from Lipfert, 1995) as a means of comparing the various pollutants and combination of pollutants. Note the absence of information on CO as a risk factor. The data for 1970 TSP (lagged about 12 years) were obtained from Lipfert (1978), assuming that Madison could represent Portage, WI. The estimates made by this procedure shown in Table 1 correspond quite closely to the figures given by Dockery et al. based on output from the Cox proportional hazards regression model. However, because there are only 6 degrees of freedom for the air quality data, the confidence limits in Table 1 are somewhat wider than those for the risk factors having individual data. As in the original paper, the relative risks are based on differences in average air pollution levels between Steubenville and Portage.

**Table 1 Estimated Relative Risks in Six U.S. Cities
Associated with a Range of Air Pollutants (from Lipfert, 1995)**

species	regr. coeff.	std error	pollution range[*]	rel. risk	95% CIs (n=6)
PM_{15}	0.0085	(0.0026)	28.3	1.27	(1.04-1.56)
$PM_{2.5}$	0.0127	(0.0034)	18.6	1.27	(1.06-1.51)
SO_4^{2-}	0.0297	(0.0081)	8.5	1.29	(1.06-1.56)
TSP	0.0037	(0.0014)	55.8	1.22	(0.99-1.53)
1970 TSP	0.0014	(0.00044)	154	1.25	(1.03-1.50)
$TSP-PM_{15}$	0.0042	(0.0032)	27.5	1.12	(0.88-1.43)
$PM_{15}-PM_{2.5}$	0.0178	(0.0098)	9.7	1.19	(0.91-1.55)
$PM_{2.5}-SO_4$	0.0255	(0.0029)	8.4	1.24	(1.16-1.32)
$PM_{15}-SO_4$	0.0121	(0.0034)	18.1	1.24	(1.05-1.48)
SO_2	0.0093	(0.0032)	19.8	1.20	(1.01-1.43)
NO_2	0.0126	(0.0046)	15.8	1.22	(1.00-1.49)

[*]range = (maximum value - minimum value) for the 6 cities

Dockery et al. noted that ozone levels varied little among the six cities; this might not have been the case if a measure of peak concentration had been used for comparison instead of the annual means. No relationship was found for aerosol acidity (H^+), but only limited data were available. Table 1 shows only small differences in the relative risks among many different pollutants, owing in part to the strong collinearity present in this data set. TSP and the coarse particle variables created through subtracting PM_{15} from TSP and $PM_{2.5}$ from PM_{15} were not significant, but the non-sulfate portion of $PM_{2.5}$ had the tightest confidence limits (to create this variable, SO_4^{2-} was multiplied by 1.2, assuming an average composition of NH_4HSO_4, before subtraction). Also, the estimated 1970 TSP variable performed better than the TSP data used by Dockery *et al.* (ca. 1982). However, all of the differences in relative risks among pollutants and their confidence limits could have occurred due to chance, given only 6 observations.

Among the six cities, Steubenville appears to be the most important. It was the only city for which the residual risk, after accounting for age, race, sex, smoking, education, body mass, and occupational exposure (however crudely), was significantly different from zero. It is also the only location on Figure 1 with incremental mortality risks not explained by increased frequency of sedentary lifestyle. Steubenville has traditionally been a city of heavy industry, primarily metalworking, and it is logical to assume that the chemical characteristics of its air pollution differ from more heterogeneous urban areas such as St. Louis, Topeka, or Watertown (Boston). For example, monthly average TSP readings as high as 800 ug/m^3 were recorded in Steubenville during the period before the study (Lipfert and Wyzga, 1991) and airborne Fe and Mn concentrations there were among the highest in the nation ca. 1980 (Lipfert et al., 1988).

Comparison of the pollution risks among the various available cohort subsets is one of the most important outcomes of a study on individuals. Such comparisons must account for the higher variability among subgroups, however, and Dockery et al. were not capable of distinguishing differences in excess risks between subgroups less than about 18% (i.e., a relative risk of 1.18 cannot be distinguished from one of 1.36, for example). Although none of these subgroup differences were statistically significant, the mortality risks associated with area of residence (and thus air pollution) were higher for females and for smokers and the risks were also higher for those occupationally exposed compared to the nonexposed. However, the personal exposures of nonsmokers and of those without occupational exposures are more likely to correlate with the outdoor monitoring data used here than those with these additional sources of exposure. It is possible that the relative risk estimates for outdoor air pollution for these lesser exposed subgroups (1.19 and 1.17, respectively) might be the most reliable estimates in the study. However, risks for these subgroups are just as likely to have been overstated by lifestyle confounding as for the others.

Dockery et al. were quite cautious in their conclusions, stating only that the results suggest that fine-particulate air pollution "contributes to excess mortality in certain U.S. cities." However, EPA has moved from this "suggestion" to a firm conclusion, which they used as a basis for new regulation.

2.2.1.2 <u>The American Cancer Society Study</u>. Pope et al. (1995) analyzed 7-year survival data (1982-89) for about 550,000 adult volunteers obtained by the American Cancer Society (ACS). This cohort is not a random sample of the U.S. population; it is 94% white and better educated than the general public, with a lower percentage of smokers than in the Six City Study. The (crude) death rate during the 7.25 years of follow-up was just under 1% per year, which is about 20% lower than expected for the white population of the U.S. in 1985, at the average age reported. In contrast, the corresponding mortality rates for the Six-Cities study discussed above tended to be higher than the U.S. average. No mention was made of residence histories for the decedents; matching was done on residence location at entry to the study. Causes of death considered included all causes, cardiopulmonary causes, lung cancer, and all other causes.

The Cox proportional hazards model was used to define individual risk factors for age, sex, race, smoking (including passive smoke exposure), occupational exposure, alcohol consumption, education, and body-mass index. The deaths, about 39,000 in all, were assigned to geographic locations using the 3-digit zip codes listed at enrollment into the ACS study in 1982. Relative risks were then computed for 151 metropolitan areas defined by these zip codes and were compared to the corresponding air quality data, ca. 1980. The sources of air quality data used were the EPA AIRS system for sulfates, as obtained from high-volume sampler filters for 1980, and the Inhalable Particulate Network for fine particles ($PM_{2.5}$). The latter data were obtained from dichotomous samplers during 1979-81; Pope et al. used the values from this data base reported by Lipfert et al. (1988) but only 50 $PM_{2.5}$ locations could be matched with the death data. The correlation between the two pollutants was 0.73. No other pollutants were considered. The authors assumed that 1979-81 pollution values would be representative of long-term cumulative exposures, in keeping with the objective of analyzing chronic effects. The sulfate values from the inhalable particle filters, which are thought to be free from artifacts, were not used by Pope et al. As in the Six City Study, all estimates of air pollution exposure were based on data obtained from central monitors in each city, rather than from personal exposures.

Pope et al. took great care with the potential confounding factors for which data were available. Several different measures of active smoking were considered, as was the time exposed to passive smoke (the only study to do so). The occupational exposure variable was specific to (any of) asbestos, chemicals/solvents, coal or stone dusts, coal tar/pitch/asphalt, diesel exhaust, or formaldehyde. The education variable was an indicator for having less than a high-school education. However, alcohol use and body-mass index were considered as linear predictors of survival, whereas other studies have indicated these effects to be non-linear (U- or J-shaped) (Gronbaek et al., 1994; Lew and Garfinkel, 1979; Doll et al., 1994). Risk factors not considered by Pope et al. include migration, income, employment status, dietary factors, drinking water hardness and physical activity levels, all of which have been shown to affect longevity (Pocock et al., 1980; Belloc, 1973; Sorlie and Rogot, 1990). Pollution coefficients were reduced by 10-15% when variables for climate extremes were added to the model.

Pope et al. found very consistent pollution risks for all-cause mortality for males and females and for ever-smokers and never-smokers. The relative risks for air pollution were slightly higher for females for cardiopulmonary causes of death. The lung cancer-sulfate association was only significant for males, except for male never-smokers. The adjusted total mortality risk ratios (computed for the range of the pollution variables) were 1.15 (95% CI = 1.09-1.22) for sulfates and 1.17 (95% CI = 1.09-1.26) for $PM_{2.5}$. When expressed as log-linear regression coefficients, both pollution measures were quite similar: 0.0070 (0.0014) per ug/m^3 for SO_4^{2-} and 0.0064 (0.0015) for $PM_{2.5}$; however, it is possible that the SO_4^{2-} results have been biased high by the presence of filter artifacts (Lipfert, 1994a). Expressed as the percentage of mortality associated with air pollution at the mean values and corrected for SO_4^{2-} filter artifact, Pope et al. found mean effects of about 5% for sulfate and 12% for $PM_{2.5}$. No significant excess mortality for the "other" causes of death was attributed to air pollution.

In some ways, the ACS study was an improvement on the Six Cities study, because of its greater care in representing some of the confounders and its increased geographic scope. However, it suffered from omitting many of the same potential regional confounders and its use of only two highly related air pollutants. Other types of particulate matter and carbon monoxide were not investigated.

<u>2.2.2 Other Long-term Health Effects.</u> There are no long-term studies of air pollution and hospitalization, in part because utilization of hospital facilities is strongly influences by supply and demand factors (Lipfert, 1994b). This imposes a severe limitation on examining coherence among long-term health endpoints.

The EPA decision document referred to Table 13-5 of the CD for other morbidity evidence supporting the need for an annual $PM_{2.5}$ standard. In that Table, only the 24-Cities Study (discussed in detail below) actually considered fine particles in this regard. The previous Six City study was also included, but its comparisons were limited to TSP and PM_{15}/PM_{10}. Data from Abbey et al. (1995) for SO_4^{2-} in California were also included with respect to bronchitis in children. Using EPA's comparison criteria for relative risks (based on arbitrary increments of air quality), the "coarse" particles (TSP, PM_{15} or PM_{10}) had an average odds ratio for children's bronchitis of 3.12 with a standard error of 0.16 (between studies), while the "fine" particles (H^+, SO_4^{2-}, $PM_{2.1}$) had an average odds ratio of only 2.26, with a standard error of 0.36. Thus, the smaller particles do not show greater toxicity for this endpoint. For decreased lung function in children, there were no significant differences across the Six Cities, and the decrements in the 24 Cities were not statistically different among any of the particle measures considered (H^+, SO_4^{2-}, $PM_{2.1}$, PM_{10}). Note that comparisons of respiratory effects due to ozone were not made in CD Table 13-5. The Six and Twentyfour Cities studies considered bronchitis prevalence, while the study of Abbey et al. (1995) considered incidence of new cases. The odds ratio for incidence was lower (1.39), but data for only one pollutant (SO_4^{2-}) were reported by EPA.

2.2.2.1 <u>The 24-Cities Study of Children's Respiratory Health.</u> In response to concerns about the toxicity of acid aerosols, a cross-sectional analysis of the respiratory health of children (ages 8-10) in 24 small towns in the U.S. and Canada was begun in 1988 and continued through 1991 (Dockery et al., 1996; Raizenne et al., 1996; Spengler et al., 1996; Brook and Spengler, 1996). The protocol involved monitoring the local air quality in each town for about a year for a wide range of species, obtaining parental responses in September and October on children's respiratory health through questionnaires, and testing the pulmonary function of the children from October to May. Roughly one third of the communities were investigated in each of the 3 school years involved. The authors of these studies reported that symptoms of bronchitis were associated with acidity and sulfates and that reduced pulmonary function was associated with particle acidity. These results were based on about 13,000 to 15,000 children, but the air quality measures were obtained from only 24 central monitoring stations. No individual exposures to air pollution were considered.

Reanalysis of the published data suggests that other interpretations may be equally or more likely. First, the design of the study has certain shortcomings. Since data were not taken simultaneously in all communities, both time and space are variables in the study. For example, 1988 was a high ozone and acidity year in much of the U.S. (Thurston et al., 1994), and it is possible that the monitoring data obtained for some of the 8 towns studied that year were not representative of long-term exposures. However, Brook and Spengler considered this problem for some (but not all) of the communities and dismissed it for the aggregate. In addition, lung function testing was performed sequentially throughout the school year and it is possible that seasonal variations may have biased the results for specific communities. The data of Spodnik et al. (1966) and of McKerrow and Rossiter (1968) show seasonal swings of a few percent, which is of the same order of magnitude as the air pollution effects reported by Raizenne et al. (1996). No mention was made of accounting for acute lung function responses to air quality differences on the days of testing. A priori, transient acute responses would be expected to exceed differences in long-term averages.

The results for respiratory symptoms in the 24 cities (Dockery et al., 1996) show that only 5 of 64 combinations of symptoms categories and pollutants were statistically significant; although not all of these comparisons are independent, this is close to what might have been expected just due to chance. In addition, the stratified sample analysis (Figure 2 of Dockery et al.) shows that the effect of acidity on bronchitis symptoms depends almost completely on the contrast between Eastern U.S. and Canadian cities. When U.S. cities were considered alone, the odds ratio was 1.0, and when only eastern cities were considered, the odds ratio exceeded 1.0 but was not significant. Thus, the acidity relationship is not robust. Also, only the first year of study (the high ozone year) was significant when the data were stratified by year. Presence of molds in the home was not mentioned as a potential (confounding) risk factor (as determined separately by Spengler et al. (1994) for the same children and by Dales et al. (1991) for Canadian children), yet use of a humidifier was shown to define the most sensitive subgroup. Regression analysis of the tabulated aggregated data for each city (n=24) did not produce plausible relationships (smoking indicated a negative effect on bronchitis prevalence, for example). It is possible that the use of parental recall to define the endpoints produced too much noise for credible findings; it is also possible that there are systematic differences between the U.S. and Canada, either involving climate or definitions or diagnoses of symptoms. All of these factors suggest that there are hypotheses that provide more reasonable alternatives to the authors' conclusion that aerosol acidity was the causal agent.

The results for lung function were more amenable to reanalysis. Figure 2 (of this review) plots the city-specific data for the average forced vital capacity (FVC, a measure of lung volume), the fraction of children who had less than 85% of the FVC predicted for that community (low FVC), after adjustment for age, sex, height, and weight, and the peak expiratory flow rates (PEFR). The data points are clustered geographically according to the authors' stratification of locations: the "SO4 belt" represents Appalachia; the "transport region" is generally downwind of these locations; west coast stations are in California, including one in a high ozone region; "background" stations are in Arkansas, South Dakota, Saskatchewan, and British Columbia. When these plots are viewed without regard to geographic clustering, they appear to support the hypothesis that acidity degrades lung function. However, for FVC, the variation within the "SO4 belt" is so large and unrelated to acidity that a regional bias factor is suggested, as an alternative to the causal acidity hypothesis. To evaluate this alternative hypothesis, the 22 aggregated observations (two of the towns refused to participate in this phase of the project) were regressed against air quality, regional dummy variables, and potential confounders such as parental smoking and education level. These regressions showed that the regional dummy variable was a more powerful predictor of FVC than any air pollutant (by deleting the dummy variables, this regression technique also reproduced the result for acidity reported by Raizenne et al. on the basis of individuals). For example, the best regression for FVC included the SO4-belt regional dummy variable (−) and education (+) (and no air pollutants) with an R^2 of 0.64; when regional dummy variables were excluded, the best regression included acidity and peak ozone and had an R^2 of 0.56. The conclusion thus follows that geography per se (i.e., long-standing regional differences) is a better predictor of these small lung function differences than air pollution. The implied mean effect of acidity on FVC was about 1.6%; of peak O_3, about 1.9%. However, for peak flow, acidity was a better predictor than the regional dummy variable (mean effect, about 4%). These increments represent very modest effects on lung function.

The lung function results should be interpreted as suggesting that Appalachian children tend to have smaller lungs (as indexed by FVC) than children elsewhere in the U.S. and Southern

Canada. This might be a result of chronic exposure to some type of air pollution, but that hypothesis seems no more likely than one related to regional ethnicity, diet or exercise, given the known regional differences in these factors (Lipfert, 1995). Particulate acidity as a long-term risk factor is an especially problematic rationale, given the low probability that ambient acidity as measured at central stations would actually be delivered to the lungs. First, ammonia in the breath will tend to neutralize inhaled acids, especially for those typically smaller particles that might be found near sources. Second, acid particles that penetrate indoors will tend to be neutralized by indoor ammonia, coming from pets, household cleansers, or personal sources. Acidity is perhaps the only air pollutant against which we have two layers of defenses, and it is difficult to rationalize its role as a cause of long-term health effects. Ozone may well be a better candidate for changes in lung function, especially since ozone has so much larger effect on acute lung function changes than acidity (EPRI, 1994).

It is also of interest to compare the results of the 24-City Study with those of its predecessor, the Harvard Six City Study. In that study, "bronchitis" was associated with TSP and with PM_{15}, and later with acidity for a subset of the cities. However, for lung function for adult never-smokers, Steubenville (the most polluted city) had a higher average lung function than would have been expected for that population, in spite of the much worse air quality that was experienced there during the 1960s and 1970s. For children, using TSP as a predictor, both St. Louis and Steubenville had lung function averages higher than expected (see Lipfert [1994b], pp. 471-3 for plots of these data). Thus, the Six City data seem to suggest that **worse** air pollution is accompanied by **better** lung function! In the Six City Study, the entire range of TSP was associated with about 3% in lung function, which seems to be a smaller incremental effect than is now seen in the 24 cities, even though pollution levels are currently much lower. Lipfert (1994b) also noted that use of long-term average ozone as a health predictor obscures the effects of peak ozone excursions; he estimated that differences in peak ozone could have predicted the Six City lung function results, with a mean effect of about 3.6%.

2.2.3. <u>Conclusions from Long-Term Supporting Studies</u>. While the prospective cohort mortality studies broke new ground in terms of cross-sectional studies, they suffered from severe data limitations, in large part because they were not designed for this purpose from the outset and thus the necessary data on each subject were not collected at entry to the studies. The regional confounding between sulfate and/or fine particles that exists in the U.S. precludes valid estimates of their long-term effects on this basis. Potential confounders include lifestyle, diet, economic and employment status, and migration. These studies thus provide no robust basis on which to estimate the existence of long-term health effects. This also precludes the use of these studies to conclude that acute mortality effects are not dominated by mortality "displacement" (subsequent decreases in daily deaths that compensate for the increases associated with the environment) or as a means of estimating the degree of life-shortening in the acute mortality studies. Failure to consider previous or cumulative exposures to air pollution overstates the effects attributed to current exposures. The facts that peak ozone and CO exposures were not considered by either study, and that the more comprehensive of the two (the ACS study) considered a very limited range of particle metrics constitute severe limitations in the usefulness of the studies.

However, there is an unexpected important implications from these studies (Lipfert, 1995): None of the population subgroups examined appeared to be significantly more sensitive to air pollution than any other (assuming equal probability of confounding). Since the relative risks were virtually unchanged by excluding subjects with hypertension and diabetes, this finding might also be extended to those with pre-existing chronic diseases. This apparent homogeneity of response has implications regarding the acceptability of population-based studies in which such stratification is not possible. For example, population-based studies such as Lipfert (1993) are unable to incorporate individual risk factors and thus must assume that all members of the population group (city, county, SMSA) are alike in these respects. The results of the prospective studies suggest that this may be a viable assumption.

The two multi-pollutant long-term morbidity studies considered by EPA failed to show any effects on children's respiratory health that could be specifically attributed to fine particles alone. If anything, there seemed to be a slight advantage to the more inclusive PM metrics (such as PM_{10}). Note that, although these particular two studies were limited in scope, they may have offered arguably the best hope of detecting such air pollution morbidity effects, if indeed they exist. They studied children, thus avoiding issues of greatly differing previous community air quality levels and issues of active smoking and occupational exposures. However, like all the other long-term studies, they suffered from lack of personal exposure data and from the potential for regional confounding. The only study to consider bronchitis incidence (as opposed to prevalence) found the lowest odds ratio of the group, implying that prevalence studies may overstate the rate of creation of new cases.

3. Recently Published Epidemiological Studies (not considered by EPA)

3.1. Daily Mortality

3.1.1 <u>Ostro (1996)</u>. In this study of mortality in Southern California (Riverside and San Bernadino Counties) daily counts from 1980-86 (mean=41 per day) were regressed against a surrogate measure of $PM_{2.5}$ that was estimated from humidity-adjusted visual range data taken at the local airport (Ontario, CA). Direct measurements of $PM_{2.5}$ were not available. Ostro's regression model controlled for day of week, weekend, year, calendar quarter, month, daily temperature and dew point, but it did not specifically control for hot days (even though the maximum temperature reached 111°F). Ostro considered several regression model forms, including ordinary least squares, Poisson, deviations from 15-day moving averages, generalized linear modeling, use of trigonometric terms to model seasonality, and stratification by season. Ozone was the only other air pollutant considered; concentrations were measured at four sites in the area; the averaging time used for O_3 was not stated, but since the mean was given as 0.14 ppm, it appears to be the peak hour for each day. The overall mean surrogate $PM_{2.5}$ (based on visibility data) was 32.5 ug/m^3, with a maximum of 190 ug/m^3. Ostro also reported that an alternative estimation method (based on local measurements) gave a mean of 42.8 ug/m^3, with a maximum of 231 ug/m^3. Ostro reported that the regression results were insensitive to the use of this alternative $PM_{2.5}$ estimation method. No other air pollutants were considered, even though data were probably available from the routine monitoring stations that recorded ozone data.

These California counties are downwind of Los Angeles and constitute one of the most polluted areas in the United States, with both O_3 and $PM_{2.5}$ levels more than twice as high as most locations in the Eastern U.S. and in frequent violation of NAAQS. Furthermore, the climate lends itself to frequent outdoor exposures, year round. The daily mortality counts here were high enough so that random noise should not have obscured the air pollution relationships. However, Ostro reported 89% of all deaths in the age 65 and over category, which is suspicious since only about 12% of the population was in that age group. Thus, a relationship between daily mortality and air pollution would be expected to be found here if it in fact existed anywhere.

However, Ostro found that surrogate $PM_{2.5}$ was significantly associated only with total and respiratory mortality, and only in the summer, not with circulatory or deaths of the elderly. This may be first time that the younger age group has been implied to be more important in such a study. Surrogate $PM_{2.5}$ was not significant for the full year for the causes of death considered. Ostro reported that his results were insensitive to the type of regression model used. He only reported ozone results for the summer and for total mortality and did not present a sensitivity analysis for O_3. When regressed as the only pollutant in the model, ozone was significant and had a mean effect of about 2.8% with an upper 95% confidence interval of about 7%. The mean effect of surrogate $PM_{2.5}$ for the same conditions was reported as 3% with an upper confidence interval of 5%. (Since Ostro reported only one digit for these results, the two pollutants cannot be compared precisely, but it appears that O_3 had the larger mean effect in separate regressions.) It is also possible that the failure to specifically consider hot day effects inflated the results for both O_3 and surrogate $PM_{2.5}$.

Considering O_3 and surrogate $PM_{2.5}$ together in a joint regression, caused both coefficients to decrease and lose significance, which is a typical result with collinear variables. The O_3-surrogate $PM_{2.5}$ correlation in these data was 0.58. In addition, possible differences in exposure error must also be considered when evaluating the results from joint regressions; when two correlated variables that have equivalent underlying effects are regressed jointly, the one with less measurement error will tend to dominate (Lipfert and Wyzga, 1995a). Ozone is strongly attenuated outdoors by local combustion sources of NO and indoors by reactions on surfaces. In contrast, $PM_{2.5}$ has a long atmospheric residence time and penetrates into buildings where it will remain suspended and may be augmented by indoor sources such as smoking or cooking. In addition, the particular $PM_{2.5}$ data used here were determined from visual range and thus are spatially integrated, which might make them actually perform better in community-wide studies than local measurements would. Thus, the near survival of ozone in the joint regressions could be interpreted as showing a stronger underlying relationship with mortality than surrogate $PM_{2.5}$. However, in this paper, the author did not interpret his results in the context of differential measurement error, but suggested that the small difference between mean surrogate $PM_{2.5}$ levels in summer (36 ug/m^3) and in winter (29 ug/m^3) was responsible for the failure of surrogate $PM_{2.5}$ to reach significance in winter (which implies a threshold). However, the range of this variable was greater in winter (180 ug/m^3) than in summer (95 ug/m^3), and the range should have been more important in the regression than the mean (unless, of course, a threshold exists).

In this paper, three variables (surrogate $PM_{2.5}$, O_3, temperature) interacted and were likely intercorrelated to varying degrees; a general correlation matrix was not presented. However, surrogate $PM_{2.5}$ was (just) significant only in summer, when its correlation with temperature was reported to be low. The failure to consider very hot days separately (as has been done by other authors, such as Schwartz and Dockery [1992]) may have overstated the ozone effects. Ostro stated that (because of intercorrelations), "disentangling the effects of temperature versus ozone is problematic". He should have extended this qualification to include the effects of $PM_{2.5}$.

Ostro noted that the results of his study were consistent with those of Kinney et al. (1991) for Los Angeles County, who also found a stronger effect for O_3 than for PM, but who also noted the difficulty of disentangling PM effects from other pollutants associated with vehicle exhaust (CO and NO_2). It is thus also likely that these problems of collinear pollutants (not considered by Ostro) also existed in Riverside and San Bernadino Counties.

The overall conclusion of this review is that Ostro's study fails to confirm the hypothesis of Schwartz et al. (1996) of an independent effect of fine particles on daily mortality, in spite of the extremely high pollutant levels encountered at the location of study. Ostro's strong results for ozone must be viewed in the context of the frequent violations of the O_3 NAAQS in this location and his inadequate treatment of very hot weather in the regression model.

3.1.2. <u>Anderson et al., 1996</u>. Daily death counts were regressed against air quality for ozone, British smoke, SO_2, and NO_2 in Greater London from 1987 to 1992 by Anderson et al. (1996). Carbon monoxide was not considered. British smoke measurements generally involve particles of less than 5 um; it thus may be considered as a specialized index of fine particles. Anderson et al.'s regression model controlled for season, secular trend, day-of-week, holidays, influenza epidemics, temperature and humidity. They considered deaths from all causes (less accidents, etc.), respiratory diseases, and cardiovascular diseases; the mean daily total death count was 175. Ozone was measured at a single station in central London and was considered as 8-hr daytime averages and daily peak values. NO_2 was measured at two urban sites and was considered as peak hour values and as 24-hr averages. SO_2 and black smoke were measured at four stations that employed the same types of monitoring systems that were used in the 1950s and 60s. the distributions of black smoke and ozone appeared to be nearly lognormal, with geometric standard deviations (GSDs) of about 1.4 and about 1.6, respectively. The GSDs were somewhat lower for SO_2 and NO_2.

The regression model was based on autoregressive log-linear Poisson regression, following prior work by Schwartz and colleagues (see Schwartz and Dockery, 1992, for example). Trigonometric terms were used for seasonal adjustment; dummy variables controlled for days of the week, holidays, and an influenza epidemic. Temperature effects were represented using a piecewise linear relationship with three segments. Lags up to two days were considered for the air pollutants; each of them was found to maximize the response to at least one of the various pollutants considered. There were 18 combinations of causes of death, pollutants, and averaging times, and each was considered separately for the entire period, the cool season, and the warm season. Thus there were 54 regression coefficients computed, only some of which were actually independent. No non-plausible combinations were considered as "controls." Anderson et al. were somewhat more thorough than many of their predecessors in terms of the range of results presented.

Regression results were reported for the entire period and by season. For the entire period, all-cause mortality was significantly associated with smoke (mean effect = 0.017) and with ozone (8-hr mean effect = 0.014; peak-hr mean effect = 0.017). The mean effects for NO_2 and SO_2 were lower and nonsignificant. No pollutants were significantly associated with cardiovascular deaths, which is perhaps the first time that this finding has been reported. Mean effects on cardiovascular deaths were all less than 0.01. Only ozone was significantly associated with respiratory deaths, with mean effects of 0.035 and 0.042 for 8-hr and peak-hr averages, respectively.

In the warm season, both ozone measures were significantly associated with all three cause of death categories, NO_2 was significant for cardiovascular deaths but negatively significant for respiratory deaths, and SO_2 was significant for all causes. Ozone was significant for respiratory deaths and smoke was significant for all causes in the cool season.

Anderson et al. also presented results for joint pollutant regressions for ozone and smoke for the same combinations of seasons and causes of death. In these regressions, coefficients for both pollutants generally increased over their separate-regression values, suggesting independent effects. However, this result is also consistent with the negative correlation between the two pollutants. The results for both O_3 and smoke were unaffected by adding SO_2 or NO_2 to the models, but the effects of SO_2 and NO_2 were sensitive to the inclusion of smoke and O_3, respectively. These results are also consistent with their respective bivariate pollutant correlations.

When compared with typical results seen in the United States, both similarities and important differences are seen. Similarities include the facts that the mean air pollution effects in London were similar for all pollutants considered, and that SO_2 and NO_2 effects were weaker than those for smoke and O_3. In addition, responses for a given pollutant, end point, and season were not reported as a function of lag, so that possible harvesting effects could not be evaluated (like most U.S. studies).

Differences with respect to U.S. studies include the strength of the O_3 effect in comparison with black smoke, although this has only been tested directly in the U.S. in Los Angeles (Kinney and Ozkaynak, 1991). The finding of ozone effects in Central London on a par with those in Los Angeles County, where concentrations are much higher and outdoor exposures may be more common, seems quite unexpected. In addition, a relationship between O_3 and respiratory disease mortality in London <u>winters</u> strains credulity. Black smoke in London may be an index of primary emissions from vehicles, especially diesel engines, whereas summer PM in the U.S. is usually dominated by secondary particles. It is unfortunate that Anderson et al. did not include CO in their study.

The conclusions of Anderson et al. should be viewed in the context of potential differential effects of measurement error. They stated that their ozone readings are undoubtedly biased low because they were obtained in the central city. Ozone values in more outlying areas may also have a more uniform temporal distribution, which could affect their correspondence with mortality peaks. Indoors, both SO_2 and O_3 will be greatly attenuated, but NO_2 and black smoke may be augmented by indoor sources, depending on the fuels used for cooking and heating. Since both CO and smoke are primary emissions from vehicles, the omission of CO from the analysis may have inflated the associations with smoke. In addition, this is one of the few time-series studies that failed to find PM associations with cardiovascular mortality. Finally, Anderson et al. were very thorough, in considering more than one averaging time for O_3 and NO_2 and several causes of death for each pollutant. Their study could have been further improved by considering different age groups and by reporting more details of the lag structures.

The results of Anderson et al. could be used to predict the outcome of a single episode that occurred in London during December 1991, in which peak-hr NO_2 reached 423 ppb and 24-h average smoke reached 228 ug/m^3, as reported by Anderson et al. (1994). The model developed from the winter results of Anderson et al. (1996) would predict about a 25% increase in all-cause mortality, almost all due to smoke; the actual observed increase was only 10%. For cardiovascular mortality, an increase of 14% was observed, but the model would have predicted no net change. For respiratory deaths, the model would have predicted an increase of about 6%; the observed increase was 22%. This unsatisfactory correspondence between predicted and observed changes in daily mortality suggests that models of this type will not be reliable in predicting future improvements in public health that might be gained by eliminating severe air pollution episodes.

3.1.4   <u>The HEI Multi-Pollutant Study of Philadelphia</u>.   The Health Effects Institute (HEI) contracted with Johns Hopkins University to replicate certain of the early acute studies of air pollution and mortality (Samet et al., 1995) and to "explore the sensitivity of findings to analytic approaches and assumptions" using data from Philadelphia (1973-88) as a basis. The first report (Samet et al. 1995) investigated six previously published data sets and generally agreed with the findings of the original investigators, using their data and methods. The report said that "it is reasonable to conclude that, in these six data sets, daily mortality from all causes combined, and from cardiovascular and respiratory causes in particular, increased as levels of particulate air pollution indexes increased". However, they also noted that data from Philadelphia (representing the most deaths under study) suggested that particles were not acting alone.

In the second report (Samet et al., 1996), they found that the results for TSP and/or SO_2 in Philadelphia were relatively insensitive to changes in regression models or to treatments of weather effects. They then added daily data on CO, O_3, and NO_2 and proceeded to a multi-pollutant analysis, starting from scratch with regression model development. They also found that air pollution associations were strongest for those aged 75 and over. In their multi-pollutant model, TSP, SO_2, CO lagged 3 days, and O_3 were significantly positively associated with total mortality, while NO_2 had a significant negative association (a fact that was not mentioned in the abstract). As a single pollutant, NO_2 was not significant (positive), so that the multi-pollutant result probably represented the effects of multicollinearity. Kinney and Ozkaynak (1991) found similar results for Los Angeles when they entered 5 pollutants simultaneously; in this case, the PM variable became negative (but not significant). Similar results were shown by Lipfert and Wyzga (1995a) with simulated data. Nevertheless, the final remarks of Samet et al. (1996) in this report bear repeating (and heeding): "We caution against using the model coefficients directly to estimate the potential consequences of lowering concentrations of the individual pollutants through regulatory measures; the pollutant concentrations are correlated and the estimates of their effects depend on modeling assumptions." It should be perhaps noted that both Samet et al. (1996) and Anderson et al. (1996) have reported significant negative mortality associations with NO_2.

3.1.3   <u>The Birmingham Replication Studies</u>.   In 1993, Schwartz published a study of daily mortality and air pollution in Birmingham, AL. This was followed in 1994 by a similar study involving hospital admissions for the elderly. In both cases, Schwartz found that a 3-day average of PM_{10} provided the maximum implied air pollution effects. Roth and Li presented a replication study of these data at a 1996 CASAC meeting, based on a large number of model variations and lag/pollutant averaging combinations. Although not all of the many combinations that Roth and Li considered were independent, their analysis implied that the claims of Schwartz were not robust and may have represented a statistical artifact.   This position was confirmed by a more formal analysis sponsored by the U.S. EPA, reported by Davis et al. (1996). They failed to find a significant association with PM_{10} and concluded: "When we use the same variables as included by Schwartz, we obtain results similar to his. But when we use alternative models we obtain different conclusions. In particular, when humidity is included among the meteorological variables (it is excluded in the analysis by Schwartz), we find that the PM_{10} effect is not statistically significant." Davis et al. went on to say "...model selection is critical in making conclusions about the effect of particulates on mortality, requiring consistent, defensible approaches to assure reliable interpretations".

Moolgavkar et al. (1997) reanalyzed the Birmingham hospital admissions data and found "little evidence of association between air pollution and hospital admissions for respiratory disease". Roth and Li had reported similar findings at the CASAC meeting. The difference between these analyses and that of Schwartz (1994) may well be in the control for differences in admissions by the day of the week. Hospital admissions tend to drop on weekends (this was even true during the major air pollution episode in London in 1952 [Lipfert, 1994]), as do emissions of many primary pollutants. Since these two phenomena are not causally related, failure to control for day-of-week will lead to spurious associations. Since ozone is often higher on weekends (because of reduced titration by NO_x), failure to account for day-of-week can lead to a spurious positive correlation for ozone for a one- or two-day lag (admissions on Mondays correlated with ozone on Saturday or Sunday). This also illustrates the flexibility that the time-series analyst has in seeking the "optimum" model with positive correlations.

3.2 Other Hospitalization and Morbidity Studies

<u>3.2.1 Hospitalization Studies.</u> Several studies have been published recently dealing with daily hospitalization for cardiac and respiratory diagnoses (Morris et al., Schwartz and Morris, Burnett et al.). The U.S. studies are limited to patients aged 65 and over, financed through Medicare. The general conclusion from these studies is that ambient CO is the best predictor for cardiac diagnoses, and O_3 is the best for respiratory diagnoses. Some PM indicators have also been implicated, notably PM_{10}, but the relative performance of these pollutants in joint regressions must be viewed in the perspective of exposure error. O_3 is strongly attenuated indoors (reference needed), and its outdoor distribution in urban areas is affected by chemical reactions (titration) with local sources of NO_x. Since CO in urban areas has only two major sources, wood burning and vehicles, its outdoor distributions tend to be much more variable than those for the smaller particles fractions of PM (reference needed). Both CO and O_3 tend to be highly correlated with PM (in winter and summer, respectively). Thus, the fact that both CO and O_3 perform better than PM in joint regressions, in spite of their typically higher exposure errors, constitute evidence that CO and O_3 are more likely to evoke health responses than PM, at current ambient levels. This conclusion follows directly from the simulation studies of Lipfert and Wyzga (1995a).

<u>3.2.2 Emergency Room Visits for Respiratory Illnesses in Montreal.</u> Delfino et al. (1997) tracked ER visits for respiratory illness for 25 hospitals in Montreal for the summers of 1992 and 1993. Only the visits that required medical treatment and monitoring on a stretcher were included in the analysis. A control group was defined consisting of non-ambulatory patients with gastrointestinal or psychiatric complaints.   Seasonal and day-of-week effects were controlled. The air pollutants considered included ozone, PM_{10}, $PM_{2.5}$, $SO_4^=$, and strong acidity (H^+); all were measured daily, ozone at 7 stations. The results showed the largest mean effect for ozone (21%), with PM_{10}, $PM_{2.5}$, $SO_4^=$, and acidity (not significant) following in order, based on single-pollutant regressions. The mean effect of $SO_4^=$ was about 6%, and the authors showed that this result was reasonably consistent with several previous studies. Only PM_{10} was significantly associated with the control admissions (at lag 0), but this raises doubts about the possibility of spurious results for the other endpoints as well.

3.2.3 A Swiss Cross-Sectional Study of Lung Function. Ackermann-Liebrich (1997) examined about 10,000 Swiss adults living in 8 different locations, including two at higher altitudes. Lung function and allergy tests were conducted and air quality data were obtained from local routine measurements (SO_2, NO_2, TSP, PM_{10}, and O_3). Analyses were done separately for by smoking status and presence of respiratory symptoms. The authors reported that "significant and consistent effects" on lung function were found for SO_2, NO_2, and PM_{10}, for all subgroups. Results were not reported for TSP, but since the correlation between TSP and PM_{10} was 0.95, one would expect the lung function associations to be similar. They declined to try to distinguish the effects of the various pollutants. However, some of the lung function differences due to air pollution lost significance when altitude was included as a variable. The magnitude of the effect attributed to PM_{10} (3.4% decrease in FVC for 10 ug/m^3 in PM_{10}) seems quite large in comparison to other studies.

4. Studies That EPA Should Have Considered in Their Proposal But Did Not.

EPA selected particular studies from the more inclusive Criteria Document for use in the Staff Paper and Decision Document. Some of the equally valid studies that present contrasting views have been discussed above. Others are discussed in the following section.

4.1 Overlooked Short-Term Mortality Studies

4.1.1 Review and Reconciliation of Results. Lipfert and Wyzga (1995b) reviewed 27 time-series mortality studies that imply that about 1-8% of daily deaths may be associated with daily variations in air quality, but the design of these studies does not allow estimates of the degree of prematurity of death. There are also uncertainties about the "responsible" pollutants, because of the fundamental collinearity among time-varying pollutants. One of the methodologies that focuses on longer-term effects is that of the cross-sectional analysis, which deals with spatial relationships. Lipfert (1995) considered and compared the two types of cross-sectional studies: prospective cohort studies, which deal with survival of individuals, and population-based "ecological" studies that analyze differences among communities. Both types relied on community-based air monitoring data and thus are "ecological" in that sense.

The published cross-sectional studies of long-term associations between air pollution and mortality identified several obstacles to precise determination of relationships. These include uncertainties as to the "correct" regression models, personal lifestyle risk factors for which appropriate data may not be available, changes in heart disease mortality (resulting in part from changes in the prevalence of these risk factors) that differ by region, collinearity among pollutants, and unavailability of appropriate data on certain pollutants such as CO. In addition, differences in historical air pollution levels that might be associated with chronic responses have usually been ignored. However, one of the outcomes from the prospective studies has been the failure to identify any particularly susceptible population subgroup that might affect the plausibility of the ecological assumption (i.e., that the behavior of individuals may be inferred from data on groups). As a result, there appear to be no reasons why the results of prospective cohort studies should not be compared to those from ecological studies.

Two such types of comparisons appear to be relevant at this time. First, all of the published cross-sectional studies that considered air pollution concurrent with mortality found significant associations; however, they differed by pollutant and by magnitude of response, and some of these differences appeared to be due to differences in the degrees to which potential confounding variables were controlled by the various models. Next, the relative risks derived by the prospective cohort studies (Dockery et al., 1993; Pope et al., 1995) appear to substantially exceed those of the daily mortality studies, which might be interpreted as either showing the existence of effects due to chronic exposures to air pollution or the presence of regional confounding. However, a portion of the "excess" relative risks reported by the prospective studies may be a result of using concurrent rather than historic or cumulative air pollution exposure levels, which are substantially higher for the most polluted locations. Substituting the higher exposure levels would bring the estimates from the prospective studies more into line with the ecological cross-sectional studies and with the daily mortality studies.

Although there are good reasons for comparing long- and short-term exposure studies, the power of such comparisons is limited by the fundamental uncertainties of both types of studies. There are lingering questions about the appropriate lag structures, seasonal adjustments, and weather variables in the daily time series studies (Wyzga and Lipfert, 1995) as well as with the duration of peak air pollution exposures. As discussed above, the long-term studies are sensitive to the regression models used, the degree of confounder control, and the lag structure assumed for long-term exposures (annual averages). In addition, there is no *a priori* reason to assume that the same pollutants would be responsible for both acute and chronic responses. The combined effects of all of these uncertainties appear to be of the same order of magnitude as the net differences between types of studies, making the establishment of chronic effects on this basis problematic.

It would thus appear that additional methods must be developed if reliable estimates of the long-term effects of air pollution are to be derived. The first requirement should be the establishment of specific disease-pollutant hypotheses to be tested, preferably on the basis of clinical theory and laboratory experiments. The relevant epidemiological data should probably then be obtained from the records of long-term pollution abatement histories. Such a space-time analysis might succeed if enough contrasting situations can be identified.

The effect on the regression coefficients of using lagged air pollution data has already been discussed. This final comment is addressed to the suggestion by Pope et al. (1995) that the ecological relationships between air pollution and age-race-sex-adjusted mortality data may be used to justify relative risks of the order of 1.25. Figure 3 is a plot of SMSA mortality against SO_4^{2-}, ca. 1980, for various adjustment procedures. The plot on the left resembles Figure 1 of Pope et al. on which this suggestion was based, although the exact correspondence between the locations considered could not be established. The plot in the center includes mortality adjustments for college education and for smoking and being overweight (state-level data for the latter two parameters). This corresponds roughly to the actual Six City analysis, and the slope is reduced somewhat. The plot on the right extends the mortality adjustments to a larger range of variables that have been shown exogenously to influence mortality; this slope is essentially zero, consistent with the regression results. Figure 4 extends this analysis to $PM_{2.5}$, for a smaller number of locations. Here the slope reduction due to introducing additional nonpollution variables is less drastic but still substantial. These plots are intended to demonstrate that relationships between longevity and air pollution based on spatial gradients cannot be established with confidence without considering a full range of possible confounders. This precept applies to prospective cohort and population-based (ecological) studies alike.

<u>4.1.2 Air Pollution and Mortality in Amsterdam</u>. Verhoef et al. (1996) studied the associations of all-cause mortality (mean daily count=19) in Amsterdam with several indices of urban air pollution from 1986-92. They considered measurements at 3 population oriented sites for smoke, PM_{10} measured every 3-4 days, SO_2, CO and O_3. Daily averages were used for all pollutants except ozone, for which peak hourly averages were used. A prediction algorithm was used to fill in the missing PM_{10} data. Weather data included temperature and relative humidity from the airport. Data on flu epidemics were also considered; day-of-week and temporal trends were also controlled. Lags up to 2 days were considered. All average air concentrations were low: smoke, 12 ug/m^3; estimated PM_{10}, 38 ug/m^3; SO_2, 5 ppb; CO, 0.85 ppm; O_3, 22 ppb. Regression results were presented for each pollutant and lag; it is not clear if the results for each lag are independent. The results were presented as relative risks for 100 ug/m^3 of each pollutant, which is very misleading because of the great disparity in mean values. Statistical significance was only obtained for smoke and ozone, but the mean effects were quite similar for all pollutants except and SO_2, which was lower. When summed over all lags, the mean effects were: smoke, 0.044; estimated PM_{10}, 0.030; SO_2, 0.014; CO, 0.03*; O_3, 0.029. When combined with gaseous pollutants in 2-pollutant regressions, the PM measures retained significance at the

------

* CO results are uncertain because only 1 digit was presented in the tabulated results.

expense of the gases. The authors concluded that black smoke was the best predictor of daily mortality in their study, in part because of particle size but also because of its chemical constituents. It should be noted that these constituents have little in common with the soluble sulfate compounds currently being emphasized in the U.S. Also, possible effects of differential measurement error should be considered here, given that a large percentage of the PM_{10} values were estimated.

4.2 Long-Term Mortality Studies

4.2.1 Prospective Cohort Study of California Seventh-Day Adventists. Abbey et al. (1991) described a prospective study of about 6000 white, non-Hispanic, nonsmoking, long-term California residents who were followed for 6-10 years, beginning in 1976. There were 845 deaths from non-external causes. The study was designed to test the use of cumulative population-oriented air quality data as an explanatory factor for disease incidence and chronic effects. Ambient air quality data (TSP, O_3) dating back to 1966 were used, and the study was restricted to those who lived within 5 miles of their current residence for at least 10 years. The follow-up analysis (Abbey et al., 1995) considered exposures to SO_4^{2-}, PM_{10} (estimated from site-specific regressions on TSP), $PM_{2.5}$ (estimated from visibility), and visibility *per se* (atmospheric extinction coefficient). All of the air quality monitors in the state were used to create individual exposure profiles (duration of exposure to specific minimum concentration levels) for each participant, by interpolating to their zip code centroids based on the three nearest monitoring stations. The Cox proportional hazards model was used, considering age, sex, past smoking, education, and presence of definite symptoms of airway obstructive disease (asthma, chronic bronchitis, or emphysema) in 1977 as individual risk factors, together with various exposure indices for air pollution, considered separately. Data on occupational exposures and history of high blood pressure were available but were not used in the mortality model. No data were available on climate, body mass, income, migration, physical activity levels or diet and separate results were not reported for mortality by gender.

Neither heart attacks or nonexternal mortality was associated with any pollutant in these studies. The authors felt that possible errors in their estimated exposures to air pollution may have contributed to the lack of significant findings. The finding of no association between long-term cumulative exposure to air pollution and mortality may be interpreted as showing the absence of chronic responses after 10 years but not necessarily the absence of (integrated) acute responses, since coincident exposures were not considered. It is also possible that the latency period for chronic effects may exceed 10 years and that additional follow-up might still reveal chronic effects. The imputed magnitudes of the other risk factors considered were not given.

4.2.2 Studies of 1980 City Mortality. The analysis by Lipfert et al. (1988) comprised a statistical analysis of 1980 spatial patterns among U.S. central cities for total mortality (all causes), evaluating demographic, socioeconomic, and air pollution factors as predictors. The advantages of studying central cities instead of SMSAs include potentially better measures of air pollution exposure because of the smaller land areas and higher population densities, and larger numbers of observations that allow analysis of subsets. Lipfert et al. found sulfate* and iron particles to be significant predictors of all-cause mortality in about 180 cities; TSP was considered only cursorily. If the elasticities for SO_4^{2-} were corrected to account for the filter artifacts, they would be reduced by about 50% in this analysis (Lipfert, 1994a). The data on PM_{15} and $PM_{2.5}$ were only available for 68 cities; neither pollutant was statistically significant in this data set but their elasticities were in the same range range found for other pollutants (0.013-0.05). Lipfert et al. also tested lagged pollution data as a means of attempting to

------

* The sulfate data (along with data on ambient SO_2 and NO_x) that yielded the best results were obtained from a long-range transport model (Shannon, 1981), instead of from local measurements. This lack of measurement error may have contributed to the success of these three variables, which were virtually indistinguishable from one another. However, Shannon has since reported in a personal communication the discovery of a systematic error in the meteorological data used as input to the model. This error biases some of the results high. For this reason, the findings of Lipfert et al. (1988) with respect to SO_4^{2-}, SO_2, and NO_x may be unreliable.

distinguish acute from chronic responses; using ca. 1970 TSP and SO_4^{2-} data to predict 1980 city mortality was slightly less effective than using ca. 1980 data for these pollutants as predictors. This result does not suggest the presence of chronic effects.

<u>4.2.3 Studies of 1980 SMSA Mortality.</u> Ozkaynak and Thurston (1987) analyzed 1980 total mortality in 98 SMSAs against EPA AIRS data on TSP and SO_4^{2-} and data on PM_{15} and $PM_{2.5}$ from the EPA inhalable particle (IP) monitoring network for 38 of these locations. The sulfate measurements that Ozkaynak and Thurston used were probably affected by artifacts from the high-volume sampler filters (Lipfert, 1994a). The sulfate data from the IP network were not used by Ozkaynak and Thurston.

Ozkaynak and Thurston ranked the importance of the various pollutants mainly by relative statistical significance in separate regressions. They concluded that the results were "suggestive" of an effect of particles on mortality decreasing with particle size, although in the basic model only SO_4^{2-} was significant. In some of the other models, $PM_{2.5}$ was also significant and PM_{15} was nearly so. However, if the effects are judged by elasticities rather than significance levels, SO_4^{2-}, $PM_{2.5}$, and PM_{15} would be judged as equivalent, with TSP ranking somewhat lower. The indicated effect of SO_4^{2-} would be reduced from an elasticity of 0.086 to about 0.05 by accounting for filter artifacts. Ozkaynak and Thurston also used source apportionment techniques to estimate that particles from coal combustion and from the metals industry appeared to be the most important.

The coefficients and significance levels obtained for TSP by Ozkaynak and Thurston may be the result of the data they used, which were based on a single monitoring station in each SMSA and thus are unlikely to be fully representative of average population exposures to TSP in areas as large as SMSAs. As a result, alternative interpretations of these findings are certainly possible.

In addition, because smoking, diet, and other socioeconomic or lifestyle variables were not considered in the regression model, the pollution coefficients may have been biased. Finally, Ozkaynak and Thurston did not specifically address the question of acute vs. chronic responses by exploring lagged pollution variables.

Lipfert (1993) used data from up to 149 metropolitan areas (mostly SMSAs) in a study of the relationships between community air pollution and "excess" mortality due to various causes for the year 1980. Several socioeconomic models, including the model proposed by Ozkaynak and Thurston (1987), were used in cross-sectional multiple regression analyses to account for non-pollution effects. Cause-of-death categories analyzed include all causes, nonexternal causes (ICD-9 0-800), major cardiovascular diseases (ICD-9 390-448), and chronic obstructive pulmonary diseases (COPD) (ICD-9 490-96). The spatial patterns for the first three groupings were quite similar but differed markedly from the patterns of COPD mortality, which tend to be higher in the Western U.S. Regressions were performed for these cause-of-death groupings as annual mortality rates ("linear" models) and as their logarithms ("log-linear" models).

Two different sources of measured air quality data were utilized. Data from the EPA AIRS database included TSP, $SO_4^=$, Mn, and ozone (obtained from a long-term average isopleth map). Data from the inhalable particulate network included PM_{15}, $PM_{2.5}$ and $SO_4^=$ from the teflon IP filters for 63 locations. All particulate data were averaged across all the monitoring stations available for each SMSA; the TSP data were restricted to the year 1980 and were based on an average of about 10 sites per SMSA. This represents a substantial improvement over previous air quality representations.

The associations between mortality and air pollution were found to be dependent on the socioeconomic factors included in the models, the specific locations included in the data set, and the type of statistical model used, as was the case with a previous analysis of 1970 data (Lipfert, 1978). An important variable that interacts with indices of industrial air pollution is the rate of population change, which is negatively correlated with mortality and with sulfur air pollution.

Stepwise regressions were run for each mortality variable, and a "parsimonious" model was developed that had statistically significant coefficients for the non-pollution variables. Most of these coefficients also agreed with exogenous estimates of the "correct" magnitudes of the risks. Using these models, statistically significant associations were found between TSP and mortality due to non-external causes with log-linear models, but not with linear models. Sulfates, manganese, inhalable particles (PM_{15}), and fine particles ($PM_{2.5}$) were not significantly ($p > 0.05$) associated with mortality with any of the parsimonious models, although $PM_{2.5}$ and Mn were close with linear models ($p = 0.07$) and statistical significance may have been affected by the use of smaller data sets for these species. $PM_{2.5}$ was a significant predictor of heart disease mortality only when the regression model was restricted to the variables used by Ozkaynak and Thurston (1987). Lipfert showed that $PM_{2.5}$ was the "strongest" particulate variable with linear models, but that TSP performed better in log-linear models. In that sense, the findings for PM were not robust. Scatter plots and quintile analyses suggested that a TSP threshold might be present for COPD mortality, at around 65 ug/m^3 (annual average).

Lipfert's results provided some support for previous findings of associations between TSP and premature mortality but not for fine particles. They also (indirectly) supported the hypothesis that improving the accuracy of pollutant exposure data tends to increase statistical significance. Similarly, the lack of significance for $SO_4^=$ may partly relate to the flawed measurement methods used at the time. The ambiguity between linear and log-linear models probably reflects the effects of influential observations. However, the study failed to consider other socioeconomic and lifestyle variables (such as diet and exercise habits) that could have been confounded with air pollution.

4.2 Overlooked Morbidity Studies

<u>4.2.1 The NHANES II Data on Lung Function.</u> Schwartz's (1989) observational study is in many ways a paradigm for deriving estimates of the effects of air pollution on lung function. The lung function data were obtained as part of the second National Health and Nutrition examination Survey (NHANES), which was intended to be a random, national sample of the civilian, noninstitutionalized population of the United States. The NHANES II spirometry data were limited to children and young adults (ages 6-24). Care was taken to ensure standardization in the collection and processing of the data, which were acquired from 1976 to 1980. Schwartz based his analysis on the best trial for each subject and deleted data that did not meet spirometry standards. The lung function data (FVC, FEV_1, and PEF) were fitted to regression models based on race, age, sex, number of cigarettes smoked per day, and several physiological measures. The residuals from these regressions were then regressed against air pollution, one pollutant at a time, as a 2-stage procedure. The possibility of neighborhood clustering* was accounted for by using a nested random effects model similar to that of Ware et al. (1986).

Air pollution data were obtained from the EPA's air quality data base (now called AIRS, Aerometric Information Retrieval System), using only those monitors that were located within 10 miles of the population centroid of the census tract of the subject's residence. This condition resulted in different numbers of acceptable cases for each pollutant considered, as shown in Table 11-2. Temporal averaging consisted of the 365 days preceding the examination. The ozone regressions were based on the annual average of daylight hours (8-hour averages). This statistic is about 50 to 100% greater than the annual average of all hours.

Schwartz found statistically significant negative associations between each of the three lung function metrics and NO_2, O_3, and TSP, but not for SO_2. (One of the problems with national data on SO_2 may be the variability among measurement methods in use, especially during this

----

* Cohen (1980) and others have explored the extent to which a genetic deficiency contributes to impaired pulmonary function. To the extent that close relatives tend to live in the same neighborhood, community, or region, confounding could thus occur between local environmental effects and the effects of shared genetic defects. This phenomenon could also help explain the apparent geographic clustering in the 24 Cities Study discussed above.

period.) The strongest associations were for ozone and FVC. Robustness of the results was established by re-estimating the relationships with non-linear pollution transforms, including thresholds; by excluding subjects with chronic respiratory conditions, smokers, subjects not residing in the state of their birth; by including factors for family socioeconomic status, urbanization, region of the country; by using 2-year means for pollution data; and by extending the allowable monitor radius to 20 miles. Excluding subjects with chronic respiratory conditions reduced all of the pollution coefficients; excluding smokers increased them. Increasing the radius for the pollution monitors reduced only the TSP coefficients. All of these findings are consistent with a causal hypothesis. Schwartz also noted that linear dose-response models for the air pollutants appeared to be superior to logarithmic transforms.

The nonlinear dose-response relationships suggested a threshold for ozone of about 0.04 ppm for the daylight average (0.02-0.025 ppm annual average); this is a relatively high value (90th percentile in Schwartz's data), which might correspond to average daily 1-hour peaks around the current NAAQS of 0.12 ppm (235 ug/m^3). For TSP, a threshold was suggested around 90 ug/m^3, which exceeds the current standard of 75 ug/m^3. The NO_2 relationship was more nearly linear, with an increase in slope at about 0.04 ppm (75 ug/m^3), which is below the current federal standard. These results suggest that most of the "signal" in Schwartz's data set was derived from the locations above the 90th percentile in air pollution, for all species. The lack of significant findings for SO_2 is also noteworthy, since the 90th percentile for SO_2 was 0.019 ppm (50 ug/m^3) and there are relatively few urban locations in the United States which now exceed this value (see Table 2-8, for example). Association with $SO_4^=$ aerosol was not evaluated, but regional dummy variables were investigated that may provide some insight, given the regional distribution of this pollutant. No dummy variable was significant for FEV_1 or FVC, but peak flow was about 5% lower in the Northeast. Schwartz also reported decrements in lung function associated with central city residence, apart from the specific air pollution relationships studied. Such a relationship would imply a primary pollutant (if any), rather than a secondary species like SO_4^{2-} or $PM_{2.5}$.

4.2.2 <u>Acid Aerosols and Respiratory Symptoms</u>. Schwartz et al. (1989) followed a cohort of about 300 children selected from the Six Cities Study, whose families kept diaries for about a year, beginning in the fall of 1984. Respiratory symptoms were recorded and each family was called biweekly and mailed their diary to the investigators monthly. Daily measurements of SO_2, NO_2, O_3, PM_{10}, $PM_{2.5}$, atmospheric visibility, $SO_4^=$, sulfuric acid, and aerosol strong acidity were made. 48-hr air quality averages were used in the regressions. For logistical reasons, the study extended over four school years (1984-88). This paper emphasized "lower respiratory illness", defined as presence of at least two symptoms among cough, chest pain, phlegm, or wheeze. An event was defined as an occurrence of new symptoms. The authors reported significant results for all pollutants except the acids. The ranking of the significant pollutants by mean effect was, in decreasing order: PM_{10} and O_3 (tied for first), $PM_{2.5}$, $SO_4^=$, visibility, and SO_2 (distant last). A dose-response plot for O_3 showed approximate linearity, with an approximate doubling of the incidence of lower respiratory illness (from 0.07% to 0.14%) as ozone increased from about 20 ppb to about 60 ppb. Temperature was also found to be an important predictor of respiratory symptoms.

4.2.3 <u>A Study of Respiratory Outpatient Visits in Alaska</u>. Gordian et al. (1996)* analyzed outpatient visits to both physicians' offices and hospital emergency rooms from May 1992 to March 1994, in Anchorage, Alaska. Air quality measurements came mainly from a central site for PM_{10}. CO was measured during winter months at five sites. The main sources of suspended particulates were reported to be crustal materials and volcanic ash. The data were adjusted for day-of-week effects and filtered to control for seasonality. The regression results showed a mean effect of PM_{10} of 13-27% on asthma visits and 4-13% for upper respiratory visits. The authors concluded that "the coarse fraction of PM_{10} may affect the health of working people".

-----

* A preliminary version of this paper was made available to EPA during preparation of the CD.

However, CO was found to be associated with visits for bronchitis and for upper respiratory conditions, which raises questions about biological plausibility for this and other similar studies. The direct effects of CO are thought to be limited to the oxygen-carrying capacity of the blood so that associations with respiratory symptoms must reflect some sort of surrogate effects. All automobile-related pollutants (CO, aldehydes, soot, NO_x, VOCs) tend to be highly correlated in time, so that any of these species could be the actual causal agent for respiratory effects, as also noted in the acute mortality study of Ozkaynak and Kinney (1991) in Los Angeles. By the same token, findings of associations between any one of these species and cardiac complaints should be interpreted as applying to any member of the group. This reinforces the ambiguity between PM and CO that has been noted throughout this review.

Finally, it should be noted that there is no evidence in the Alaska study implicating sulfates, aerosol acidity, or any other soluble fine particle with respiratory morbidity.

5. Important Technical Issues Affecting the Scientific Validity of EPA's Proposal

5.1 Issues in Time-Series Studies

5.1.1. <u>Seasonal Adjustment and Lag Selection</u>. In addition to daily variations in the timing of health-related events, there are also large seasonal variations that must be removed or "controlled", as well as differences that occur by day of the week and on holidays.* There are many ways of trying to do this, some better than others. After such treatment, the air quality and health data can be plotted as series of peaks and valleys oscillating about a horizontal line. The extent to which the health and pollution peaks match up defines their implied statistical relationship. A certain amount of matching is to be expected just due to chance, and if the lag between exposure and response is free to vary, it becomes much easier to find such a match (Pickles, 1981). Averaging the pollution data over more than one day also improves the probability of matching.

During the major air pollution episodes of past decades (which form the paradigm for time-series analyses), health responses appeared to track pollution excursions throughout the periods of poor air quality. By selecting an appropriate lag between pollution excursion and response, a near-linear dose-response function could be hypothesized that included both increases and decreases over several days. This kind of matching gives rise to correlations around 0.5 or more. In contrast, the vast majority of current situations involve peaks for single days only which are thus shaped as triangles; the shapes or temporal profiles of the excursions thus play little or no role in the analysis. This means that the correlations between excursions and health responses depend on the numbers of triangles that appear to coincide with appropriate lags; the shapes or durations of episodes are no longer a factor. These correlations are much lower (ca. 0.1) and depend on a large number of repeating cycles to achieve significance. Examination of episodes of high pollution and of high mortality in contemporary atmospheres shows little or no correspondence; this can only be explained by either substantial measurement error (people are not exposed to the highest concentrations) or by (undefined) statistical artifacts in the analysis.

----

* Such effects can give rise to spurious correlations between health and air pollution, because of unrelated factors that intervene or confound. Mortality has an annual cycle peaking in winter with a low in summer, in all climates (Lipfert, 1991); this may be due to increased presence of viral agents or due to the tendency to spend more time indoors in close proximity to other people in inclement weather. Ambient air quality has annual cycles resulting from seasonal differences in emissions (space heating, air conditioning) and dispersion (mixing height, ventilation). These cycles could give the appearance of relationships between health and air quality (both positive and negative, depending on the pollutant), whereas in reality extraneous factors are responsible. Similarly, fewer people die on Sundays and more on Mondays (for reasons that are not dully understood), which could be spuriously associated with reduced air pollution on weekends resulting from reduced commercial activities.

5.1.2. <u>Pollutant and Weather Collinearity</u>. All air pollutants that are generated locally fluctuate similarly to changes in weather, such as with wind speed.   Pollutants that emanate from a common source may share even more characteristics; examples are the particles and gases released as combustion products.   However, monitoring networks are not the same for all pollutants and these uncertainties will tend to mask the degree of collinearity actually present among pollutants.   For example, in Philadelphia, we find that SO_2, NO_2 and CO track PM nearly perfectly (Figures 5-10, based on data from Wyzga and Lipfert, 1995) after averaging to reduce measurement variability.   Daily change in temperature also tracks PM in this way (Figure 11). Regional pollutants, such as ozone, sulfates, and fine particles, only partially track the other pollutants because of their dependence on photochemistry; however, these pollutants also tend to track weather phenomena, but differently than primary pollutants. The ability to distinguish these collinear pollutants from one another depends on their relative measurement errors.

EPA and others seem to be relying upon spatial contrasts among time-series results to address the pollutant collinearity issue. EPA cites the finding of similar associations between PM and daily mortality in locations with both high and low levels of sulfur oxides as evidence that PM is the agent and not SO_x, for example. However, there are important gaps in this line of reasoning. The most common co-pollutant that has been considered with PM is SO_2; these two species tend to be highly correlated and, since SO_2 has a greater exposure uncertainty (because it is highly attenuated indoors), PM is usually the "winner" in a joint regression.   Carbon monoxide is a much more likely confounder of PM and has only been considered in a few instances. Kinney and Ozkaynak (1991) found that they could not distinguish among PM, CO or NO_2 in Los Angeles, for example. CO has been found to be a more important predictor of cardiac hospital admissions than PM in several studies (Morris et al., Schwartz and Morris; Burnett et al.). the Pope et al. studies of air pollution and health in Utah Valley are an example of a situation in which CO should have been considered along with PM, because there is a common source (traffic, steel production) and because the ambient CO standard was violated there. Communities where residential wood burning is a major source of PM are another such situation; adding CO to the analysis of Fairley ( ) for Santa Clara, CA, results is a drastic attenuation of the association between daily mortality and PM.   These factors must be considered in the light of generally more spatial variability in CO than in PM, which would suggest that PM ought to be the "winner" in joint regressions. Those situations in which CO does in fact prevail must be regarded as powerful evidence that the underlying toxicity of low concentrations of CO is greater than those of PM.

Finally, the "de minimus" argument for ignoring "low" concentrations of copollutants is inconsistent with the parallel belief in the nonexistence of pollutant thresholds. If the "new" epidemiology is to be taken at face value (which we do not recommend), that it becomes impossible to state a priori which concentrations are "too low" to warrant serious consideration. Viewed in the context of concentration levels at which serious health effects have been rigorously demonstrated, all of today's concentrations are "too low."

5.1.3. <u>Exposure Errors</u>.   The studies use outdoor temperature data from airports to represent weather effects, so that the levels of adverse heat or cold people are actually exposed in conjunction with air pollution are unknown.   Uncertainties in exposures to weather stresses thus may be just as important as those for exposure to air pollution, which means that the division of responses between these two types of environmental stresses is also uncertain. For example, the failure to find robust effects of PM on daily mortality in Birmingham, AL, might be attributed to heavier use of air conditioning there, or to adaptation of the population to heat stress (and interactions or collinearity between heat and air pollution).

Relevant sources of exposure uncertainty include:

1. <u>Instrumental error</u>. Measurement systems are not always repeatable and vary in their conformity with standards. Coarse particle measurements are among the least reliable.

2. <u>Spatial variability in the ambient</u>. Any epidemiological study must rely on a finite number of air monitors in a given city. Concentrations of air pollutants vary considerably spatially (including vertically), because of the local effects of pollutant sources and sinks within the city.

3. <u>Temporal variability</u>. It is not always clear which pollutant averaging time should be used; peak concentrations tend to vary more in space than averages. Studies that seek to describe chronic responses should use some measure of long-term, cumulative exposures.

4. <u>Personal exposures vs. ambient air quality</u>. Different pollutants have been shown to vary considerably in this regard, depending on activity patterns and the balance between sources and sinks in indoor environments and on rates of penetration of outdoor air.

5. <u>Other factors</u>. Other factors include human heterogeneity with respect to breathing rates and metabolic uptake of inhaled pollutants that control doses to target organs.

Table 2 is a subjective assessment that compares the air pollutants of interest in this regard. Tradeoffs are involved, for example, because some of the most accurately measured pollutants tend to vary more spatially, and vice versa.

These concepts are also extended to ambient temperature, which is a common confounder in many studies. Airport data are usually used, but virtually no one is actually exposed to those measured levels. Heat-wave victims, for example, are often found to have been exposed to much higher temperatures.

TABLE 2
SUBJECTIVE JUDGMENTS ABOUT EXPOSURES TO DIFFERENT POLLUTANTS

species	instrument error	spatial variability	personal exposures	human variability
TSP	overestimate	fair	unknown	unknown
PM_{10}	underestimate	fair	variable	unknown
$PM_{2.5}$	random	small	> ambient	high (smoking)
CP	underestimate	fair (?)	unknown	unknown
SO_4^{2-}	(depends on filter)	small	unbiased	unknown
H^+	small	high	< ambient	high
SO_2	nil	high	<< ambient	high
O_3	nil	high (urban)	< ambient	high
NO_2	nil	fair (?)	< or > ambient	high (smoking)
CO	nil	high	< or > ambient	high (smoking)
temperature	nil	small	< or > ambient	high

The entry "smoking" refers to the heterogeneity that is expected when the population of interest includes both smokers and non-smokers. Source: Lipfert and Wyzga, 1996.

5.1.4 <u>Basis for Selecting "Responsible" Agents</u>. Lipfert and Wyzga (1995b) found that almost all of the common air pollutants display a certain degree of matching with health events, with about the same mean effects. Data on personal exposures are lacking for all pollutants. As a result, identification of the pollutants most closely associated with health responses (and thus to be controlled) may or may not be correct, especially since medical science has no rational explanations for the epidemiological findings.

5.1.5. <u>Studies Provide No Clues on Mechanisms</u>. The fact that studies of acute responses to PM in various locations tend to find similar results, even when the physical and chemical natures of the particles are quite different (because of the geographic differences and presence of different types of sources) has important implications with respect to the implied health effects. If the particles are fundamentally different but are still associated with the same outcomes (premature cardiovascular mortality, for example), then the health responses must be non-specific, rather than pathological. For example, an irritation response would be more likely than a chemical reaction to specific substances. This, together with the rapid response times that have been indicated (less than one day), would seem to rule out most deep lung injury processes.

5.2 Issues in Long-Term (Cross-Sectional) Studies

Mortality rates or probabilities of survival may differ by location for any of a number of reasons:

> 1. geographic variations in the presence of specific endemic diseases, such as malaria or influenza.

> 2. geographic variations in risk factors, which may either be associated with the location *per se* or with the people who tend to live there, such as climate or the quality of the environment, the age and racial distributions of the population, its dietary and smoking habits, or occupational factors associated with local industry.

> 3. geographic variations in the frequencies of acute events, such as heat waves, accidents, or natural disasters.

Epidemiological studies of the long-term effects of community air pollution on mortality are most likely to be concerned with the second and third categories, since (aside from occupational lung disease) there are no diseases specific to these pollutants.

The second category of geographic factors relates to chronic effects on health, and it is the task of the analyst to distinguish the effects of air pollution from those of other risk factors in much the same way that it is necessary to adjust for seasonal trends in time-series studies. Long-term health risk factors may be further subdivided into factors that relate to the population of a given place (age, race, education, lifestyle, for example) and factors that relate to the physico-chemical environment of that place (climate, air and water quality). There are also likely to be interactions between these two subcategories, since places with desirable environments may attract as in-migrants that portion of the population that is better off economically while the disadvantaged part of the population may be forced to remain in less desirable locations and in those with depressed economies.

The third category above is included in long-term studies because annual mortality rates must also reflect the net sum of the acute events that took place that year (Evans *et al.*, 1984a). If the increases in daily death rates associated with acute events are not subsequently canceled by decreases (a phenomenon referred to as "harvesting"), annual rates will indicate the history of these acute effects. Thus, differences in long-term mortality rates associated with air pollution are likely to reflect some combination of acute and chronic effects. Although both types of information are useful contributions to the overall understanding of the health effects of air pollution, their distinction may be difficult if based on statistical criteria alone.

5.2.1 <u>Typical Geographic Patterns in Disease</u>. Spatial patterns of U.S. mortality rates show some well-defined trends that have existed for decades (Lipfert, 1994b). Such patterns are sometimes called the "geography of disease." In general, heart disease is higher east of the Mississippi and ischemic heart disease shows even sharper gradients and peaks in the Northeast (part of this gradient could be due to differences in diagnostic practices, although cold weather has also been implicated [Lloyd, 1991]). Cancer death rates tend to be higher in the Northeast, but not exclusively in industrialized states (Vermont, New Hampshire and Maine are relatively high). Gorham *et al.* (1990) argue that breast cancer is higher in northern latitudes because of reduced intake of vitamin D; such patterns are also seen in the former Soviet Union, for example. Lung cancer deaths are more evenly distributed and include some of the states with high tobacco use (Nevada, Kentucky, Virginia, but not North Carolina). Pneumonia and influenza deaths are distributed across the country but tend to be higher north of about the 36th parallel. The "stroke belt" has been defined as a broad east-west stripe across the southern part of the country.

Spatial trends in air pollution have both local and regional patterns. Local patterns within cities reflect the presence of primary pollutants from local sources (CO from traffic, particles from industrial operations, SO_2 and NO_2 from combustion sources, for example). There are also multi-state regional patterns in secondary pollutants, such as sulfates and other fine particles in Appalachia and the East North Central "rust belt," and ozone in Southern California and along the Northeast corridor. Collinearity among pollutants results from common spatial patterns of their major sources.

Associations between air pollution and health may be suggested by coincident spatial patterns; the challenge to the epidemiologist is to establish whether such associations are causal or merely circumstantial, since there are also many lifestyle parameters that vary spatially, such as smoking, diet, exercise and employment status. Furthermore, some of these factors have much stronger effects on mortality than air pollution: for example, a difference of 11 years in life expectancy according to whether all the "good" health practices are followed (Belloc, 1973), and 55% higher mortality for unemployed white men, aged 25-64 (Sorlie and Rogot, 1990). (See Yeager *et al.* (1995) for a recent showing of the significant correlation between sedentary lifestyle and coronary heart disease mortality at the state level that remained after controlling for smoking and hypertension; the relative risk for sedentary lifestyle was about 1.8.) Unfortunately, suitable data on these potential confounders are not always available. The possibilities for confounding by regional factors vary with the scale of the analysis; comparisons within regions may thus be less susceptible than comparisons across the whole country. For this reason, consistency between different types of studies becomes very important in considering causality. As stated by Schwartz (1993), "the strength of association within study is best assessed by the stability of the regression coefficient to multiple model specifications..." (as opposed to the p-value or t statistic). Thus, relative magnitudes of the imputed effects are more important than statistical significance per se.

5.2.2 <u>Regional Confounding</u>. The fundamental problem with cross-sectional studies is the difficulty in accounting for all of the relevant factors that characterize geographic differences in health. There are regional differences in smoking habits, diet, lifestyle, and climate that affect health, some of which also vary systematically with regional air pollutants. There are local differences among places that may relate to the types of industry present and the air pollutants that they produce. For example, persons of lower socioeconomic status may tend to live closer to the industries in which they work, so that they are exposed on the job (to high pollution levels) and at home (to community air pollution levels). Omitting such confounding factors from the statistical analysis tends to shift the blame onto poor ambient air quality; the same result can occur when inadequate surrogates are used in attempts to describe these complex socioeconomic phenomena.

5.2.3 <u>Defining Appropriate Exposure Metrics for Long-term Studies</u>. When cross-sectional studies are used to study the incidence (or prevalence) of chronic disease, the latency period between exposure and full-blown disease development must be considered. Such periods may run from 10 to 30 years, and in order to test the hypothesis that exposure to air pollution was responsible or contributed to the observed health response, exposures must be considered for appropriate periods, not just for the period of the survey of outcomes.

The air pollution exposure data must also be consistent with the nature of the disease or condition under study. If repeated acute exposures are suspected of implication (as might be the case for ozone), then statistics on the frequencies of specified concentration levels are needed, not just the annual average concentration. When more than one pollutant is of interest and the pollutants tend to be collinear (which is usually the case), then data on their relative measurement errors will be needed. Measurement artifacts that may vary geographically, such as particles formed due to gaseous reactions on filters, must be considered.

5.2.4 <u>Other Potential Geographic Confounders</u>. Geographic differences in climatic factors that affect exposure to outdoor air pollution must also be considered. Heavy use of air conditioning will result in less exposure to outdoor air pollution for example. Drinking water quality, especially hardness, has been implicated in geographic studies of heart disease, and selenium deficiency may be a factor in cancer prevalence.

Additional geographic variability may come from differences in access to medical care and the quality of that care. New therapies and treatment techniques may appear first at the large metropolitan teaching hospitals first and then diffuse to the hinterland later. Several studies have shown that the remarkable decline in heart disease mortality that began in the late 1970s started on both coasts and was experienced much later in the Southeast, for example.

5.2.5. <u>Assessment of Extant Cross-Sectional Studies</u>. None of the extant cross-sectional mortality studies meets these expectations. EPA has singled out two prospective cohort studies (Dockery et al., 1993; Pope et al., 1995) as supportive of their proposed new standard for fine particles. Since these were studies of individuals, adjustments for age, race, or smoking status were not at issue, as they can be for studies of whole communities. However, the amount of additional information about these individuals was meager at best. Data on educational attainment and body mass index were included in both studies, as linearized variables; Pope et al. also included data on alcohol consumption. Occupational exposures were poorly characterized by Dockery et al.; the categories were much too broad. Pope et al. did a better job here and also considered climatic factors, inclusion of which reduced the estimated air pollution effects by about 10%. Both studies relied upon ambient air quality taken during the periods of study; no consideration was given to prior or cumulative exposures. Neither study included data on migration (deaths in locations other than that at entry to the study were given no special consideration), diet, physical activity, income, unemployment, or access to medical care. Possible contributions from ozone were not considered adequately and CO was neglected altogether; these are two of the pollutants whose biological mechanisms are relatively well understood and for which relatively robust associations have been shown with hospital admissions. Finally, there is good evidence that regional differences in lifestyle (as exemplified by statistics on lack of exercise) could have explained a good part of the excess mortality that was assigned to fine particle air pollution.

The above considerations provide doubt as to whether these studies can provide any reliable information about long-term effects of air quality on health. However, even if these faults were overlooked, the studies do not convincingly discriminate among pollutants. Pope et al. considered only fine particles (about 50 cities) and sulfate data obtained from glass-fiber filters (about 150 cities); no other measures of PM were considered and no gaseous pollutants were considered. Multiple-pollutant regressions were not performed and thus it is not possible to partition the observed responses either among the pollutants that were studied or those that were not. It should be noted that it is not possible to produce "combustion-related" particles (which the authors blamed for the associations that they found) without simultaneously producing "combustion-related" gaseous air pollutants. The Six-Cities study only had 6 different levels of air quality and thus had little power to discriminate among pollutants. Steubenville, OH, had the highest concentrations of almost all pollutants, and most of these were higher still during the years before the study began (NAAQS were violated there).

6. Conclusions

This review has shown that there is a lack of specific statistical support for an ambient air quality standard for fine particles, in addition to the existing standard for PM_{10}. While the studies do not exclude the existence of associations between fine particles and health responses, they fail to show responses that are specific to this class of particles or to any of their specific chemical constituents. Arguments have been made by others in favor of hypotheses involving specific PM constituents, either by particle size or chemistry; adopting the proposed EPA standards for $PM_{2.5}$ would almost certainly foreclose further research on these and other hypotheses. If only $PM_{2.5}$ data were collected in the future, only $PM_{2.5}$ effects could be studied. The preponderance of evidence from the current epidemiological data base supports a broader research approach, especially given the lack of appropriate biological mechanisms and the dearth of information on actual personal exposures to all air pollutants. This broader research approach should include a full range of particle sizes and chemical constituents, as well as other air pollutants that are correlated with PM.

7. Recommendations

Given the weak scientific support for EPA's proposed revisions to the NAAQS for PM, it is recommended that the proposal be placed on hold pending completion of a comprehensive research and monitoring program. However, all particle sizes should be studied, not just $PM_{2.5}$. In addition, a retrospective long-term study of the benefits of previous improvements in air quality should be conducted. An important issue is that of the spatial differences in the improvements in heart disease mortality starting in the late 1970s.

8. References

D.E. Abbey, P.K. Mills, F.F. Petersen, and W.L. Beeson, Long-Term Ambient Concentrations of Total Suspended Particulates and Oxidants As Related to Incidence of Chronic Disease in California Seventh-Day Adventists, *Env.Health Perspect.* 94:43-50 (1991).

D.E. Abbey, M.D. Lebowitz, P.K. Mills, F.F. Petersen, W.L. Beeson, and R.J. Burchette, Long-Term Ambient Concentrations of Particulates and Oxidants and Development of Chronic Disease in a Cohort of Nonsmoking California Residents, *Inhal.Tox.* 7:19-34 (1995).

U. Ackermann-Liebrich et al. (1997), Lung Function and Long Term Exposure to Air Pollutants in Switzerland, Am.J. Respi Crit Care Med 155:122-9.

H.R. Anderson, A.P. de Leon, J.M. Bland, J.S. Bower, and D.P. Strachan (1996), Air pollution and daily mortality in London: 1987-92, Br.Med.J. 312:665-9.

R.T. Anderson, P. Sorlie, E. Backlund, N. Johnson, G.A. Kaplan (1997), Mortality Effects of Community Socioeconomic Status, Epidemiology 8:42-47.

Armstrong, B.G.; The Effects of Measurement Errors on Relative Risk Regressions, Am.J. Epidem. 132:1176-84 (1990).

N.D. Belloc, Relationship of Health Practices and Mortality, *Prev.Med.* 2:67-81 (1973).

J.E. Brody, Life in Womb May Affect Adult Heart Disease Risk, NY Times (need date).

Cohen, B.H., Chronic Obstructive Pulmonary Disease: A Challenge in Genetic Epidemiology, Am.J. Epidemiology 112:274-288 (1980).

C.J. Crespo, S.J. Keteyian. G.W. Heath, C.T. Sempos (1996), Leisure-Time Physical Activity Among U.S. Adults, Arch Intern med 156:93-98.

R.E. Dales, H. Zwanenburg, R. Burmett, and C.A. Franklin (1991), Respiratory Health Effects of Home Dampness and Molds Among Canadian Children, Am.J.Epidem. 134:196-203.

J.M. Davis, J. Sacks, N. Saltzmann, R.L. Smith, and P. Styer (1996), Airborne Particulate Matter and Daily Mortality in Birmingham, Alabama, Technical Report #55, National Institute of Statistical Sciences, Research Triangle Park, NC 27709.

R.J. Delfino, A.M. Murphy-Moulton, R.T. Burnett, J.R. Brook, and M.R. Becklake (1997). Effects of Ozone and Particulate Air Pollution on Emergency Room Visits for Respiratory Illnesses in Montreal, accepted by Am.J. Resp Crit Care Med.

D.W. Dockery and J.D. Spengler (1981), Personal Exposures to Respirable Particulates and Sulfates, J.APCA 31:153-9.

D.W. Dockery, F.E. Speizer, D.O. Stram, J.H. Ware, J.D. Spengler, and B.G. Ferris, Jr., Effects of Inhalable Particles on Respiratory Health of Children, Am.Rev.Resp.Dis. 139:587-594 (1989).

D.W. Dockery et al., An Association Between Air Pollution and Mortality in Six U.S. Cities, N.Eng.J.Med. 329:1753-9 (1993).

D.W. Dockery et al. (1996), Health Effects of Acid Aerosols on North American Children: Respiratory Symptoms, Envir. Health Perspect. 104:500-5.

D.W. Dockery, J. Schwartz, and J.D. Spengler (1992), Air Pollution and Daily Mortality: Associations with Particulates and Acid Aerosols, Envir. Res. 59:362-373.

R. Doll, R. Peto, E. Hall, K. Wheatley, and R. Gray, Mortality in relation to consumption of alcohol: 13 years' observations on male British doctors, Brit.Med.J. 309:911-8 (1994).

Dzubay, T.G.; Barbour, R.K. "A Method to Improve the Adhesion of Aerosol Particles on Teflon Filters," J.APCA 1983, 33:962-5.

Electric Power Research Institute (EPRI) (1994), Health Effects of Acid Aerosols, Scientific Perspective, Palo Alto, CA.

Gordian, M.E., Ozkaynak, H., Xue, J., Morris, S.S., and Spengler, J.D. (1996), Particulate Air Pollution and Respiratory Disease in Anchorage, Alaska, Envir.Health Perspect. 104:290-7.

Gronbaek, M., Deis, A., Sorensen, T.I.A., Becker, U., Borch-Johnsen, K., Muller, C., Schnohr, P., and Fjensen, G. (1994), Influence of sex, age, body-mass index, and smoking on alcohol intake and mortality, Brit.Med.J. 308:302-6.

R. Hanzlick (1996), Protocol for Writing Cause-of-Death Statements for Deaths Due to Natural Causes, Arch Intern Med 156:25-6.

E.A. Lew and L. Garfinkel, Variations in Mortality by Weight Among 750,000 Men and Women, J.Chron.Dis. 32:563-76 (1979).

F.W. Lipfert, "The Association of Human Mortality with Air Pollution: Statistical Analyses by Region, by Age, and by Cause of Death", Ph.D. Dissertation, Union Graduate School, Cincinnati, Ohio. Available from University Microfilms (1978).

F.W. Lipfert, Air Pollution and Mortality: Specification Searches Using SMSA-based Data, J.Env.Econ.&Mgmt. 11:208-243 (1984).

F.W. Lipfert, Community Air Pollution and Mortality: Analysis of 1980 Data from U.S. Metropolitan Areas I. Particulate Air Pollution, BNL 48446-R, prepared for U.S. Department of

Energy, Brookhaven National Laboratory, August 1993.

F.W. Lipfert (1994a), Filter artifacts associated with particulate measurements: Recent evidence and effects on statistical relationships, *Atm. Envir.* 28:3233-49.

F.W. Lipfert, Air Pollution and Community Health, Van Nostrand Reinhold, New York (1994b).

F.W. Lipfert, Estimating Air Pollution-Mortality Risks from Cross-Sectional Studies: Prospective vs. Ecologic Study Designs, in Particulate Matter: Health and Regulatory Issues, AWMA Publ. VIP 49, Proc. of International Specialty Conference, Pittsburgh, PA April 1995. pp. 78-102.

F.W. Lipfert, R.G. Malone, M.L. Daum, N.R. Mendell, and C.-C. Yang, (1988), A Statistical Study of the Macroepidemiology of Air Pollution and Total Mortality, BNL Report 52122, Brookhaven National Laboratory, Upton, NY.

F.W. Lipfert and R.E. Wyzga (1995a), Uncertainties in Identifying "Responsible" Pollutants in Observational Epidemiology Studies, Inhalation Tox. 7:671-89.

F.W. Lipfert and R.E. Wyzga (1995b), Air Pollution and Mortality: Issues and Uncertainties, J.AWMA 45:949-66.

F.W. Lipfert and R.E. Wyzga (1996), The Effects of Exposure Error on Environmental Epidemiology, Proc. 2nd Colloquium on Particulate Air Pollution and Human Health, ed. by J. Lee and R. Phalen, Park City, UT, May 1996, p. 4-295 to 4-302. Also submitted to Water, Soil, and Air Pollution.

F.W. Lipfert and R.E. Wyzga (1997), Daily Mortality and Size-Fractionated Particulate Matter in Six U.S. Metropolitan Areas: The Implications of Measurement and Modeling Uncertainties (accepted by J. AWMA).

McKerrow, C.B., and Rossiter, C.E. (1968), An annual cycle in the ventilatory capacity of men with pneumoconiosis and of normal subjects, Thorax 23:340-349.

G.B.M. Mesink, M.Deketh, M.D.M. Mul, A.J. Schuit, and H. Hoffmeister (1996), Physical Activity and Its Association with Cardiovascular Risk Factors and Mortality, Epidemiology 7:391-7.

S.H. Moolgavkar, E.G. Luebeck, and E.L. Anderson (1997), Air Pollution and Hospital Admissions for Respiratory Causes in Minneapolis-St. Paul and Birmingham, accepted by Epidemiology.

B. Ostro (1995), Fine Particulate Air Pollution and Mortality in Two Southern California Counties, Envir.Res. 70:98-104.

H. Ozkaynak and G. Thurston, Associations Between 1980 U.S. Mortality Rates and Alternative Measures of Airborne Particle Concentration, *Risk Analysis* 7: 449-462 (1987).

J.H. Pickles (1981), 'Peak Matching' in Time Series Analysis of London Daily Mortality Data, 1958-59 and 1959-60, memorandum LM/PHYS/242 from Central Electricity Research Laboratories, Leatherhead, Surrey, UK.

J.H. Pickles (1982), Air Pollution Estimation Error and What It Does to Epidemiological Analysis, Atmos. Envir. 16:2241-5.

S.J. Pocock, A.G. Shaper, D.G. Cook, R.F. Packham, R.F. Lacey, P. Powell, and P.F. Russell, British regional heart study: geographic variations in cardiovascular mortality, and the role of water quality, *Brit.Med.J.* 24 May 1980, 1243-9.

C.A. Pope, III, M.J. Thun, M.M. Namboodiri, D.W. Dockery, and J.S. Evans, Particulate air pollution as a predictor of mortality in a prospective study of U.S. adults, Am.J.Resp.Crit. Care Med. 151:669-74 (1995).

M. Raizenne et al. (1996), Health Effects of Acid Aerosols on North American Children: Pulmonary Function, Envir. Health Perspect. 104:506-14.

E. Rogot, P.D. Sorlie, N.J. Johnson, and C. Schmitt, A Mortality Study of 1.3 Million Persons, NIH Publication No. 92-3297, National Heart Lung and Blood Institute, Bethesda, MD. (1992).

Samet, J.M., Zeger, S.L., and Berhane, K. (1996), Particulate Air Pollution and Daily Mortality: Replication and Validation of Selected Studies, Phase I Report of the Particle Epidemiology Evaluation Project, Health Effects Institute, Cambridge, MA.

Samet, J.M., Zeger, S.L., Kelsall, J.E., and Xu, J. (1996), Air Pollution, Weather, and Mortality in Philadelphia, 1973-88, Report to the Heath Effects Institute on Phase 1B: Particle Epidemiology Evaluation Project, Johns Hopkins University, Baltimore, MD.

L. Sandvik *et al.*, Physical Fitness as a Predictor of Mortality Among Healthy, Middle-Aged Norwegian Men, *N Engl J Med* 328:533-7 (1993).

Schwartz, J. Lung Function and Chronic Exposure to Air Pollution: A Cross-Sectional Analysis of NHANES II, Env.Res. 50:309-321 (1989).

J. Schwartz, Air Pollution and Daily Mortality in Birmingham, AL, Am.J. Epidemiol. 137:1136-1147 (1993).

J. Schwartz, Air Pollution and Hospital Admissions for the Elderly in Birmingham, AL, Am.J. Epidemiol. 139:589-598 (1994).

J. Schwartz et al. (1989). Acute Effects of Acid Aerosols on Respiratory Symptom Reporting in Children, AWMA paper 89-92.1, presented at the Annual Meeting of the Air&Waste management Association, Anaheim, CA.

J. Schwartz and D. Dockery, Particulate Air Pollution and Daily Mortality in Steubenville, Ohio, Am.J.Epidemiol.135:12-29 (1992).

Schwartz, J.; Dockery, D.W.; Neas, L.M. "Is Daily Mortality Associated Specifically with Fine Particles?" J.AWMA **1996**, 46: 927-39.

Science (1995), Wood Smoke Fires Infections, Vol. 267, p. 1771.

J.V. Selby et al. (1996), Variation Among Hospitals in Coronary-Angiography Practices and Outcomes after Myocardial Infarction in a Large Health Maintenance Organization, N.Engl.J.Med. 335:1888-96.

Shannon, J.D., A Model of Regional Long-term Average Sulfur Atmospheric Pollution, Surface Removal, and Wet Horizontal Flux, Atm.Env. 15:689-701 (1981).

P.D. Sorlie, and E. Rogot, Mortality by Employment Status in the National Longitudinal Mortality Study, *Am.J.Epidem.* 132:983-92 (1990).

Spengler, J.D.; Briggs, S.L.K.; Ozkaynak, H. "Relationships Between TSP Measurements and Size-Fractionated Particle Mass Measurements in Six Cities Participating in the Harvard Air Pollution Health Study," report prepared for the U.S. Environmental Protection Agency, Office of

Air Quality Planning and Standards (1986).

J. Spengler et al. (1994), Respiratory Symptoms and Housing Characteristics, Indoor Air 4:72-82.

J.D. Spengler, P. Koutrakis, D.W. Dockery, M. Raizenne, F.E. Speizer (1996), Health Effects of Acid Aerosols on North American Children: Air Pollution Exposures, Envir. Health Perspect. 104:492-99.

Spodnik, M.J., Jr., Cushman, G.D., Kerr, D.H., Blide, R.W., and Spicer, W.S., Jr., (1965), Effects of Environment on Respiratory Function, Arch.Env. Health 13:243-254.

Thurston, G.G., Ito, K., Hayes, C.G., Bates, D.V., and Lippmann, M. (1994), Respiratory Hospital Admissions and Summertime Haze Air Pollution in Toronto, Ontario: Consideration of the Role of Acid Aerosols, Envir.Res. 65:271-90.

U.S. Environmental Protection Agency (1996), Air Quality Criteria for Particulate Matter, EPA/600/P-95/001aF, Washington, DC.

A. P. Verhoeff, G. Hoek, J. Schwartz, and J.H. van Wijnen (1996), Air Pollution and Daily Mortality in Amsterdam, Epidemiology 7:225-30.

R.E. Wyzga and F.W. Lipfert, Temperature-Pollution Interactions with Daily Mortality in Philadelphia, in Particulate Matter: Health and Regulatory Issues, AWMA Publ. VIP 49, Proc. of International Specialty Conference, Pittsburgh, PA April 1995.
pp. 3-42.

X. Xu and L. Wang (1993), Association of Indoor and Outdoor Particulate Level with Chronic Respiratory Illness, Am.Rev.Respir.Dis. 148:1516-22.

Yoshimura, I.; The Effect of Measurement Error on the Dose-Response Curve, Env.Health Persp. 87:173-8 (1990).

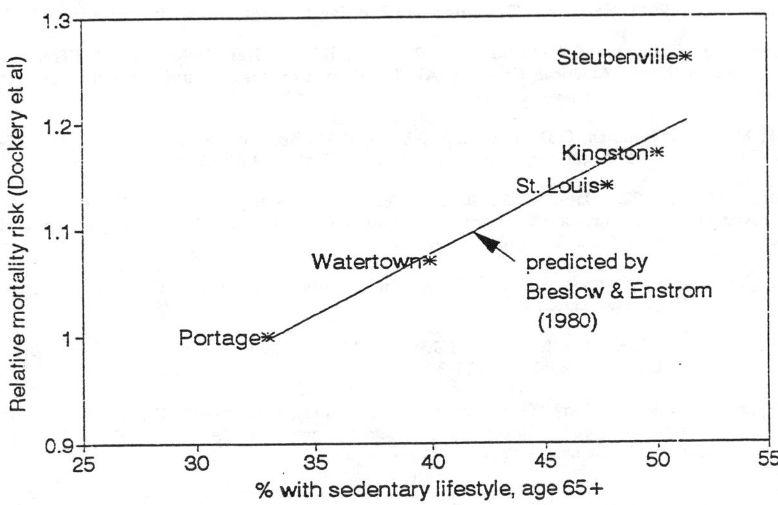

Figure 1. Relative mortality risks in 5 of the Harvard Six Cities as a function of percentage of population with sedentary lifestyles.   Source: Lipfert, 1995.

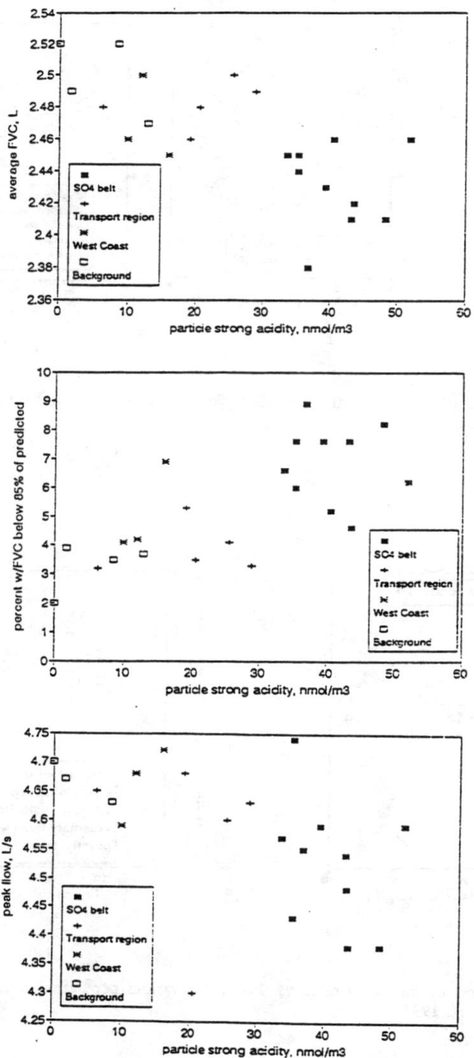

Figure 2. Plots of lung function data from the 24 Cities Study. Data sources: Spengler et al., 1996; Raizenne et al., 1996.

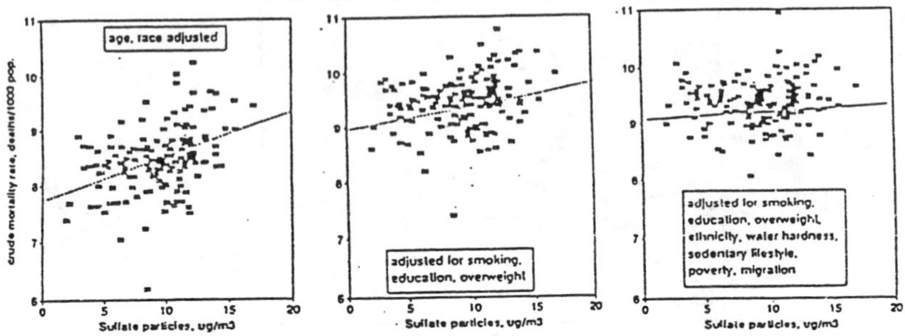

Figure 3. Scatter plots of mortality rates vs. SO_4^{2-} for various population-based models for 148 SMSAs. Source: Lipfert, 1995.

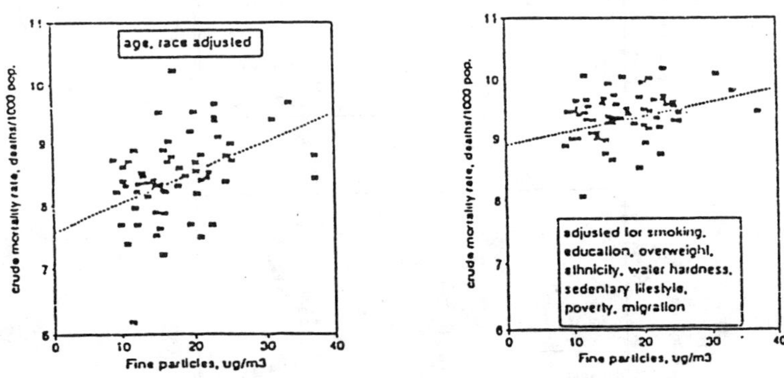

Figure 4. Scatter plots of mortality rates vs. $PM_{2.5}$ for various population-based models for 62 SMSAs. Source: Lipfert, 1995.

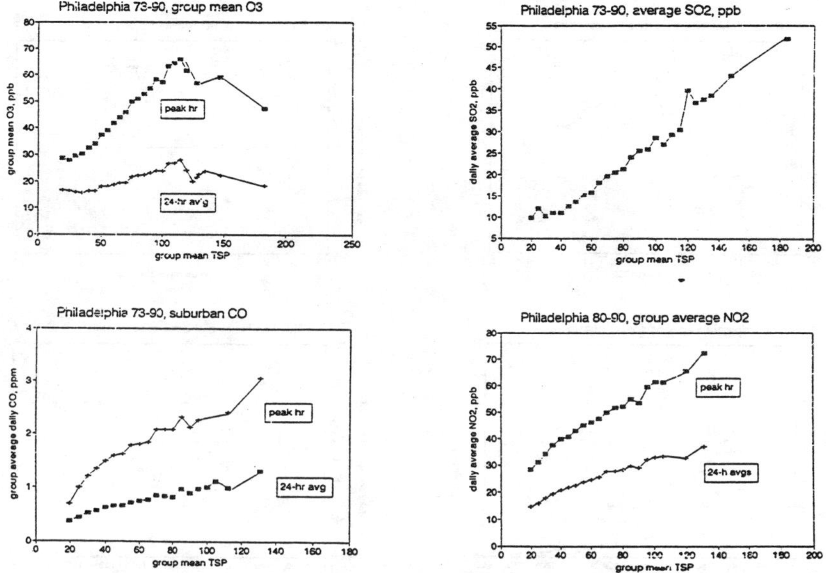

Figure 5. Cross-plots of air quality data from Philadelphia, aggregated according to average TSP levels, 1973-90.

45

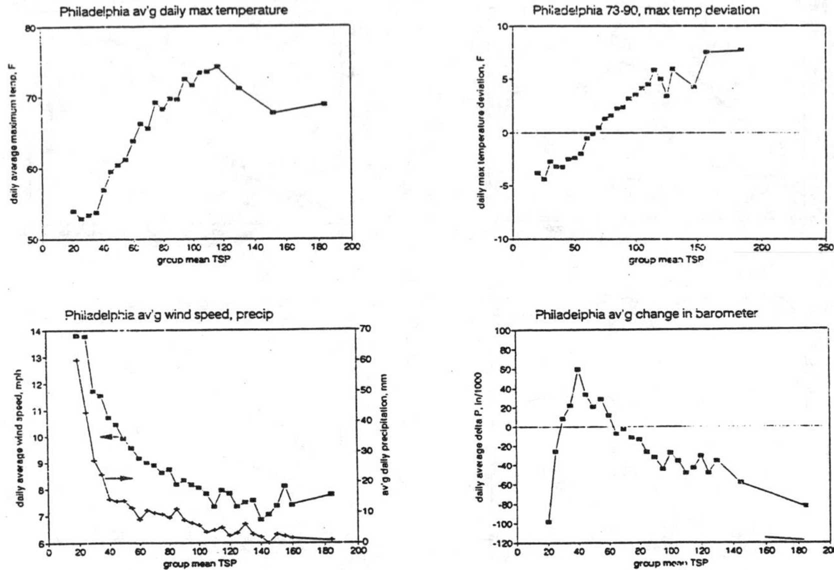

Figure 6. Cross-plots of meteorological data from Philadelphia, aggregated according to average TSP levels, 1973-90.

Appendix A

Summary Comments on EPA's Selected Epidemiological Studies and Justifications for New Regulations

This section presents a capsule view of EPA's overall rationales and justifications for the new NAAQS for $PM_{2.5}$ that they proposed. The first table compares features and findings for the seven epidemiological studies that EPA presented in Table 13-5 of the Criteria Document, with respect to their support for a $PM_{2.5}$ standard. The column on the right presents alternative interpretations of the findings of each of these studies. In no case does an unequivocal justification for the proposed new standard maintain.

The next two tables are based on two summary tables of the EPA Proposal, as published in the Federal Register (FR 61:65638, Dec. 13, 1996). In "Interpretation of Individual Study Results", EPA discusses some of the problems that have been pointed out during the process of developing their regulations and rationalizes them in support of their decision. The right-hand side of this table presents alternative conclusions on each topic. Question 4 was not posed by EPA but should have been. This format is also followed in "Consistency and Coherence Arguments", and again, the key question was not posed.

COMPARISON CHART FOR EPIDEMIOLOGICAL STUDIES OF FINE PARTICLES

Authors year endpoint	Significant pollutants	Study strengths	Study weaknesses	Alternative conclusions from the study

A. Acute studies selected by EPA

Authors year endpoint	Significant pollutants	Study strengths	Study weaknesses	Alternative conclusions from the study
Dockery et al. 1982 daily mortality	PM10	considered wide range of PM species including PM elemental composition	time too short; CO, peak O3 not included areas too big for a single monitor errors in coarse particle measurements	coarse particles implicated
Schwartz et al. 1996 daily mortality	PM10, PM2.5 SO4	size-classified PM data	CO, peak O3 not included areas too big for a single monitor errors in coarse particle measurements	no difference by particle size when meas. errors are considered.
Thurston et al. 1994 resp.hosp.admiss.	O3, H+ PM2.5,PM10	size-classified PM, acidity data	inadequate seasonal, hot day control significance used as criterion 1-way significance tests	O3, PM10 have biggest mean effects

B. Long-term studies selected by EPA

Authors year endpoint	Significant pollutants	Study strengths	Study weaknesses	Alternative conclusions from the study
Dockery et al. 1993 long-term mortality	PM15,FM2.5 SO4	indiv. data on smoking,education,body mass	likely regional confounding (no data on diet, lifestyle, migration) no data on individual or prior exposures too few locations poor occupational exposure data	prior exposure likely important and reduces PM coefficients SO4,PM2.5 effects likely confounded OK to use population groups instead of individuals no difference by particle size
Pope et al. 1995 long-term mortality	SO4, PM2.5	indiv. data on smoking,education,body mass, alcohol use, occupational exposure	likely regional confounding (no data on diet, lifestyle, migration) no data on individual or prior exposures too few pollutants (no PM10)	prior exposure likely important and reduces PM coefficients SO4,PM2.5 effects likely confounded OK to use population groups instead of individuals no data on particle size effect
Dockery et al. 1996 child resp.sympt.	H+,SO4,SO2	wide range of pollutants	symptoms based on 1-yr recall by parents too few locations likely regional confounding	effects could be due to chance or regional bias U.S.-Canada difference is crucial
Raizenne et al. 1996 child lung function	H+,SC4 PM10, PM2.1	wide range of pollutants	may be confounded by season, acute effects too few locations likely regional confounding	effects could be due to regional bias

EPA's "Consistency and Coherence" Arguments (from Proposal, FR)	Alternative rationales.
1. Consistent positive significant associations in many different places by many investigators.	1. (a) Many of these investigators are colleagues or collaborators. (b) All of the reanalyses that considered alternative models found different results than the originals. (c) It is difficult to publish negative findings, since failure to find an effect doesn't necessarily mean that it doesn't exist. (d) Many of the "consistent" results are achieved by considering different lags and/or averaging times; some of the results may be consistent but the models are often not. (e) The results may actually be more consistent than should be expected, given the variability in populations, PM composition, and exposures. Perhaps the consistency comes from the structures of the studies/models.
2. Co-pollutant effects inconsistent.	2. (a) SO_2 is the most frequent co-pollutant investigated; it is not expected to be important because it is strongly absorbed indoors. (b) CO was rarely investigated for mortality and was highly significant for cardiac hospitalization. (c) The relative performance of co-pollutants depends on their relative measurement errors. (d) In a no-threshold model, low concentration is not a criterion for unimportance. All pollutant concentrations are "low" according to toxicity criteria. (e) The ACS study did not consider an adequate range of copollutants. (f) The most recent studies don't find effects that can be unequivocally assigned to PM.
3. Coherence across time scales and endpoints.	3. (a) Coherence in time-series studies derives in part from their nearly identical structures. However, there are many more deaths than would be expected from the small numbers of hospital admissions. (b) There is less evidence of coherence for long-term studies (no hospitalization studies). Long-standing regional differences in diet and lifestyle could explain both mortality and morbidity patterns

(c) The acute effects of PM on lung function are much smaller than those that would be expected to cause premature death.

(d) Daily and annual mortality effects cannot be compared because of the large uncertainties for both.

4. Are there unique effects that can be assigned only (or mainly) to $PM_{2.5}$?

(This question was not posed by EPA!)

4. No. Mean health responses are consistently higher for PM_{10} than for $PM_{2.5}$. Sulfate is a distant third. A separate $PM_{2.5}$ standard is thus not justified.

EPA's "interpretation of Individual Study Results" (FR)

Alternative interpretations.

1. Is PM a surrogate or a general air pollution index, or does it have specific health effects of its own? In any event, will reducing PM improve health?

1. There are no biological mechanisms for PM as an unspecified mixture. Since the responses are quite similar in different places where PM composition varies, it must be acting as a surrogate. It could be a surrogate for gaseous pollutants such as CO or NO2, and controlling only PM would not yield health benefits in these instances.

2. Are the indicated PM health effects robust to model changes, treatment of weather, or co-pollutants?

2. No. Certainly models, lags, and averaging times can be found that will optimize the indicated response to almost any pollutant. Since the associations are weak and all models tend to fit equally poorly, it is impossible to define the "best" model. PM effect has been the criterion.

3. What are the implications of poor correlations between personal exposures and the outdoor ambient data used in the epi studies?

3. The theory that slopes can only be biased towards the null (i.e., attenuated) only applies to the classical situation of a single independent variable whose error is normally distributed with zero mean. This situation is never encountered in practice. Biasing the slope downward also biases the threshold to the left and makes the function appear to be linear. Poorer correlations increase the bias. The problem is more serious than EPA acknowledges; they cite nonrepresentative situations (nonsmoking non carpeted Japanese homes) to try to minimize the problem.

Personal exposures to PM tend to be higher than outdoor data, lower for O_3 and SO_2. Personal exposures to CO and NO_2 depend on proximity to sources (vehicles, gas stoves).

4. What are the causal implications of the varying lag structures found by different investigators? (question not discussed by EPA)

4. The author's freedom to search for the lag that optimizes the pollutant signal constitutes multiple comparisons that should be reflected in the significance tests but that are not. It is difficult to rationalize why different lags should be found in different places and times (as they are). It is also difficult to rationalize how death or admission to hospital can occur on the same day as the pollution excursion, given the tendency of some pollutants to peak in mid-afternoon.